GROWTH FACTORS AND CYTOKINES IN HEALTH AND DISEASE

A Multi-Volume Treatise

Volume 1B • 1996

GROWTH FACTORS

GROWTH FACTORS AND CYTOKINES IN HEALTH AND DISEASE

A Multi-Volume Treatise

GROWTH FACTORS

Editors: DEREK LEROITH
Diabetes Branch
NIDDK
National Institutes of Health
Bethesda, Maryland

CAROLYN BONDY
Developmental Endocrinology Branch
NICHD
National Institutes of Health
Bethesda, Maryland

VOLUME 1B • 1996

JAI PRESS INC.

Greenwich, Connecticut *London, England*

55 Old Post Road, No. 2
Greenwich, Connecticut 06836

JAI PRESS LTD.
38 Tavistock Street
Covent Garden
London WC2E 7PB
England

ISBN: 0-7623-0091-4

Transferred to digital printing in 2008.

Printed in United States of America

CONTENTS

LIST OF CONTRIBUTORS vii

PREFACE xi
Derek LeRoith and Carolyn Bondy

TGFα: A MEMBER OF THE EGF RECEPTOR LIGAND FAMILY
David C. Lee 277

TRANSFORMING GROWTH FACTOR βS: BIOCHEMISTRY AND BIOLOGICAL ACTIVITIES *IN VITRO* AND *IN VIVO*
Wei Cui and Rosemary J. Akhurst 319

THE TRANSFORMING GROWTH FACTOR β SUPERFAMILY
Alexandra C. McPherron and Se-Jin Lee 357

THE TGFβ RECEPTORS AND SIGNALING PATHWAYS
Michael B. Datto, Craig H. Bassing, and Xiao-Fan Wang 395

THE VASCULAR ENDOTHELIAL CELL GROWTH FACTOR FAMILY AND ITS RECEPTORS: MOLECULAR AND BIOLOGICAL PROPERTIES
Denis Gospodarowicz 433

THE HEPATOCYTE GROWTH FACTOR RECEPTOR (MET): AN UNCONVENTIONAL TRANSDUCER OF MITOGENIC AND MOTOGENIC SIGNALS
Paolo M. Comoglio 465

BOMBESIN/GRP AND VASOACTIVE INTESTINAL PEPTIDE/PACAP AS GROWTH FACTORS
Terry W. Moody and Robert T. Jensen 491

ROLE OF SPHINGOLIPID METABOLITES IN THE ACTIONS OF GROWTH FACTORS AND CYTOKINES
Sarah Spiegel and Sheldon Milstien 537

INDEX 565

LIST OF CONTRIBUTORS

Andrew Baird	Department of Cell Biology The Scripps Research Institute La Jolla, California
Mariano Barbacid	Department of Molecular Biology Bristol-Myers Squibb Pharmaceutical Research Institute Princeton, New Jersey
Craig H. Bassing	Department of Pharmacology Duke University Medical Center Durham, North Carolina
Gerhard Baumann	Center for Endocrinology, Metabolism and Molecular Medicine Department of Medicine Northwestern University Medical School Chicago, Illinois
Carolyn Bondy	Developmental Endocrinology Branch NIHCD National Institutes of Health Bethesda, Maryland
Cristin Carter–Su	Department of Physiology University of Michigan Medical School Ann Arbor, Michigan
Paolo M. Comoglio	Department of Biochemical Sciences University of Torino School of Medicine Torino, Italy
Wei Cui	Department of Medical Genetics University of Glasgow Duncan Guthrie Institute Yorkhill Hospitals Glasgow, United Kingdom

Michael B. Datto — Department of Pharmacology
Duke University Medical Center
Durham, North Carolina

Wendy J. Fantl — Chiron Corporation
Emeryville, California

Denis Gospodarowicz — Laboratory of Cellular Chemistry
Chiron Corporation
Emeryville, California

Gordon Guroff — Section on Growth Factors
National Institute of Child Health and Human Development
National Institutes of Health
Bethesda, Maryland

Carl-Henrik Heldin — Ludwig Institute for Cancer Research
Biomedical Center
Uppsala, Sweden

Yoko Hirata — Section on Growth Factors
National Institute of Child Health and Human Development
National Institutes of Health
Bethesda, Maryland

Robert T. Jensen — Digestive Diseases Branch
National Institute of Diabetes and Digestive and Kidney Diseases
National Institutes of Health
Bethesda, Maryland

David C. Lee — Lineberger Comprehensive Cancer Center and Department of Microbiology and Immunology
School of Medicine
University of North Carolina
Chapel Hill, North Carolina

Se-Jun Lee — Department of Molecular Biology and Genetics
Johns Hopkins University
School of Medicine
Baltimore, Maryland

Ann Logan	Molecular Biology Research Group The Wolfson Research Laboratories Queen Elizabeth Medical Centre Edgbaston, Birmingham, United Kingdom
Derek LeRoith	Diabetes Branch NIDDK National Institutes of Health Bethesda, Maryland
Alexandra C. McPherron	Department of Molecular Biology and Genetics Johns Hopkins University School of Medicine Baltimore, Maryland
Sheldon Milstien	Laboratory of Neurochemistry National Institute of Mental Health National Institutes of Health Bethesda, Maryland
Terry W. Moody	Biomarkers and Prevention Research Branch National Cancer Institute Rockville, Maryland
Salil K. Niyogi	Biology Division Oak Ridge National Laboratory Oak Ridge, Tennessee
Mari Oshima	Section on Growth Factors National Institute of Child Health and Human Development National Institutes of Health Bethesda, Maryland
Arne Östman	Ludwig Institute for Cancer Research Biomedical Center Uppsala, Sweden
Kevin G. Peters	Duke University Medical Center Durham, North Carolina
Lisa S. Smit	Department of Physiology University of Michigan Medical School Ann Arbor, Michigan

Sarah Spiegel — Department of Biochemistry and Molecular Biology
Georgetown University Medical Center
Washington, D.C.

Douglas K. Tadaki — Naval Medical Research Institute
Immune Cell Biology Program
Bethesda, Maryland

Xiao-Fan Wang — Department of Pharmacology
Duke University Medical Center
Durham, North Carolina

Bengt Westermark — Department of Pathology
University Hospital
Uppsala, Sweden

Lewis T. Williams — Chiron Corporation
Emeryville, California

PREFACE

Advances in molecular technology in recent years have catalyzed an explosive growth of information about intercellular peptide messengers and their receptors. For example, ten years ago the only neurotrophin characterized at the molecular level was nerve growth factor (NGF) and the only recognized neurotrophin receptor was the p75 NGF receptor. At present, the number of described neurotrophic peptides approaches 30 and the number of receptors is increasing apace. Just six years ago, the characterized interleukins numbered about three while now there are at least 16. Because many of these new peptide ligands and receptors were identified by "reverse genetic" techniques the understanding of their biological roles lags behind the knowledge of their molecular structures. Over the past few years, however, a new era of functional studies has begun because recombinant proteins have become available for clinical studies. In addition, animal models have been and are being developed using recombinant DNA techniques. Both the clinical studies and studies of transgenic and target deleted mice will allow for further physiologic elucidation of the biological roles of these messenger peptides and their receptors.

This series on Growth Factors and Cytokines is divided into three main sections: Growth Factors (Volume I), Cytokines (Volume II) and Systems (Volume III). Although volumes I and II are separate the distinction between "growth factors" and "cytokines" is probably more historical or pragmatic than indicative of differences in function. The term "growth factors" refers to a wide variety of locally or systemically produced proteins with pleiotropic actions on tissue growth and differentiation. The term "cytokines" describes a group of proteins identified primarily within the immune and hematopoietic systems, although it is likely that such a narrow view of cytokines will not survive for long. For example it appears that some interleukins and interleukin receptors are expressed by neuroepithelial cells *in vivo* suggesting that these interleukins may have intrinsic roles within the nervous system. Furthermore, tumor necrosis factor (TNF) has been identified as a potential adipose tissue regulatory factor which is both produced and acts locally. The third volume entitled Systems deals more directly with the role of these factors in both normal physiology and the disease processes resulting from the deficiency or excess of growth factors/cytokines and their receptors.

The first volume deals with peptide growth factors and their receptors. Here too there is an arbitrary division of ligands and their receptors. In some instances (e.g., insulin-like growth factors) the proteins and their corresponding receptors are discussed in the same chapter, whereas in other cases, for example, NGF and platelet-derived growth factor they are discussed separately. While we have attempted to be as comprehensive and inclusive as possible, there will always be some regrettable omissions. At the publishing date we recognize that a few growth factors and cytokines have not been included in this review. These new discoveries will for certain be reviewed in similar pages in the future.

Derek Le Roith
Carolyn Bondy

TGFα: A MEMBER OF THE EGF RECEPTOR LIGAND FAMILY

David C. Lee

I.	Introduction	278
II.	Historical Background	278
III.	The EGF-R and Related Proteins	279
IV.	The Mammalian EGF Superfamily	280
V.	Viral and Nonvertebrate EGF Family Homologues	282
VI.	Structure/Function Analyses of Mature TGFα	283
VII.	The EGF-R Ligands and Differences in Biological Activity	284
VIII.	ProTGFα and Its Processing	284
IX.	ProTGFα: A Model for Membrane-Anchored Growth Factors	287
X.	Expression of TGFα/ProTGFα in Normal Tissues	288
XI.	Expression by Neoplastic Cells and Tissues	291
XII.	TGFα and Transformation	292
XIII.	Chimeric TGFα Proteins as Therapeutic Agents	295
XIV.	The TGFα Gene	296
XV.	The TGFα Promoter	297
XVI.	Regulation of TGFα Gene Expression	298
XVII.	Development of TGFα Knockout Mice	300

Growth Factors and Cytokines in Health and Disease
Volume 1B, pages 277–318.

ISBN: 0-7623-0091-4

XVIII. Conclusions . . . 303
Acknowledgments . . . 303
References . . . 303

I. INTRODUCTION

Since its discovery 20 years ago, transforming growth factor α (TGFα) has been one of the most intensively studied growth factors. This high level of interest stems from its implied role in multistage carcinogenesis and its early identification as an example of a membrane-anchored receptor ligand. In writing this review, I have attempted to summarize the impressive body of knowledge accumulated about this growth factor without losing sight of the numerous, fundamental questions that remain to be answered.

II. HISTORICAL BACKGROUND

Todaro and colleagues observed in the late 1970s that several retrovirally-transformed cell lines displayed reduced surface binding of epidermal growth factor (EGF) compared to their normal cellular counterparts (Todaro et al., 1976; Todaro et al., 1977). This led to the hypothesis that sarcoma viruses encode an EGF-related protein that can compete for binding to the EGF receptor (EGF-R). EGF-R-binding activities that were not recognized by anti-EGF antibodies were subsequently identified in media conditioned by sarcoma virus-transformed fibroblasts (De Larco and Todaro, 1978). These trypsin- and dithiothreitol-sensitive, heat-stable activities of 25, 12, and 7 kilodaltons (kDa) were shown to promote a phenotypic transformation of normal cultured fibroblasts; treated cells lost the property of contact inhibition and acquired the ability to grow in soft agar. Initially referred to as "sarcoma growth factor (SGF)", they were collectively renamed "transforming growth factor (TGF)" when similar activities were shown to be produced by both tumor-derived (Todaro et al., 1980) and chemically-transformed cell lines (Roberts et al., 1980). Subsequently, Roberts and colleagues (1981) described the isolation from non-neoplastic tissues of a new class of transforming growth factors whose ability to stimulate the anchorage-independent growth of normal fibroblasts was potentiated by exogenous EGF. The significance of this observation was further clarified when it was shown that transforming growth factor activity from sarcoma virus-transformed cells could be fractionated by high-performance liquid chromatography (HPLC) into two, synergistically acting subsets: a component, renamed TGFα, whose actions were mediated through the EGF-R (Massagué, 1983; Carpenter et al., 1983), and a second activity, named TGFβ, that potentiated the EGF- or TGFα-dependent growth of fibroblasts in soft agar (Anzano et al., 1982, 1983). Determination of the amino acid sequence of these proteins confirmed that TGFα is structurally and functionally related to EGF (Marquardt et al., 1983, 1984) and revealed that TGFβ is a structurally unrelated protein (Derynck et al., 1985). The discovery of TGFβ provided satisfying explanations for differences in the activities of early SGF preparations compared to those of pure EGF, as well as the misleading identification of a novel receptor that reportedly bound TGF but not EGF (Massagué et al., 1982). Many intervening studies have since gone on to show that TGFβ displays diverse activities including regulation of immune function, extracellular matrix production, and cell growth (reviewed by Roberts and Sporn, 1990). Although TGFβ stimulates the anchorage-independent growth of fibroblasts in a manner consistent with its discovery, it is a potent inhibitor of epithelial cell proliferation.

III. THE EGF-R AND RELATED PROTEINS

Work from many laboratories has documented TGFα's ability to regulate cell proliferation, migration, and differentiation in culture, and, generally, to mimic the actions of EGF (reviewed by Carpenter and Wahl, 1990). In addition, TGFα has been implicated in fundamental biological processes including wound healing (Schultz et al., 1987; Rappolee et al., 1988b), liver regeneration (Mead and Fausto, 1989), angiogenesis (Schreiber et al., 1986), bone remodeling (Ibbotson et al., 1985), development (Lee et al., 1985a; Wilcox and Derynck, 1988a), and tumorigenesis (Sporn and Todaro, 1980). All of these actions are apparently mediated through the EGF-R because no other receptor has been identified. Binding of the growth factor rapidly induces receptor dimerization, activation of the receptor's intrinsic tyrosine kinase activity, and autophosphorylation of the receptor itself. As a direct result, various signaling molecules are recruited to phosphorylated tyrosines in the receptor's cytoplasmic domain (Koch et al., 1991; Carpenter, 1992), and this, in turn, leads to the activation of any of several different signaling pathways, with the specific response depending, in part, on the nature of the cell type. Signaling systems activated by EGF-R include the ras/raf/MAP-kinase cascade, a phospholipase C-γ-(PLC-γ)-mediated pathway that activates Ca^{2+}-dependent kinases, and a newly discovered pathway leading to the activation of latent transcription factors (so-called STAT proteins) that, then, translocate from the cytoplasm to the nucleus. While the recent elucidation of these pathways stands as a major accomplishment of modern cellular biology, many details of the way these signals are integrated and coupled to molecular regulation of proliferation, migration, and differentiation are still unclear.

EGF-R (or HER/erbB) is a member of a family of structurally related proteins that includes HER2/neu/erbB2, HER3/erbB3, and HER4/erbB4 (reviewed in Plowman et al., 1993). These 170–185 kDa proteins are characterized by the conserved presence of two cysteine-rich elements in their ectodomains and an uninterrupted tyrosine kinase homology in their intracellular regions. The tyrosine kinase sequence of HER3 is most diverged, and a recent study indicates that it has little if any activity *in vitro* (Guy et al., 1994). Instead, HER3 may function only as a receptor and docking protein. HER3 and HER4 are binding receptors for heregulin/neu differentiation factor (NDF) (Holmes et al., 1992; Peles et al., 1992; Wen et al., 1992; this human/rat ligand contains a motif related to but distinct from the EGF-like sequence (described below). A direct binding ligand for HER2/neu has not yet been identified, although HER2/neu can apparently heterodimerize with either HER3 or HER4 to form a high-affinity, heregulin/NDF-binding complex. Indeed, the ability of the four EGF-R family members to form various combinations of heterodimers has been clearly established, at least in transfected cells (Carraway III and Cantley, 1994). Illustrating that heterodimerization could have important functional consequences, a recent report described evidence to support the notion that EGF-induced PI3-kinase activation is mediated through the formation of an EGF-R/HER3 heterodimer (Soltoff et al., 1994). Whether EGF-R's propensity to heterodimerize is variably affected by the binding of different EGF family ligands (described below) is unknown.

IV. THE MAMMALIAN EGF SUPERFAMILY

EGF and TGFα belong to a family of five structurally and functionally related proteins that all bind and activate the EGF-R. The other members of this family include amphiregulin (Shoyab et al., 1988, 1989), heparin-binding EGF (Higashiyama et al., 1991), betacellulin (Shing et al., 1993) [A sixth member, epiregulin, was reported while this manuscript was in press (Toyoda et al., 1995).]. A potential seventh member, schwannoma-derived growth factor (SDGF) (Kimura et al., 1990) is almost certainly a species variant of amphiregulin because it is highly homologous and maps to the same chromosomal locus (Pathak et al., 1995). These various ligands all bind to the EGF-R with high affinity via a conserved motif referred to as the EGF-like sequence:

$$CX_7CX_{2-3}GXCX_{10}CXCX_3YXGXRC$$

Although only formally proven for EGF and TGFα, disulfide bonds formed between the six similarly spaced cysteines in this conserved sequence are presumed to impart a characteristic three-loop structure on each of the various proteins that is required for EGF-R binding activity. [Note that while cripto (Ciccodicola et al., 1989) is sometimes referred to as an EGF-like protein, a comparison of its corresponding sequence reveals considerable deviation in the positioning of the six cysteines relative to the sequence shown above, strongly arguing that it does not function as an EGF-R ligand.] In the case of the five known cellular EGF-R ligands, this EGF-like sequence resides in the ectodomain of an integral membrane precursor and is released in the context of variably sized larger peptides by proteolytic processing. For example, the EGF-like sequence of the fully processed 50 amino acid TGFα is flanked by seven amino acids on both the NH_2 and COOH termini, and mature EGF is similar in size. In contrast, the secreted forms of amphiregulin, HB-EGF, and betacellulin are characterized by larger NH_2-terminal-extending peptides (Figure 1) that are rich in hydrophilic amino acids and confer the ability to bind to heparin on this subset ligands.

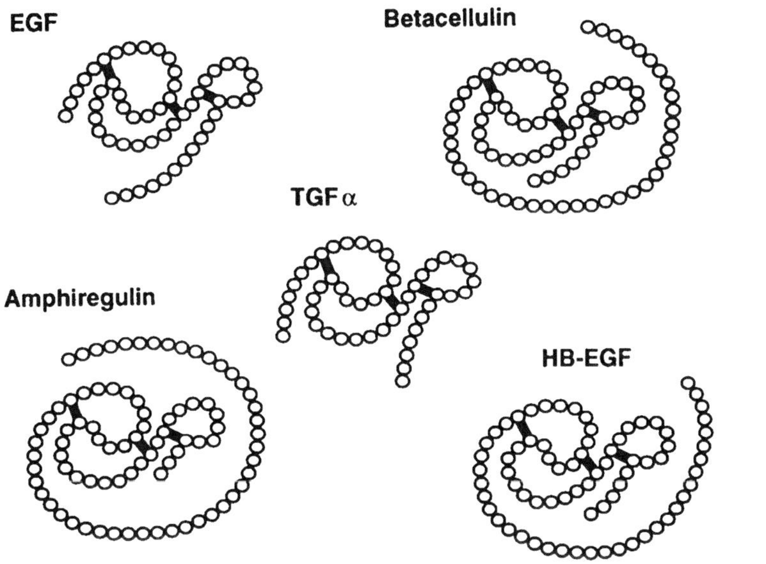

Figure 1. The mammalian EGF-R ligands. A schematic representation of the predominant secreted forms of the five EGF-like ligands is shown with disulfide bonds (black bars) formed between the first and third, second and fourth, and third and sixth cysteines, as has been experimentally established only for EGF and TGFα. Note the conservation of the central three-loop structure, and the fact that amphiregulin, heparin-binding (HB) EGF, and betacellulin are all distinguished by the presence of extended amino-terminal sequences. Note also that an alternate form of secreted amphiregulin lacking the amino-terminal, six amino acids has been described.

As previously noted, soluble forms of the five EGF family members are all derived from integral membrane precursors. Amongst these, proEGF is distinguished by its large size (it contains approximately 1200 amino acids) and eight EGF-related sequences in its extracellular domain, none of which are thought to function as EGF-R ligands (Scott et al., 1983; Gray et al., 1983). Integral membrane precursors to the other EGF family members are considerably smaller proteins, each containing only a single EGF-like sequence in its ectodomain. It is noteworthy that the EGF-like sequence is the only apparent homology among the five integral membrane precursors.

The five cellular EGF-R ligands are, in turn, part of a superfamily of cellular proteins all containing a six-cysteine motif related to the EGF-like sequence. These include heregulin/NDF, the previously described HER3/HER4 ligand, and a large number of membrane-associated or secreted proteins that have diverse functions but are not ligands for protein kinase receptors (summarized in Carpenter and Wahl, 1990). Proteins in this latter group, which include Notch, the low density lipoprotein receptor, and serum clotting factors, often contain multiple EGF-like motifs. Although the EGF-related sequence of heregulin/NDF is distinguished from the consensus sequence of EGF-R ligands by a greater separation of the third and fourth cysteines (thus, producing a larger second loop) and the substitution of phenylalanine for a conserved tyrosine, the EGF-related sequence of the nonreceptor-ligand proteins diverges more significantly. A possible function for this more distantly related motif is suggested by findings that many of its host proteins are dependent on Ca^{2+} for activity and that some variations of the motif contain a consensus signal for aspartate-β-hydroxylation, a modification known to contribute to the formation of high-affinity Ca^{2+} binding sites (Handford et al., 1991; Huang et al., 1991; Selander--Sunnerhagen et al., 1992).

V. VIRAL AND NONVERTEBRATE EGF FAMILY HOMOLOGUES

Pox family viruses encode EGF-related proteins that bind and activate EGF-R. The best characterized of these is vaccinia virus growth factor (VGF) (Blomquist et al., 1984; Brown et al., 1985; Stroobant et al., 1985). The VGF precursor is a 140 amino acid integral membrane protein containing a single EGF-like sequence. Soluble VGF appears to be derived by a single cleavage event that releases the entire 77 amino acid extracellular domain. Expressed during the early phase of viral replication, it may be responsible for the observed hyperproliferation of adjacent uninfected cells (Buller et al., 1988). Related proteins are also encoded by the Shope fibroma (Chang et al., 1987; Chang et al., 1990) and myxoma (Upton et al., 1987) viruses. Interestingly, these appear to lack membrane domains and, hence, they could be directly secreted.

Non vertebrate homologues of the EGF-like ligands have been identified as well. Interestingly, genetic analyses suggest that their roles may be more related to the control of cellular differentiation rather than proliferation per se. The *torpedo (top)/DER* gene of *Drosophila* is a homologue of the vertebrate EGF-R (Price et al., 1989; Schejter and Shilo, 1989); its expression is required for normal oogenesis, embryogenesis, and imaginal disc formation (Clifford and Schupbach, 1992; Raz and Shilo, 1992). Possible ligands for *top/DER* include products of the *gurken* (*grk*) (Neuman–Silberberg and Schupbach, 1993) and *spitz* (*spi*) (Rutledge et al., 1992) genes. *Grk*, a predicted transmembrane protein of 294 amino acids with a single TGFα-like motif in its ectodomain, is asymmetrically localized in the oocyte. *Spi* is a 230 amino acid transmembrane protein with one EGF-like peptide in its extracellular region. Despite its structural similarity to mammalian EGF-R ligands, *spi* may not be a top/DER ligand because the phenotypes of *spi* mutants most closely resemble those of rhomboid, a putative membrane receptor with several membrane-spanning domains but no discernible signaling motif (Bier et al., 1990).

EGF-related molecules have also been implicated in the development of the nematode *Caenorhabditis elegans* (*C. elegans*). The *C. elegans* vulva forms from six precursor cells stimulated to differentiate in response to a signal emanating from a so-called anchor cell in the adjacent gonadal tissue (Hill and Sternberg, 1992). This signal appears to be mediated by interactions between lin-3, a transmembrane protein of 438 or 423 amino acids that contains an extracellular EGF-like sequence, and the EGF-R homologue let-23. Because the inductive signal acts over a distance, lin-3 could be released as a diffusible ligand.

VI. STRUCTURE/FUNCTION ANALYSES OF MATURE TGFα

Sequence-specific ^{1}H NMR resonance assignments for EGF and TGFα have been interpreted to suggest that, in each case, the EGF-like sequence folds into two independent domains that together form a mitten-like structure (Kohda et al., 1989; Montelione et al., 1989; Kline et al., 1990). According to this model, the palm of the mitten would be formed from the first and second disulfide loops, and the thumb corresponds to the third disulfide loop. This predicted structure suggests that TGFα and the four related ligands might all bind EGF-R as a mitten would hold an object—a suggestion that, in turn, implies a requirement for all three disulfide loops. An early report that a peptide corresponding to the third disulfide loop of TGFα had significant biological activity (i.e., bound and activated the EGF-R) (Nestor et al., 1985) has not been reproduced (Darlak et al., 1988; Defeo-Jones et al., 1988). Moreover, the destruction of any of the three disulfide bonds caused by the substitution of alanines for cysteines yielded biologically inactive TGFα proteins (Defeo-Jones et al., 1988). In contrast, deletion of residues in the short amino-terminal leader peptide that precedes the first cysteine of the EGF-like sequence had relatively little impact on TGFα's biological activity.

A comparison of the five mammalian EGF-R ligands shows that, in addition to the six cysteines, only four other amino acids are conserved throughout the family. The individual mutation of three of these residues, Gly^{19}, Gly^{40}, and Arg^{42} (numbered relative to the first residue of the mature 50 amino acid TGFα), yielded TGFα proteins with little or no activity (Feild et al., 1992; Defeo-Jones et al., 1989; Lazar et al., 1989). The effects of mutating Tyr^{38}, on the other hand, depended on the nature of the substitution. With regard to those residues that are partially conserved (or conservatively replaced) amongst family members, mutation of Phe^{15}, His^{18}, Asp^{47}, or Leu^{48} all inhibited TGFα's biological activity, although the extent of inhibition varied from moderate to marked (Defeo-Jones et al., 1988, 1989; Lazar et al., 1988, 1989; Feild et al., 1992). On the other hand, substituting the relatively nonconserved amino acids or deleting the carboxy-terminal $Leu^{49}Ala^{50}$ dipeptide of TGFα (Yang et al., 1990), produced only small to moderate declines in biological activity.

VII. THE EGF-R LIGANDS AND DIFFERENCES IN BIOLOGICAL ACTIVITY

The five EGF family members all bind and activate EGFR, and no other receptors have yet been identified for any of these proteins. Nevertheless, quantitative and, possibly, qualitative differences in biological response have been ascribed to the different ligands. TGFα is generally more potent than EGF, and in some contexts, its effects have been described as stronger and/or more prolonged (Barrandon and Green, 1987; Schreiber et al., 1986). Other differences have also been noted. For example, HB-EGF was 40-fold more potent than EGF in stimulating the proliferation of smooth aortic muscle cells (Higashiyama et al., 1991). Although amphiregulin was equipotent to EGF in stimulating the proliferation of BALB/MK cells, it (in contrast to EGF or TGFα) couldn't induce anchorage-independent growth of NRK cells in the presence of TGFβ (Shoyab et al., 1989). Some of these distinctions in biological response could reflect differences in the manner these various ligands bind EGF-R. Unfortunately, comparisons of ligand binding affinity have often been confounded by the mixing of ligands and receptors from different species. The potential significance of this point is suggested by the finding that chicken EGF-R showed a 100-fold greater affinity for human TGFα compared to murine EGF (Lax et al., 1988). Using homologous components, Ebner and Derynck (1991) found that EGF and TGFα bound EGF-R with comparable affinities, indicating that, in this case, distinctions in biological activity could not be accounted for by differences in receptor binding. On the other hand, amphiregulin displaced only a portion of the ^{125}I-EGF bound to EGF-R on the surface of human A431 cells (Shoyab et al., 1989), suggesting that this particular ligand may bind only a subset of EGF-R molecules. The fact that, in contrast to EGF and TGFα, amphiregulin, HB-EGF and betacellulin all bind to heparin could also contribute to differences in biological response because presentation of a ligand/heparin complex might modify receptor activation. Finally, distinctions in biological response could also derive from differences in the trafficking of various ligand/receptor complexes. Ebner and Derynck (1991) found that in the case of an internalized TGFα/EGF-R complex, both ligand and receptor were preferentially recycled back to the cell surface, but, when an EGF/EGF-R complex was internalized in the same cell type, both components were efficiently degraded. The investigators suggested that this potentially important distinction in fate might be due to observed differences in the pH stability of the two complexes. Finally, indicating that such differences in processing could be cell-type specific, preferential degradation of TGFα versus EGF has been observed with other cell lines (Korc and Finman, 1989; Garcia et al., 1989).

VIII. PROTGFα AND ITS PROCESSING

The cloning of human (Derynck et al., 1984) and rat (Lee et al., 1985b) cDNAs led to predictions of a primary TGFα translation product of 160 or 159 amino acids,

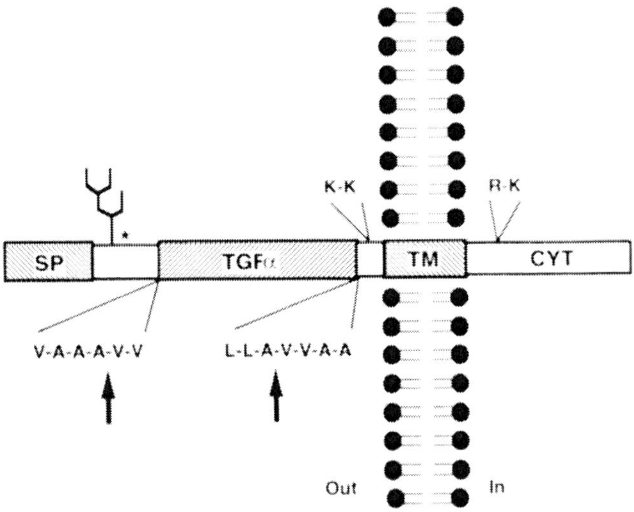

Figure 2. Schematic representation of preproTGFα. The precursor molecule is shown embedded in the plasma membrane. The signal peptide (SP), 50 amino acid EGF-like sequence (TGFα), transmembrane (TM), and cytoplasmic (CYT) domains are indicated. The branched structure indicates the location of an N-glycosylation site, and the asterisk denotes the presence of an amino acid variably present as a result of microheterogeneous RNA splicing. Amino acids flanking the alanine/valine cleavage sites are indicated; arrows mark the dipeptide bonds that must be cleaved to release the fully processed, 50 amino acid TGFα. Lysine–lysine and arginine–lysine bonds that could potentially serve as substrates for proteases with basic residue specificity are indicated in the extracellular juxtamembrane and cytoplasmic domains, respectively. Amino acid abbreviations: V, valine; A, alanine; K, lysine; R, arginine.

respectively, that contains sequence motifs characteristic of an integral-membrane glycoprotein (Figure 2). This precursor (preproTGFα) was predicted to include a putative signal peptide of 22 amino acids followed by a 16 or 17 amino acid sequence containing an N-linked glycosylation consensus site, the mature 50 amino acid TGFα, a juxta-membrane region of 9 amino acids, a hydrophobic transmembrane domain of 23 amino acids flanked by Lys-Lys and Arg-Lys bonds on the amino and carboxy termini, respectively, and a cytoplasmic sequence of 39 amino acids, 7 of which are cysteines. Interestingly, a comparison of the human and rat proTGFα sequences shows that the transmembrane and cytoplasmic domains are the most highly conserved with only a single conservative amino acid substitution in the two regions (Todaro et al., 1985). Using domain-specific antibodies, the predicted characteristics of preproTGFα were confirmed by showing that an appropriately sized glycoprotein was preferentially associated with plasma membrane fractions derived from cells harboring appropriate expression vectors (Gentry et al., 1987; Bringman et al., 1987) and inserted with the correct orientation into microsomal vesicles present during *in vitro* translation of TGFα mRNA (Teixido et al., 1987).

Release of the mature, 50 amino acid TGFα from proTGFα (the precursor lacking the signal peptide) requires the cleavage of Ala-Val bonds at both termini (Marquardt et al., 1983). Moreover, the two Ala-Val dipeptides are each flanked by similar alanine-, valine- or leucine-rich sequences (/ indicates the cleaved bond):

NH_2- ... Val-Ala-Ala-Ala-/-Val-Val ... mature TGFα ... Leu-Leu-Ala-/-Val-Val-Ala-Ala ... -COOH

This specificity resembles that of elastase enzymes. Some members of this chymotrypsin family will, in fact, convert larger forms of TGFα to species comigrating with the 6-kDa mature protein (Ignotz et al., 1986; Luetteke et al., 1988; Mueller et al., 1990). Although their apparent sequence similarity suggests that the two Ala-Val dipeptides could be hydrolyzed by the same enzyme, the two sites can, nevertheless, be cleaved at markedly different rates. Pulse-chase analyses of transfected Chinese hamster ovary cells showed that, whereas cleavage of the amino-terminal site occurred rapidly (i.e., within 1 hr), the carboxyl-terminal cleavage required for release of soluble TGFα was a much slower event (i.e., it required up to 4 hr) (Teixido et al., 1990). On the other hand, cleavage of the amino-terminal site must be rate limiting in some transfected or transformed lines because these cells secrete larger (typically, 18–21 kDa) forms of TGFα; the latter correspond to the mature, 50 amino acid sequence linked to amino-terminal precursor sequences containing the N-glycosylation site (Ignotz et al., 1986; Bringman et al., 1987; Teixido and Massagué, 1988; Luetteke et al., 1988). These cell-specific differences in cleavage rates could indicate that the two sites are cleaved by distinct enzymes that are differentially expressed or activated in various cell lines. Alternatively, cleavage at the amino-terminal site could be variably influenced by the nearby addition of both N- and O-linked sugars, the extent and form of which is known to vary between cell types. In fact, the impact of glycosylation on the expression and processing of proTGFα is presently uncertain. Although production of soluble TGFα by an O-glycosylation-defective CHO clone was comparable to that of wild-type CHO cells (Teixido et al., 1990), secretion of mature TGFα and accumulation of larger species were both diminished in the presence of the N-glycosylation inhibitor tunicamycin (Bringman et al., 1987; Teixido and Massagué, 1988). Finally, soluble TGFα species of 30–40 kDa have also been described (Sherwin et al., 1983; Linsley et al., 1985). It is not known whether these include additional regions of the precursor or are more extensively glycosylated. However, in the context of this discussion it is important to note that the variably sized, larger secreted forms of TGFα are all biologically active (Ignotz et al., 1986; Bringman et al., 1987; Luetteke et al., 1988).

As noted, release of soluble TGFα from transfected CHO cells is normally slow. However, it can be rapidly stimulated by exposing these cells to phorbol esters (e.g., TPA), crude serum factors, or calcium ionophores (Pandiella and Massagué, 1991a,b). Activation of processing by these pleiotropic agents, which appears to involve both protein kinase C-dependent and -independent pathways, is not specific to proTGFα; release of soluble c-*kit* ligand (KL) from its integral membrane precursor was also rapidly induced by these same agents. Nevertheless, different enzymes may be involved because activated processing of these two precursors showed distinct sensitivities to a panel of protease inhibitors (Pandiella et al., 1992). Studies of TPA-induced processing have produced two additional observations. The first is that only those proTGFα molecules on the cell surface are targets for activated processing. This conclusion is based on the findings that activated processing was not observed when cells were treated with brefeldin A, resulting in the accumulation of precursor in the Golgi, but was observed with cytosol-deficient cells that had been permeabilized by exposure to streptolysin O (Bosenberg et al., 1992; Bosenberg et al., 1993). The second, more surprising finding is that activated cleavage depends on the presence of a valine residue at the carboxy terminus of proTGFα's cytoplasmic domain. Whereas deletion or nonconservative substitution of this valine prevented the activation of proTGFα processing, internal deletions of cytoplasmic sequence had relatively little effect as long as the resulting cytoplasmic fragments still retained a C-terminal valine residue (Bosenberg et al., 1992). Moreover, activated processing was observed with a chimeric precursor containing the ectodomain of proTGFα linked to the cytoplasmic domain of proKL; the latter contains a C-terminal valine, though according to a recent study, this residue is not required to process this precursor (Cheng and Flanagan, 1994).

The phenomenon of activated cleavage has suggested a model in which proTGFα is processed by a specialized protease (or proteolytic system)—one that acts at the cell's external surface but is regulated via determinants and signaling in the cytoplasm. Whether the same enzyme or system is responsible for constitutive processing of proTGFα in transformed cells remains to be determined. Inasmuch as soluble TGFα was also rapidly released by cells exposed to low doses of UV light, Brown and colleagues (1993) suggested that induced processing reflects a generalized stress response involving the global activation of ectopeptidase activity. Further characterization of recently described candidate proTGFα processing enzymes, including one that is membrane-associated and phorbol ester-inducible (Harano and Mizuno, 1994), may shed further light on this issue.

IX. PROTGFα: A MODEL FOR MEMBRANE-ANCHORED GROWTH FACTORS

Although the processing of membrane-associated precursors may be an important regulatory step, it does not appear required for biological activity. Two laboratories independently altered amino acids that flank the cleavage sites of proTGFα and, thereby, generated forms that were not cleaved or were cleaved at dramatically reduced rates relative to the wild-type precursor (Wong et al., 1989; Brachmann et al., 1989). When cells harboring these cleavage-resistant forms of proTGFα were cocultured with cells expressing EGF-R, rapid autophosphorylation of the receptor protein occurred, and this was followed by a transient rise in free intracellular Ca^{2+} and the induction of c-fos mRNA in the target cell. These results were, subsequently, extended to show that (1) in response to expression of noncleavable forms of proTGFα, transfected NRK cells exhibited transformed-like growth *in vitro* (Brachmann et al., 1989; Blasband et al., 1990a) and formed tumors in nude mice (Blasband et al., 1990a) and (2) episomal expression of wild-type proTGFα in a bone marrow stromal cell line allowed these cells to adhere to an EGF-R-expressing, EGF/IL-3-dependent hematopoietic cell line in culture (Anklesaria et al., 1990). Inasmuch as a portion of the adhered progenitor cells were stimulated to undergo DNA replication, it would appear that interactions between proTGFα and EGF-R could have multiple consequences (e.g., the sequential stimulation of adherence and proliferation). These various findings, which have been reinforced by studies of other membrane-associated ligands (Mroczkowski et al., 1989), collectively support the hypothesis that regulated processing of membrane-anchored growth factors could provide an important switch from localized autocrine or juxtacrine actions to more distal paracrine actions requiring soluble growth factor. Although physiological roles for proTGFα (or any of the related EGF family precursors) have not been confirmed, it is relevant to note that defects in melanogenesis, gametogenesis, and hematopoiesis are all associated with a mutation (Sl^d) of the proKL gene that eliminates the integral membrane precursor but not the soluble form of this ligand (Flanagan et al., 1991; Brannan et al., 1991).

Finally, an additional possibility regarding membrane-anchored growth factors is that they have a dual role both as ligand and receptor. This possibility, first suggested for proEGF by Pfeffer and Ullrich (1985), is consistent with the previously noted conservation of proTGFα's cytoplasmic domain, as well as the fact that proHB-EGF is an internalizing receptor for diphtheria toxin (Naglich et al., 1992). An intriguing observation that may also be relevant is that proTGFα was recently coimmunoprecipitated from transfected CHO cells in a complex with proteins of 86 and 106 kDa (Strum et al., 1994). Conceivably, these unidentified proteins, at least one of which appeared to have protein kinase activity, could be involved in reverse signaling following the binding of membrane-anchored growth factor to its cognate receptor. Alternatively, they could be part of the regulated protease system proposed above.

X. EXPRESSION OF TGFα/ProTGFα IN NORMAL TISSUES

TGFα is expressed throughout development, at least, in rodents. Early reports of TGFα-like activity in extracts from midgestation fetuses and/or placenta (Twardzik et al., 1982a, 1985) were extended by (1) the detection of TGFα transcripts in preimplantation blastocysts by RT-PCR (Rappollee et al., 1988a), (2) the *in situ* localization of TGFα transcripts to various tissues of the midgestation (days 9 and 10) mouse fetus including the otic vesicle, oral cavity, pharyngeal pouch, first and second branchial arches, and developing kidneys (Wilcox and Derynck, 1988a), and (3) the detection of TGFα antigen in tissues of late-gestation rat fetuses (Brown et al., 1990; Strandjord et al., 1994). Finally, TGFα expression was also localized to regions of the rodent uterus during the peri-implantation period (days 1–4) (Tamada et al., 1991), and to the decidua (Han et al., 1987) and placenta (Wilcox and Derynck, 1988a) during midgestation.

TGFα protein and/or mRNA have also been detected in numerous adult tissues. Developmentally regulated expression is widespread throughout the rodent *central nervous system* with regional variability in the absolute levels (Wilcox and Derynk, 1988b; Kudlow et al., 1989; Seroogy et al., 1991). Expression is prominent in some neurons, but localization of transcripts to regions of white matter (Seroogy et al., 1993) suggests that glial cells also express the growth factor. Although their role in the CNS has not been clarified, EGF-R transcripts are also present in proliferative zones of the postnatal rat brain (Seroogy et al., 1995), raising the possibility that TGFα/EGF-R interactions serve a mitogenic function. Additionally, TGFα could provide trophic support for neurons. Such a role in the nigrostriatal system would be consistent with the finding of relatively high levels of TGFα and EGF-R mRNAs in the striatum (Seroogy et al., 1993; Kornblum et al., 1994) and midbrain dopaminergic neurons (Seroogy et al., 1994), respectively. Finally, yet other roles have been suggested, for example, the possibility that TGFα might regulate the activity of hypothalamic cells controlling sexual maturation (Ma et al., 1992).

Expression of TGFα is prominent in cells of the *integument*. For example, cultured skin keratinocytes express the growth factor (Coffoy et al., 1987a) and respond to it with increased proliferation and motility (Barrandon and Green, 1987; Coffey et al., 1988). *In vivo* expression of TGFα has been localized to the stratified epidermis, and its deregulated expression has been associated with psoriasis and other hyperproliferative skin lesions (Gottlieb et al., 1988; Elder et al., 1989). Moreover, TGFα has been implicated in skin wound healing (Rappollee et al., 1988b; Schultz et al., 1987, 1992), although it is noteworthy that repair of full-thickness skin incisions was not compromised in the absence of this growth factor (Luetteke et al., 1993b). Recently, TGFα transcripts were localized to the inner root sheath of active hair follicles (Luetteke et al., 1993b), and the loss of the growth factor associated with anomalies in the organization and function of these structures (Luetteke et al., 1993b; Mann et al., 1993). TGFα mRNA in mouse adipose tissue (Crandall et al., 1992; Luetteke et al., 1993a), and overexpression of TGFα in transgenic mice produced significant reductions in adipose tissue weight and cellularity (Luetteke et al., 1993a). These two findings are consistent with a possible role for the growth factor in regulating fat deposition. Finally, TGFα transcripts have been detected in virgin rodent and human mammary glands and in pregnant rodent mammary glands (Liscia et al., 1990; Snedeker et al., 1991). Whether it is also expressed in lactating glands is less clear. Although Liscia and co-workers (1990) found transcripts in cells of both nulliparous and parous human mammary glands, Snedeker and colleagues (1991) did not detect expression in pregnant mouse glands by RT-PCR. Because inappropriate levels of TGFα interfere with normal mammary gland development (Vonderhaar, 1987; Sandgren et al., 1995), the growth factor could have a physiological role in this process.

Expression of TGFα is prominent in tissues of the *digestive tract.* TGFα mRNA and protein have been detected in adult rodent liver at higher levels during the fetal and neonatal periods (Mead and Fausto, 1989; Brown et al., 1990). Expression was most prominent in hepatocytes but was also localized to oval and Ito cells (Evarts et al., 1992). Significantly, the levels of TGFα mRNA and protein were induced after partial hepatectomy (Mead and Fausto, 1989; Evarts et al., 1993), suggesting that TGFα could be a component of the mitogenic stimulus to liver regeneration. Pointing to the possible physiological relevance of observations that the growth factor is a potent regulator of gastric acid secretion (Tam, 1985), TGFα transcripts and protein have been localized to gastric mucosa (Beauchamp et al., 1989; Polk et al., 1991). Additionally, expression has been detected throughout the various segments of the gut (Barnard et al., 1991; Thomas et al., 1992; Miettinen, 1993). Because this expression was most prominently associated with the nonproliferating cellular compartments, the growth factor may have a role in inducing and/or maintaining the differentiated phenotype. Finally, as is the case with EGF, TGFα is found in salivary glands and at relatively high concentrations in saliva (Humphreys–Beher et al., 1994).

Numerous other tissues have been reported to contain TGFα antigen or mRNA. These include the *endocrine* organs pituitary (Kobrin et al., 1987), hypothalamus, thyroid, parathyroid, adrenal cortex, adrenal medulla, and pancreatic islets (Driman et al., 1992), and tissues of the *urinary* (kidney) (Gomella et al., 1989; Brown et al., 1990), *respiratory* (lung) (Brown et al., 1990; Liu et al., 1990; Seroogy et al., 1991) and *reproductive* tracts. In particular, several laboratories have described the presence of TGFα in ovary (Kudlow et al., 1987; Lobb et al., 1989). Whether the growth factor is a so expressed in testes has been controversial (Skinner et al., 1989; Teerds et al., 1990). Finally, transcripts or protein have been detected in the vascular smooth muscle cells of arterioles (Mueller et al., 1990), and in cultured, activated macrophages (Madtes et al., 1988) and those recovered from wound sites (Rappollee et al., 1988b).

It is apparent from the preceding discussion that TGFα should not be considered an "oncofetal" growth factor, but rather a widely disseminated ligand for the EGF-R. Its expression appears to be more widespread than that of EGF. Efforts to understand how its activities mesh with those of the other EGF-R ligands are presently confounded by the lack of information about the expression of the other EGF family members.

XI. EXPRESSION BY NEOPLASTIC CELLS AND TISSUES

Deregulation of autocrine or paracrine growth loops, resulting from either the overproduction of growth factors or the elevated synthesis and/or mutation of their receptors, is a common theme in multistep tumorigenesis. The overproduction of TGFα by human and rodent tumors and by transformed or tumor-derived cell lines is an example. An early broad sampling revealed elevated expression of TGFα mRNA in a variety of primary and metastatic tumors that included squamous carcinomas of the head, neck, and lung, renal and mammary carcinomas, adenocarcinomas of the stomach and endometrium, large cell carcinomas of the lung, and melanomas (Derynck et al., 1987). This initial sampling was subsequently extended by reports from numerous laboratories including one that described the detection of diffuse TGFα immunostaining in the epithelium of 80 of 84 human pancreatic adenocarcinomas (Barton et al., 1991). The potential significance of this observation is underscored by findings that cultured pancreatic cancer cells produce and utilize TGFα as an autocrine growth factor (Smith et al., 1987; Ohmura et al., 1990).

Increased levels of TGFα mRNA and protein have been associated with chemically induced transformation of liver cells *in vitro* (Liu et al., 1988) and *in vivo* (Luetteke et al., 1988), and the tumorigenic index of a series of clonally derived, transformed rat liver epithelial cells correlated with the combined overproduction of TGFα and c-myc (L.W. Lee et al., 1991). Additionally, the proliferation of a transformed rat liver epithelial cell line was inhibited by expression of a TGFα antisense gene (Laird et al., 1994). Consistent with these observations, Yeh and colleagues (1987) found that the average urinary TGFα concentration was 4–5-fold higher in a group of 31 hepatocellular carcinoma (HCC) patients compared to 33 age- and sex-matched controls. As a result, it was suggested that the growth factor might be a useful complementary tumor marker for the management of HCC patients who display low levels of the commonly used marker, α-fetoprotein.

Increased levels of immunoreactive TGFα have also been associated with colorectal neoplasias: Liu and co-workers (1990) found that the average TGFα content of 29 primary colonic tumors was 4-fold higher than that of normal mucosa. However, when normal and tumor samples from the same individual were compared, the levels of TGFα protein were increased from 4 to 260-fold in 12 of the 22 colonic tumors for which matched samples were available. Tanaka and colleagues (1991) also reported that immunoreactive TGFα is commonly present in colorectal neoplasias. Moreover, increased production of the growth factor has been reported for a number of human colon carcinoma cell lines (Coffey et al., 1987b; Hanauske et al., 1987; Anzano et al., 1989), some of which also overexpress EGF-R (Untawale et al., 1993), and its production correlated with the tumorigenicity of clones of the SW613-S human colon carcinoma cell line (Modjtahedi et al., 1992).

Other studies have implicated TGFα in mammary tumorigenesis. An early report described elevated TGFα levels in urine from patients with disseminated breast cancer compared to controls (Stromberg et al., 1987). Although a recent report questioned whether increased production of the growth factor was consistently associated with mammary tumorigenesis, it also suggested that concurrent EGF-R/TGFα immunostaining might characterize a clinically aggressive subset of breast carcinomas (Castellani et al., 1994). Estrogen-regulated production of TGFα by human breast cancer cells has been well documented (Dickson et al., 1987; Liu et al., 1987; Bates et al., 1988), and EGF-R blocking antibodies inhibited the growth of some breast cancer cells in culture (Arteaga et al., 1988a; Ahmed et al., 1991). Finally, the proliferation of an estrogen-inducible breast cancer cell line was inhibited by expression of a transfected TGFα antisense gene (Ready et al., 1994).

Production of TGFα-like activity by lung carcinoma cell lines has been described (Siegfried et al., 1987), and primary lung tumors were found to express variably increased levels of the growth factor relative to normal parenchyma (Liu et al., 1990; Liu and Tsao, 1993). In a study of 131 primary human lung adenocarcinomas, 87 were immunohistochemically positive for TGFα. For those patients whose EGF-R-positive tumors displayed high versus low levels of TGFα staining, the 5-year survival rates were 36% and 85%, respectively (Tateishi et al., 1990).

Finally, elevated production of TGFα has also been demonstrated for endocrine organ (Driman et al., 1992), kidney (Gomella et al., 1989; Mydlo et al., 1989) and ovarian (Sherwin et al., 1983; Arteaga et al., 1988a; Morishige et al., 1991a) tumors. And, in a survey of several different tumor types, TGFα protein was detected by radioimmunoassay at levels from 1.6–50 ng/ml in tumor effusions from 13 of 34 breast cancers, 12 of 24 lung cancers, and 13 of 31 ovarian cancers (Arteaga et al., 1988a). In contrast, it was undetectable in 14 of 17 controls and was present at concentrations less than 2 ng/ml in the three positive samples.

In summary, the results of these and other studies strongly suggest that induction of TGFα frequently accompanies the neoplastic progression of a variety of solid tumors. A caveat to this interpretation is that very few large scale studies have been performed, and the results of cell culture studies must be viewed with a degree of skepticism.

XII. TGFα AND TRANSFORMATION

Several laboratories have reported that deregulated production of TGFα results in the phenotypic transformation of mammalian cell lines. For example, transfected Rat-1 cells, which expressed TGFα under the control of the SV40 early promoter, formed overgrown foci in monolayer and large colonies in soft agar (Rosenthal et al., 1986). Similarly, normal rat kidney (NRK) cells, which expressed either a retrovirally-encoded secreted form of TGFα (Watanabe et al., 1987) or a plasmid-encoded noncleavable proTGFα (Blasband et al., 1990a), displayed transformed-like growth in culture and formed tumors in nude mice. In contrast, Finzi and co-workers (1987 and 1990) found that TGFα did not transform some fibroblast lines. The latter observations were clarified by Di Marco and colleagues (1989), who described a quantitative requirement for EGF-R in the transformation of several rodent cell lines. While overexpression of TGFα alone did not directly transform any of the mouse or rat fibroblast lines examined, the combined expression of retrovirally-encoded TGFα and EGF-R genes produced clones that exhibited the growth characteristics of fully transformed cells. Finally, the ability of TGFα to transform cells *in vitro* is apparently not restricted to fibroblasts because mouse mammary epithelial cell lines have also been transformed by the introduction of TGFα expression vectors (Shankar et al., 1989; McGeady et al., 1989).

Although these various results support the view that deregulated expression of TGFα contributes to neoplastic growth, the possibility that established cell lines may be already partially transformed and, hence, more susceptible to the actions of a growth factor, led several laboratories to examine the consequences of deregulated TGFα expression *in vivo* using transgenic mouse models. The results of these studies further confirmed the importance of TGFα in multistep tumorigenesis, but also showed that the consequences of its deregulated production depend very much on the tissue context. Thus, both Sandgren and colleagues (1990) and Jhappan and co-workers (1990) found that broad expression of a TGFα transgene under the control of the metal-inducible metallothionein-1 (MT) promoter produced a spectrum of tissue-specific phenotypes. For example, regions of the gut generally displayed enhanced growth with the preservation of normal tissue architecture, although the pancreas exhibited abnormal growth that produced tissue lesions but no frank tumors. Expression of MT-TGFα transgenes in liver, on the other hand, was associated with enhanced growth and focal lesions and the development of malignant tumors with an incidence that varied considerably depending on the genetic background of the host mouse strain (Takagi et al., 1992; G.–H. Lee et al., 1992; Sandgren et al., 1993). Finally, expression of the MT-TGFα transgene also produced mammary gland tumors with 100% incidence in postlactational female mice (Sandgren et al., 1990). A similar result was coincidentally reported for mice expressing a mouse mammary tumor virus (MMTV)-TGFα transgene (Matsui et al., 1990; Halter et al., 1992), although in contrast to findings with MT-TGFα mice, tumors were also occasionally observed in male and virgin female mice. Presumably this difference reflects the fact that the MMTV promoter is active in the virgin mammary gland but the MT promoter is not.

In addition to the aforementioned studies, specific targeting of TGFα expression to mouse skin under the control of the keratin promoters has produced abnormalities in tissue architecture and the frequent appearance of benign papillomas, particularly, in areas of irritation or wounding (Vassar and Fuchs, 1991; Dominey et al., 1993). Neoplastic progression of these papillomas has been induced by topical application of phorbol esters. Interestingly, progression was not characterized by the spontaneous mutation of c-Ha-ras that frequently accompanies the progression of chemically induced skin papillomas (Vassar et al., 1992). This finding suggests that deregulated production of TGFα eliminates the need for ras mutations in this context.

The TGFα-mediated induction of both liver and mammary tumors has been further characterized. Interestingly, despite an already high level of transgene expression, most liver tumors overexpressed TGFα transgenes relative to surrounding normal parenchyma (Takagi et al., 1992; Sandgren et al., 1993). A significant fraction of TGFα-induced liver tumors also overexpressed the endogenous c-myc and insulin-like growth factor II genes, but not the EGF-R gene. None of the liver tumors examined contained mutations in either the Ha-ras or Ki-ras genes (Takagi et al., 1992). The finding of c-myc overexpression seems particularly significant given the synergisms between TGFα and c-myc transgenes, described below, and the aforementioned finding that the tumorigenicity of a series of clonally derived transformed rat liver epithelial cells correlated with combined overexpression of the endogenous TGFα and c-myc genes (L.W. Lee et al., 1991).

To further characterize TGFα-mediated mammary tumorigenesis, transgene expression was specifically targeted to the lactating gland using the whey acidic protein (WAP) promoter (Sandgren et al., 1995). This resulted in the appearance of mammary tumors in 100% of postlactational transgenic females. TGFα-induced tumors were relatively differentiated in appearance, the majority being either adenomas or adenocarcinomas. Most of the tumors displayed transplantable growth in syngeneic mice, although metastases to lung were rarely observed. In contrast to TGFα-induced liver tumors, most mammary tumors did not overexpress the transgene, and PCR-SSCP analysis did not reveal evidence of common p53 mutations. Interestingly, a majority of the TGFα-induced tumors contained significantly higher levels of cyclin D1 mRNA than present in nontumorigenic mammary tissue. This observation is of interest because the cyclin D1 gene is commonly amplified and overexpressed in human breast cancers, and its amplification correlates with poor prognosis (Schuuring et al., 1992). Finally, efforts to characterize the genesis of TGFα-induced mammary tumors revealed that deregulated expression of the growth factor induced precocious alveolar development during pregnancy, and also inhibited glandular involution following the cessation of lactogenic stimulation (Sandgren et al., 1995). Because involution, which occurs through an apoptotic pathway (Strange et al., 1992), was similarly inhibited with each successive pregnancy, the mammary glands of multiparous TGFα transgenic females accumulated large numbers of glandular-like structures termed hyperplastic alveolar nodules (HANs). Supporting the hypothesis that HANs might represent preneoplastic lesions in a TGFα-mediated progression pathway are findings that HANs (1) displayed a capacity for growth as evidenced by the fact that some enlarged to the point of becoming confluent, (2) showed occasional evidence of mild dysplasia, (3) bore a striking similarity to the glandular-type tumors that most commonly arose in TGFα transgenic females, and (4) closely resembled structures that have been implicated as preneoplastic lesions in carcinogen-induced rodent mammary cancer. If this hypothesis is correct, TGFα might initiate tumorigenesis in the mammary gland by promoting the inappropriate survival of mammary epithelium. If so, its actions in this context could be likened to those of Bcl-2.

The availability of transgenic models has also allowed the demonstration that deregulated TGFα can dramatically accelerate tumorigenesis induced by other agents, including viral and cellular oncogenes and chemical carcinogens. For example, tumors coexpressing a TGFα transgene and a transgene encoding either SV40 T-antigen or c-myc grew significantly faster than those expressing any one of the transgenes alone (Sandgren et al., 1993; Murakami et al., 1993; Sandgren et al., 1995). This dramatic acceleration of tumor growth was observed in tissues that are directly transformed by TGFα (i.e., liver and mammary gland) and those that are not (i.e., pancreas). It was accompanied by a consistent decrease in the average survival time of dual-transgene versus single-transgene-positive mice. TGFα overexpression also dramatically accelerated chemically induced hepatocarcinogenesis in transgenic males but not females (Takagi et al., 1993). Both genotoxic and nongenotoxic chemicals produced malignant tumors in TGFα transgenic mice with a high incidence several months prior to the time when tumors developed in either untreated TGFα or carcinogen treated nontransgenic, control mice.

Although TGFα's ability to cooperate in tumorigenesis likely reflects its potent growth promoting activity, a variety of evidence suggests that its influence is not restricted to enhanced proliferation. For example, liver tumors in bitransgenic c-myc x TGFα mice were more malignant in appearance, and they displayed a higher transplant rate than those induced by either transgene alone (Sandgren et al., 1993). Moreover, coexpression of WAP-c-myc and WAP-TGFα transgenes produced mammary tumors in virgin and male mice although expression of the WAP-TGFα alone did not (Sandgren et al., 1995). Finally, histological analysis revealed the presence of a greater number of preneoplastic lesions in livers and pancreases of dual transgenic mice (Murakami et al., 1993; Sandgren et al., 1993), suggesting that TGFα might influence early events in neoplastic progression.

Taken together, the results of these various animal studies strongly implicate deregulated TGFα/EGF-R loops in multistep tumorigenesis, and they also provide evidence of a potent collaboration between TGFα and c-myc in neoplastic progression.

XIII. CHIMERIC TGFα PROTEINS AS THERAPEUTIC AGENTS

The fact that many tumor cells overexpress EGF-R has spurred the development of a novel therapeutic approach that is based on the creation of chimeric TGFα/toxin fusion proteins. Chaudhary et al. (1987) synthesized a novel gene encoding mature TGFα sequences fused to a modified *Pseudomonas* toxin lacking its usual cell-recognition domain. The resulting chimera retrained EGF-R-binding activity and efficiently killed human tumor-derived cells harboring high (10^5–10^6) numbers of receptors on their cell surface. In contrast, it did not kill cells with relatively low ($<10^4$) numbers of surface EGF-R molecules. Following this early observation, it was subsequently shown that delivery of a TGFα/*Pseudomonas* exotoxin fusion protein doubled the survival time of nude mice injected with EGF-R-positive tumor cells, but did not affect the survival of those injected with EGF-R-negative tumor cells (Heimbrook et al., 1990). Additionally, the fusion protein delayed the growth of tumors derived from implanted epidermoid and prostate carcinoma cells (Pai et al., 1991). More recent studies have emphasized the design of different TGFα/toxin variants, including one whose ability to kill tumor cells was reportedly enhanced by the addition of a heparin-binding domain (Mesri et al., 1993).

XIV. THE TGFα GENE

The TGFα gene has been localized to a linkage group on the short arm of human chromosome 2 (p11–p13 region) (Brissenden et al., 1985; Tricoli et al., 1986) that is at least partially conserved in the 35-cM region of mouse chromosome 6 (Fowler et al., 1993; Barrow et al., 1994). Genes included within this linkage group are *Igk-Efl, Igk-J, Ly-2, Ly-3, FABPI, Egr4* and *Anx4*. Recent studies have suggested a weak association between the human TGFα locus and genetic susceptibility to cleft lip that presents either with or without cleft palate (Chenevix–Trench et al., 1992; Stoll et al., 1993; Sassani et al., 1993; Farrall et al., 1993). Additionally, the position of the human TGFα gene is close to the breakpoint of the Burkitt's lymphoma t(2:8) variant, but no significance has been ascribed to this observation.

The TGFα gene gives rise to a 4.5–4.8 kilobase (kb) mRNA (Derynck et al., 1984; Lee et al., 1985b). The proTGFα coding sequence is contained within the 5′-most 700 base pairs (bp) and is preceded by a short, G/C-rich 5′ untranslated (UT) sequence. The bulk of the mRNA is comprised of a large 3′ UT that contains an ACA repeat and multiple A/T motifs of unknown significance. An infrequently used alternate polyadenylation site in the middle of the 3′ UT gives rise to a truncated rat cDNA of 2.3 kb that was fortuitously cloned by Lee and colleagues (1985b). The 4.5-kb mRNA is derived from six rat exons that together span 85–90 kb of genomic DNA (Blasband et al., 1990b) (Figure 3). Exon 1 encodes the 5′ UT and the first part of the signal peptide; exon 2 specifies the remainder of the signal peptide; exon 3 encodes the first two disulfide-bonded loops of the mature growth factor; exon 4 encodes the third disulfide loop together with the transmembrane domain; exon 5 specifies the cytoplasmic sequence; and exon 6 encodes the entire 3′ UT. Interestingly, the exonic organization of the EGF-like sequence (namely, the fact that the three disulfide loops are encoded by two exons, with the second exon encoding the third loop together with the transmembrane domain) is characteristic of all five EGF-R ligands; it distinguishes them from EGF-related sequences in proteins that do not serve as receptor ligands because these are variably encoded by one or more exons (Bell et al., 1986).

Two other features of the TGFα gene are notable. First, the presence of a tandem CAG donor sequence at the 3′ end of the second intron results in microheterogeneous splicing and the production of both 159 and 160 amino acid forms of proTGFα (Kudlow et al., 1989; Blasband et al., 1990b). However, no functional distinctions have been ascribed to the two forms. Second, exon 6 of the human and rat TGFα genes encodes a small antisense transcript of 0.2–0.35 kb (Jakowlew et al., 1988; Blasband et al., 1990b). The rat transcript includes an open reading frame of 51 amino acids that lacks an apparent signal peptide but contains an unusual asn-gln-$(asn)_7$ sequence and a putative N-glycosylation site. The human RNA was found to be inversely expressed relative to TGFα mRNA in a limited tissue and cell survey, leading to speculation that it might function as a natural antisense regulator of TGFα expression (Jakowlew et al., 1988). In contrast, the rat transcript is constitutively expressed in cells and tissues that contain markedly different levels of TGFα mRNA.

XV. THE TGFα PROMOTER

Like those of so-called housekeeping genes, the TGFα promoter lies within a conserved region of high guanine/cytosine G/C content that lacks recognizable TATA and CAAT motifs (Jakobovits et al., 1988; Blasband et al., 1990b). Under the influence of this promoter, transcription of the human and rat genes starts at a prominent site located around -60 (all positions are relative to the first base of the initiating ATG codon). However, 5′ ends of rat TGFα mRNA have also been mapped to numerous sites in the region from -300 to -150 by both primer extension and S1 nuclease analyses (Blasband et al., 1990b). Although most of these appear to correspond to minor initiation sites, a second prominent start site is located within the cluster at approximately -200.

A variety of evidence implicates the transcription factor Sp1 in the regulation of TGFα promoter activity (Chen et al., 1992, 1994). These include findings that (1) a CAT reporter gene under the control of the promoter was efficiently expressed in *Drosophila* Schneider cells only when those cells were cotransfected with an Sp1 expression vector, (2) oligonucleotides corresponding to putative Sp1 elements in the promoter region displayed similar gel-shift activity in the presence of either crude cell nuclear extract or purified Sp1 protein, (3) *in vivo* footprinting methodology revealed protein binding at some Sp1-like sites in transformed cells that actively express the TGFα gene, and (4) site-directed mutagenesis of individual Sp1 elements reduced transcription from the major start sites *in vitro*, with the mutation of multiple sites having a combinatorial effect. Findings consistent with these have been reported for the human promoter by Shin and co-workers (1992) who focused on the mechanism(s) by which 5-azaC treatment induced expression of a TGFα promoter/reporter gene in transfected cells. They showed that 5-azaC induction correlated with enhanced formation of a promoter DNA/protein complex containing immunoreactive Sp1 protein and that DNase 1 footprinting of this complex revealed a footprint indistinguishable from that produced by addition of pure Sp1 protein. Moreover, extracts from 5-azaC-treated cells produced Sp1-like gel-shifts with promoter fragments containing putative Sp1 recognition sites. Because 5-azaC treatment did not alter the levels of Sp1 mRNA or protein, the investigators suggested that the drug may enhance the accessibility of Sp1 protein to sites on the DNA.

Two facts related to the aforementioned findings are noteworthy. First, Sp1 is part of a family of related proteins that have either stimulatory or inhibitory activity (Kingsley and Winoto, 1992; Majello et al., 1994); hence, the possibility exists that other members of the family have roles in regulating TGFα gene transcription. Second, other G/C box binding proteins may also have roles. For example, episomal expression of GC factor (GCF), a protein originally implicated in repression of the EGF-R promoter, inhibited expression of both endogenous TGFα mRNA and a cotransfected TGFα promoter/reporter construct (Kitadi et al., 1993).

XVI. REGULATION OF TGFα GENE EXPRESSION

Numerous reports have described increased levels of TGFα protein and/or mRNA in transformed cells compared to those in counterpart normal cells. These include cells transformed by retroviruses (Twardzik et al., 1982c, 1983) and chemicals (Luetteke et al., 1988; Liu et al., 1988; L.W. Lee et al., 1991). Additionally, increased expression of TGFα has been associated with transformation induced by direct introduction of activated oncogenes (Ciardiello et al., 1990c). In particular, the results of several investigations have implied a link between ras activation and induced expression of TGFα. For example, cells transformed with a temperature-sensitive Kirsten murine sarcoma virus were found to secrete TGFα only at the permissive temperature (Kaplan et al., 1982). In addition, the levels of endogenous TGFα mRNA were increased in Rat-1 cells that were induced to express a stably transfected MT-c-Ha-ras gene (Godwin and Lieberman, 1990). Evidence linking ras and the induction of TGFα expression has also been provided for mammary (Salomon et al., 1987; Ciardiello et al., 1988) and intestinal (Filmus et al., 1993) epithelial cells. It may be noteworthy that the 5′ flanking region of the rat TGFα gene contains the sequence GTGACTAA at -460 because this sequence has been implicated as a ras response element within the polyoma virus enhancer (Imler et al., 1988). However, its function in the context of the TGFα promoter has not been established.

Inductions of TGFα mRNA have been widely observed in response to treating various cell types with growth factors, phorbol esters (e.g., TPA), and steroid hormones. Among the growth factors that regulate expression of this gene are members of the EGF-R ligand family including TGFα itself which apparently autoregulates its own production (Coffey et al., 1987a; Pittelkow et al., 1989; Mueller et al., 1989; Raymond et al., 1989; Bjorge et al., 1989a). However, phorbol ester has produced some of the most dramatic reported inductions of TGFα mRNA in cell culture. For example, transcript levels were increased 8 to 10-fold in a transformed rat liver epithelial cell line (GP6ac) 12 hrs after the addition of 100-nM TPA (Raymond et al., 1989). Nuclear run-on analyses have shown that the induction in some cell lines results from increased gene transcription, although stabilization of the mRNA may also play a role (Raymond et al., 1989; Coffey et al., 1992; Gaido et al., 1992). With some cell lines, inhibition or downregulation of protein kinase C (PKC) has blunted the induction by both TPA and EGF or TGFα (Bjorge et al., 1989b), although, in one case, only the TPA induction was inhibited (Klein et al., 1992). Other supporting evidence for a PKC-dependent pathway includes findings that TGFα mRNA levels were enhanced in both human keratinocytes exposed to diacylglycerol (Pittelkow et al., 1989) and GP6ac cells treated with hormones (e.g., angiotensin, epinephrine, and bradykinin) that are known to activate PKC in this cell type (Raymond et al., 1989). RNA and protein synthesis inhibitors have also been used to further characterize the mechanism of induction; however, the effects of these inhibitors are difficult to summarize because they have apparently depended on both the timing of their addition (relative to that of the inducer) and the cell type examined (Pittelkow et al., 1989; Raymond et al., 1989; Glinsmann–Gibson and Korc, 1991; Coffey et al., 1992). Finally, Raja et al. (1991) have, at least, partially reproduced TPA regulation with a transfected TGFα promoter/reporter construct, though the response element(s) has not yet been clearly defined. In summary, these data indicate that, although there may be cell-type differences in the mechanisms, production of TGFα can be regulated and autoregulated through PKC-dependent pathways. Additionally, they suggest that tumor-promoting agents could contribute to neoplastic progression in part by inducing expression of this growth factor. Finally, the phenomenon of auto- or cross-induction is not restricted to TGFα because the expression of other EGF-R ligands is similarly regulated (Barnard et al., 1994).

The list of polypeptide growth factors that also influence TGFα expression includes TGFβ, which has been shown to increase or decrease TGFα mRNA levels depending on the cell type (Mueller and Kudlow, 1991; Zipfel et al., 1993; Lynch et al., 1993). Paradoxically, TGFβ has simultaneously induced expression of the growth factor and inhibited cellular proliferation. The TGFβ responsiveness of TGFα promoter/reporter constructs in transfected human colon adenocarcinoma cells has been localized to sequences in the -201 to -77 region. In addition to TGFβ, other growth regulators, that reportedly influence TGFα expression or promoter activity, include keratinocyte growth factor (KGF) (Dlugosz et al., 1994), tumor necrosis factor-α (Schmiegel et al., 1993; Kalthoff et al., 1993), interleukin-1 (S.W. Lee et al., 1991), and γ-interferon (Hamburger and Pinnamaneni, 1993).

The regulation of TGFα expression by steroid hormones has been widely observed. A number of laboratories have described severalfold inductions of TGFα protein and/or mRNA in human breast cancer cells (Dickson et al., 1987; Arteaga et al., 1988a; Bates et al., 1988; Murphy and Dotzlaw, 1989; Musgrove et al., 1991; Murphy et al., 1991) or human endometrial adenocarcinoma cells (Gong et al., 1991) treated with estrogen or progesterone, and more dramatic inductions have been observed in estrogen-treated uterus (Nelson et al., 1992) and anterior pituitary gland (Borgundvaag et al., 1992). These observations have been interpreted to suggest that TGFα might be a local mediator of estrogen's growth-promoting actions. An 1140-bp fragment of the human TGFα 5′ flanking region was shown to direct a 2 to 10-fold increase in CAT activity (or 30 to 300-fold increases in luciferase activity) in estrogen-responsive, but not unresponsive, cell lines (Saeki et al., 1991). Subsequent deletion analysis further localized estrogen-responsive sequences to the region from -343 to -77.

Although less well studied, regulation of TGFα expression or TGFα promoter activity has been observed in certain cell types in response to high cell density (Pittelkow et al., 1993) or the addition of HIV-1 *tat* protein (Nabell et al., 1994), glucose/glucosamine (McClain et al., 1992), or dioxin (Choi et al., 1991).

To summarize, although it appears to be constitutively expressed in a wide array of tissues and cell types, production of TGFα can nevertheless be influenced by a variety of agents. Of particular concern, from a clinical perspective, are the inductions that frequently accompany neoplastic progression. In the light of the attention focused on this growth factor for the past several years, it is surprising that so little is yet known about the mechanisms that regulate the production of its mRNA.

XVII. DEVELOPMENT OF TGFα KNOCKOUT MICE

A prominent, recent development has been the derivation in two independent laboratories of knockout mice that do not express any functional TGFα protein (Luetteke et al., 1993b; Mann et al., 1993). Despite the widespread expression of this EGF-R ligand in developing and adult rodent tissues (see the preceding section), its loss did not adversely affect the viability or fertility of these mice. Nevertheless, two phenotypes were associated with the homozygous null mutation. The most conspicuous was a pronounced waviness of the whiskers and fur most evident during the first hair growth cycle but persisting throughout the life of the mouse. All of the hair types (guards, awls, etc.) were affected; individual hairs were curled and kinky with frequent irregularities in both the caliber of the hair shaft and the internal deposition of pigmented medullary cells. Underlying these effects on hair morphology was a dramatic disorganization of active hair follicles. In contrast to follicles in wild-type skin, which were precisely aligned at a consistent depth in the dermis, hair follicles in targeted mice were positioned throughout the fat layer. Occasionally, follicles with erupting hair shafts were seen growing next to, and parallel with, the subcutaneous muscle layer; some of these follicles displayed dramatic bends towards the epidermis. Moreover, hair follicles in TGFα-deficient skin were often variably oriented with respect to one another, with some sections containing adjacent follicles in oblique, transverse, and longitudinal sections. However, it should be noted that, despite their deviant shape and patterning, the internal structure of the hair follicles was not obviously altered in homozygous null skin.

The preceding description implies that TGFα could have distinct roles in regulating hair growth, on the one hand, and the three-dimensional organization of hair follicles on the other. Inasmuch as hair growth is thought to be regulated by interactions between the inner and outer root sheaths of the hair shaft, it appears significant that expression of TGFα (Luetteke et al., 1993b) and EGF-R (Green et al., 1984; Luetteke et al., 1994) have been localized to the inner and outer root sheaths, respectively. In the case of TGFα, transcripts were present at highest levels in the differentiation zone and were not detectable in the proliferative bulb of the follicle. EGF-R transcripts, on the other hand, were distributed throughout the follicle. These data suggest a model in which interactions between TGFα and EGF-R facilitate the controlled upward movement of the hair shaft. Additionally, the aberrant shape of some follicles could also contribute to irregularities in hair shaft calliber and septulation. How TGFα might regulate the positioning of hair follicles is less clear, but formation of these structures during development is dependent on a series of inductive signals passed between epithelial and mesenchymal cells. That TGFα might have a role in this signaling is an intriguing possibility.

In addition to the skin phenotype, which exhibited 100% penetrance in homozygous null mice, loss of TGFα also produced eye problems of variable incidence and severity. Of the homozygous null mice initially characterized by Luetteke and colleagues (1993b), 40% displayed some form of ocular anomaly that ranged from mild corneal opacity to reduced globe size (microphthalmia) and partially or completely shut eyes. Closer examination revealed frequent corneal scarring, neovascularization, and abundant exudate. Affected eyes were characterized histologically by a spectrum of defects that included corneal ulceration, anterior segment dysgenesis, the absence of a lens, and retinal dysplasia evidenced in part by incomplete differentiation of the iris. Additionally, the eyelids of affected mice were frequently malformed; they lacked organized, differentiated structures and showed evidence of marked epithelial hyperplasia. Although the eye anomalies of homozygous null mice were generally more severe, 13% of heterozygous null mice also displayed eye problems. Moreover, the severe pathology displayed by homozygous mice was often unilateral. Currently, there are no explanations for these findings or the observation that eye anomalies were incompletely penetrant among homozygous null mice. However, it may be significant that those mice destined to develop severe eye problems were first recognized by the phenotype of open eye at birth. Recent unpublished work in the author's laboratory has shown that eyelid fusion, which is normally complete 3–4 days prior to birth, is significantly delayed in TGFα homozygous null mice. Although the majority of affected eyes still close prior to birth, a subset does not, perhaps, as a result of the aforementioned eyelid defects. Hence, the spectrum of eye problems observed in adult mice may represent primary and secondary sequella to premature aeration, the lack of lubricating fluids, etc.

The conclusions from the TGFα knockout mice have been unexpectedly corroborated by the recent elucidation of defects underlying two mutant mouse phenotypes first described over 60 years ago. Both waved-1 (wa-1) and waved-2 (wa-2) mice display hair and eye defects that are strikingly similar to those of TGFα knockout mice. Adult mice from both lines are characterized by pronounced waviness of the hair and whiskers, and some wa-1 and wa-2 pups are born with open eyes (Crew, 1933; Keeler, 1935; Bennett and Gresham, 1956). Significantly, the wa-1 and wa-2 mutations have been colocalized to the same regions of mouse chromosomes 6 and 11 that harbor the TGFα and EGF-R genes, respectively. Moreover, TGFα and wa-1 have been shown to act as alleles in matings between homozygous TGFα null and wa-1 mice (Luetteke et al., 1993b; Mann et al., 1993). Recent studies have confirmed that wa-1 mice produce abnormally low (i.e., roughly 10-fold reduced) levels of apparently normal TGFα mRNA and protein (Luetteke et al., 1993b; Mann et al., 1993). The molecular basis of this diminished production is unknown, but the nucleotide sequence of the TGFα promoter region is identical in DNAs from wild-type and wa-1 mice, and the two promoters direct comparable levels of reporter gene expression in transfected cells (unpublished data). Additionally, Southern blots of pulse-field gels have not revealed evidence of chromosomal alterations within a 1000 kilobase region of genomic DNA spanning the TGFα gene.

In contrast to the apparent basis of the TGFα defect in wa-1 mice, EGF-R expression is not diminished in tissues of wa-2 mice, and the binding of ligand (e.g., EGF) to the wa-2 receptor is not obviously altered. Instead, the ability of the wa-2 receptor to phosphorylate either itself (autophosphorylation) or synthetic substrates is greatly diminished, at least, *in vitro* (Luetteke et al., 1994; Fowler et al., 1995). Consistent with these findings, a comparison of the complete coding regions of wild-type and wa-2 EGF-R cDNAs revealed a single nucleotide discrepancy in the amino-terminal portion of the tyrosine kinase domain that would result in the substitution of a glycine for a valine residue in the wild-type protein. Significantly, a similarly positioned valine is either conserved or conservatively replaced in a large group of receptor and nonreceptor tyrosine kineses. Confirming the importance of this substitution, the analogous mutation of an otherwise wild-type protein produced an EGF-R that, when expressed in transfected cells, displayed markedly reduced tyrosine kinase activity. Interestingly, the effect of the spontaneous wa-2 mutation was less apparent when EGF-dependent phosphosphorylation of either EGF-R or downstream substrates was examined in intact tissues; inhibition of tyrosine kinase activity was overcome by relatively small increases in the dose of exogenous EGF.

Collectively, the combined characterization of deliberately targeted and spontaneously derived mutant mouse lines has established the physiological importance of TGFα/EGF-R interactions in skin and eyes. Inasmuch as TGFα and EGF-R are broadly expressed in developing and adult tissues, it is surprising that phenotypes resulting from the loss of either component are restricted to these two tissues. However, the redundancy of EGF-R ligands and the fact that EGF-R activity is only partially inhibited in wa-2 tissues offer logical explanations. In any event, the generation or identification of TGFα-deficient and EGF-R-impaired mice, together with the forthcoming development of knockout mice for other EGF-R homologs and ligands, should provide invaluable tools for elucidating the roles of these various signaling components in normal and neoplastic biology.

XVIII. CONCLUSIONS

The fifteen years since its discovery have seen an explosion of information about TGFα. Nevertheless, many fundamental questions remain to be addressed. A small sampling includes these problems: What are TGFα's precise physiological roles, and how do these compare with those of the related EGF-R ligands? Does proTGFα have distinct physiological roles, and how is its processing regulated? What are the mechanisms that regulate TGFα expression at the transcriptional and translational levels, and in particular, what factors mediate the sometimes dramatic induction of TGFα gene activity in neoplastic cells? Hopefully, with the tools and animal models now developed, the next 15 years will see some of these questions answered.

ACKNOWLEDGMENTS

I thank Kim Seroogy for helpful comments and Mary Davis for assistance in preparing the manuscript. This work was supported by grants CA43793 and CA61896 from the National Institutes of Health.

REFERENCES

Ahmed, S. R., Badger, B., Wright, C., & Manni, A. (1991). Role of transforming growth factor α (TGF-α) in basal and hormone-stimulated growth by estradiol, prolactin, and progesterone in human and rat mammary tumor cells: Studies using TGF-α and EGF receptor antibodies. J. Steroid Biochem. Mol. Biol. 38, 687–693.

Anklesaria, P., Teixidó, J., Lalho, M., Pierce, J. H., Greenberger, J. S., & Massagué, J. (1990). Cell–cell adhesion mediated by binding of membrane-anchored transforming growth factor α to epidermal growth factor receptors promotes cell proliferation. Proc. Natl. Acad. Sci. USA 87, 3289–3293.

Anzano, M. A., Rieman, D., Prichett, W., Bowen-Pope, D. F., & Greig, R. (1989). Growth factor production by human colon carcinoma cell lines. Cancer Res. 49, 2898–2904.

Anzano, M. A., Roberts, A. B., Meyers, C. A., Komoriya, A., Lamb, L. C., Smith, J. M., & Sporn, M.B. (1982). Synergistic interaction of two classes of transforming growth factors from murine sarcoma cells. Cancer Res. 42, 4776–4778.

Anzano, M. A., Roberts, A. B., Smith, J. M., Sporn, M. B., & De Larco, J. E. (1983). Sarcoma growth factor from conditioned medium of virally transformed cells is composed of both type α and type β transforming growth factors. Proc. Natl. Acad. Sci. USA 80, 6264–6268.

Arteaga, C. L., Coronado, E., & Osborne, C. K. (1988b). Blockade of the epidermal growth factor receptor inhibits transforming growth factor α-induced but not estrogen-induced growth of hormone-dependent human breast cancer. Mol. Endocrinol. 2, 1064–1069.

Arteaga, C. L., Hanauske, A. R., Clark, G. M., Osborne, C. K., Hazarika, P., Pardue, R. L., Tio, F., & Von Hoff, D. D. (1988a). Immunoreactive α transforming growth factor activity in effusions from cancer patients as a marker of tumor burden and patient prognosis. Cancer Res. 48, 5023–5028.

Barnard, J. A., Gravesdeal, R., Pittelkow, M. R., Dubois, R., Cook, P., Ramsey, G. W., Bishop, P. R., Damstrup, L., & Coffoy, R. J. (1994). Auto- and cross-induction within the mammalian epidermal growth factor-related peptide family. J. Biol. Chem. 269, 22817–22822.

Barnard, J. A., Polk, W. H., Moses, H. L., & Coffey, R. J. (1991). Production of transforming growth factor α by normal rat small intestine. Am. J. Physiol. 261, C994–C1000.

Barrandon, Y. & Green, H. (1987). Cell migration is essential for sustained growth of keratinocyte colonies: The roles of transforming growth factor α and epidermal growth factor. Cell 50, 1131–1137.

Barrow, L. L., Simin, K., Jones, J. M., Lee, D. C., & Meisler, M. H. (1994). Conserved linkage of early growth response 4, annexin 4, and transforming growth factor α on mouse chromosome 6. Genomics 19, 388–390.

Barton, C. M., Hall, P. A., Hughes, C. M., Gullick, W. J., & Lemoine, N. R. (1991). Transforming growth factor α and epidermal growth factor in human pancreatic cancer. J. Pathol. 163, 111–116.

Bates, S. E., Davidson, N. E., Valverius, E. M., Freter, C. E., Dickson, R. B., Tam, J. P., Kudlow, J. E., Lippman, M. E., & Salomon, D. S. (1988). Expression of transforming growth factor α and its messenger ribonucleic acid in human breast cancer: Its regulation by estrogen and its possible functional significance. Mol. Endocrinol. 2, 543–555.

Beauchamp, R. D., Barnard, J. A., McCutchen, C. M., Cherner, J. A., & Coffey, R. J. (1989). Localization of transforming growth factor α and its receptor in gastric mucosal cells. J. Clin. Invest. 84, 1017–1023.

Bell, G. I., Fong, N. M., Stempien, M. M., Wormsted, M. A., Caput, D., Ku, L., Urdea, M. S., Rall, L. B., & Sanchez–Pescador, R. (1986). Human epidermal growth factor precursor: cDNA sequence, expression *in vitro* and gene organization. Nucleic Acids Res. 14, 8427–8446.

Bennett, J. H. & Gresham, G. A. (1956). A gene for eyelids open at birth in the house mouse. Nature 178, 272–273.

Bier, E., Jan, L. Y., & Jan, Y. N. (1990). *rhomboid*, a gene required for dorsoventral axis establishment and peripheral nervous system development in *Drosophila melanogaster*. Genes & Dev. 4, 190–203.

Bjorge, J. D., Kudlow, J. E., Mills, G. B., & Paterson, A. J. (1989b). Inhibition of stimulus-dependent epidermal growth factor receptor and transforming growth factor α mRNA accumulation by the protein kinase C inhibitor staurosporine. FEBS Lett. 243, 404–408.

Bjorge, J. D., Paterson, A. J., & Kudlow, J. E. (1989a). Phorbol ester or epidermal growth factor (EGF) stimulates the concurrent accumulation of mRNA for the EGF receptor and its ligand transforming growth factor α in a breast cancer cell line. J. Biol. Chem. 264, 4021–4027.

Blasband, A. J., Gilligan, D. M., Winchell, L. F., Luetteke, N. C., Rogers, K. T., & Lee, D. C. (1990a). Expression of the TGFα integral membrane precursor induces transformation of NRK cells. Oncogene 5, 1213–1221.

Blasband, A. J., Rogers, K. T., Chen, X., Azizkhan, J. C., & Lee, D. C. (1990b). Characterization of the rat transforming growth factor α gene and identification of promoter sequences. Mol. Cell. Biol. 10, 2111–2121.

Blomquist, M. C., Hunt, L. T., & Barker, W. C. (1984). Vaccinia virus 19-kilodalton protein: Relationship to several mammalian proteins, including two growth factors. Proc. Natl. Acad. Sci. USA 81, 7363–7367.

Borgundvaag, B., Kudlow, J. E., Mueller, S. G., & George, S. R. (1992). Dopamine receptor activation inhibits estrogen-stimulated transforming growth factor α gene expression and growth in anterior pituitary, but not in uterus. Endocrinology 130, 3453–3458.

Bosenberg, M. W., Pandiella, A., & Massagué, J. (1992). The cytoplasmic carboxy-terminal amino acid specifies cleavage of membrane TGFα into soluble growth factor. Cell 71, 1157–1165.

Bosenberg, M. W., Pandiella, A., & Massagué, J. (1993). Activated release of membrane-anchored TGFα in the absence of cytosol. J. Cell Biol. 122, 95–101.

Brachmann, R., Lindquist, P. B., Nagashima, M., Kohr, W., Lipari, T., Napier, M., & Derynck, R. (1989). Transmembrane TGFα precursors activate EGF/TGFα receptors. Cell 56, 691–700.

Brannan, C. I., Lyman, S. D., Williams, D. E., Eisenman, J., Anderson, D. M., Cosman, D., Bedell, M. A., Jenkins, N. A., & Copeland, N. G. (1991). Steel–Dickie mutation encodes a c-kit ligand lacking transmembrane and cytoplasmic domains. Proc. Natl. Acad. Sci. USA 88, 4671–4674.

Bringman, T. S., Lindquist, P. B., & Derynck, R. (1987). Different transforming growth factor α species are derived from a glycosylated and palmitoylated transmembrane precursor. Cell 48, 429–440.

Brissenden, J. E., Derynck, R., & Francke, U. (1985). Mapping of transforming growth factor α gene on human chromosome 2 close to the breakpoint of the Burkitt's lymphoma t(2;8) variant translocation. Cancer Res. 45, 5593–5597.

Brown, J. P., Twardzik, D. R., Marquardt, H., & Todaro, G. J. (1985). Vaccinia virus encodes a peptide homologous to epidermal growth factor and transforming growth factor. Nature 313, 491–492.

Brown, P. I., Lam, R., Lakshmanan, J., & Fisher, D. A. (1990). Transforming growth factor α in developing rats. Am. J. Physiol. 256–260.

Brown, S. B., Krause, D., & Ellem, K. A. O. (1993). Low fluences of ultraviolet irradiation stimulate HeLa cell surface aminopeptidase and candidate "TGFαase" activity. J. Cell. Biochem. 51, 102–115.

Buller, R. M. L., Chakrabarti, S., Moss, B., & Fredrickson, T. (1988). Cell proliferative response to vaccinia virus is mediated by VGF. Virology 164, 182–192.

Carpenter, G., Stoscheck, C. M., Preston, Y. A., & DeLarco, J. E. (1983). Antibodies to the epidermal growth factor receptor block the biological activities of sarcoma growth factor. Proc. Natl. Acad. Sci. USA 80, 5627–5630.

Carpenter, G. & Wahl, M. I. (1990). The epidermal growth factor family. In: Peptide Growth Factors and Their Receptors I (Sporn, M. B. & Roberts, A. B., eds.), Springer-Verlag, Berlin, pp. 69–171.

Carpenter, G. (1992). Receptor tyrosine kinase substrates: *src* homology domains and signal transduction. FASEB J. 6, 3283–3289.

Carraway, K. L., III & Cantley, L. C. (1994). A neu acquaintance for ErbB3 and ErbB4: A role for receptor heterodimerization in growth signaling. Cell 78, 5–8.

Castellani, R., Visscher, D. W., Wykes, S., Sarkar, F. H., & Crissman, J. D. (1994). Interaction of transforming growth factor α and epidermal growth factor receptor in breast carcinoma. Cancer 73, 344–349.

Chang, W., Macaulay, C., Hue, S–L., Tam, J. P., & McFadden, G. (1990). Tumorigenic poxviruses: Characterization of the expression of an epidermal growth factor-related gene in shope fibroma virus. Virology 179, 926–930.

Chang, W., Upton, C., Hue, S–L., Purchio, A. F., & McFadden, G. (1987). The genome of the Shope fibroma virus, a tumorigenic poxvirus, contains a growth factor gene with sequence similarity to those encoding epidermal growth factor and transforming growth factor α. Mol. Cell. Biol. 7, 535–540.

Chaudhary, V. K., Fitzgerald, D. J., Adhya, S., & Pastan, I. (1987). Activity of a recombinant fusion protein between transforming growth factor type α and *Pseudomonas* toxin. Proc. Natl. Acad. Sci. USA 84, 4538–4542.

Chen, X., Azizkhan, J. C., & Lee, D. C. (1992). The binding of transcription factor Sp1 to multiple sites is required for maximal expression from the rat transforming growth factor α promoter. Oncogene 7, 1805–1815.

Chen, X., Wright, K. L., Berkowitz, E. A., Azizkhan, J. C., Ting, J. P.–Y., & Lee, D. C. (1994). Protein interactions at Sp1-like sites in the TGFα promoter as visualized by *in vivo* footprinting. Oncogene 9, 3179–3187.

Chenevix–Trench, G., Jones, K., Green, A. C., Duffy, D. L., & Martin, N. G. (1992). Cleft lip with or without cleft palate: Associations with transforming growth factor α and retinoic acid receptor loci. Am. J. Hum. Genet. 51, 1377–1385.

Cheng, H–J. & Flanagan, J. G. (1994). Transmembrane kit ligand cleavage does not require a signal in the cytoplasmic domain and occurs at a site dependent on spacing from the membrane. Mol. Biol. Cell 5, 943–953.

Choi, E. J., Toscano, D. G., Ryan, J. A., Riedel, N., & Toscano, W. A., Jr. (1991). Dioxin induces transforming growth factor α in human keratinocytes. J. Biol. Chem. 266, 9591–9597.

Ciardiello, F., Kim, N., Hynes, N., Jaggi, R., Redmond, S., Liscia, D. S., Sanfilippo, B., Merlo, G., Callahan, R., Kidwell, W. R., & Salomon, D. S. (1988). Induction of transforming growth factor α expression in mouse mammary epithelial cells after transformation with a point-mutated c-Ha-*ras* protooncogene. Mol. Endocrinol. 2, 1202–1216.

Ciardiello, F., Valverius, E. M., Colucci–D'Amato, G. L., Kim, N., Bassin, R. H., & Salomon, D. S. (1990c). Differential growth factor expression in transformed mouse NIH-3T3 cells. J. Cell. Biochem. 42, 45–57.

Ciccodicola, A., Dono, R., Obici, S., Simeone, A., Zollo, M., & Persico, M. G. (1989). Molecular characterization of a gene of the EGF family expressed in undifferentiated human NTERA2 teratocarcinoma cells. EMBO J. 8, 1987–1991.

Clifford, R. J. & Schüpbach, T. (1992). The *torpedo* (DER) receptor tyrosine kinase is required at multiple times during *Drosophila* embryogenesis. Development 115, 853–872.

Coffey, R. J., Derynck, R., Wilcox, J. N., Bringman, T. S., Goustin, A. S., Moses, H. L., & Pittelkow, M. R. (1987a). Production and autoinduction of transforming growth factor α in human keratinocytes. Nature 328, 817–820.

Coffey, R. J., Jr., Goustin, A. S., Soderquist, A. M., Shipley, G. D., Wolfshohl, J., Carpenter, G., & Moses, H. L. (1987b). Transforming growth factor α and β expression in human colon cancer lines: Implications for an autocrine model. Cancer Res. 47, 4590–4594.

Coffey, R. J., Jr., Graves–Deal, R., Dempsey, P. J., Whitehead, R. H., & Pittelkow, M. R. (1992). Differential regulation of transforming growth factor α autoinduction in a nontransformed and transformed epithelial cell. Cell Growth Diff. 3, 347–354.

Coffey, R. J., Sipes, N. J., Bascom, C. C., Graves–Deal, R., Pennington, C. Y., Weissman, B. E., & Moses, H. L. (1988). Growth modulation of mouse keratinocytes by transforming growth factors. Cancer Res. 48, 1596–1602.

Crandall, D. L., Gordon, G., Herzlinger, H. E., Saunders, B. D., Zolotor, R. C., Cervoni, P., & Kral, J. G. (1992). Transforming growth factor α and atrial natriuretic peptide in white adipose tissue depots in rats. Eur. J. Clin. Invest. 22, 676–680.

Crew, F. A. (1933). Waved: An autosomal recessive coat form character in the mouse. J. Genet. 27, 95–96.

Darlak, K., Franklin, G., Woost, P., Sonnenfeld, E., Twardzik, D., Spatola, A., & Schultz, G. (1988). Assessment of biological activity of synthetic fragments of transforming growth factor α. J. Cell. Biochem. 36, 341–352.

De Larco, J. E. & Todaro, G. J. (1978). Growth factors from murine sarcoma virus-transformed cells. Proc. Natl. Acad. Sci. USA 75, 4001–4005.

Defeo–Jones, D., Tai, J. Y., Vuocolo, G. A., Wegrzyn, R. J., Schofield, T. L., Riemen, M. W., & Oliff, A. (1989). Substitution of lysine for arginine at position 42 of human transforming growth factor α eliminates biological activity without changing internal disulfide bonds. Mol. Cell. Biol. 9, 4083–4086.

Defeo–Jones, D., Tai, J. Y., Wegrzyn, R. J., Vuocol, G. A., Baker, A. E., Payne, L. S., Garsky, V. M., Oliff, A., & Riemen, M. W. (1988). Structure-function analysis of synthetic and recombinant derivatives of transforming growth factor α. Mol. Cell. Biol. 8, 2999–3007.

Derynck, R., Jarrett, J. A., Chen, E. Y., Eaton, D. H., Bell, J. R., Assoian, R. K., Roberts, A. B., Sporn, M. B., & Goeddel, D. V. (1985). Human transforming growth factor—beta cDNA sequence and expression in tumor cell lines. Nature 316, 701–705.

Derynck, R., Goeddel, D. V., Ullrich, A., Gutterman, J. U., Williams, R. D., Bingman, T. S., & Berger, W. H. (1987). Synthesis of mRNAs for transforming growth factors α and β and the epidermal growth factor receptor by human tumors. Cancer Res. 47, 707–712.

Derynck, R., Roberts, A. B., Winkler, M. E., Chen, E. Y., & Goeddel, D. V. (1984). Human transforming growth factor α precursor structure and expression in *E. coli*. Cell 38, 287–297.

Di Marco, E., Pierce, J. H., Fleming, T. P., Kraus, M. H., Molloy, C. J., Aaronson, S. A., & Di Fiore, P. P. (1989). Autocrine interaction between TGFα and the EGF-receptor: Quantitative requirements for induction of the malignant phenotype. Oncogene 4, 831–838.

Dickson, R. B., Kasid, A., Huff, K. K., Bates, S. E., Knabbe, C., Bronzert, D., Gelmann, E. P., & Lippman, M. E. (1987). Activation of growth factor secretion in tumorigenic states of breast cancer induced by 17β-estradiol or v-Ha-*ras* oncogene. Proc. Natl. Acad. Sci. USA 84, 837–841.

Dlugosz, A. A., Cheng, C., Denning, M. F., Dempsey, P. J., Coffey, R. J. Jr., & Yuspa, S. H. (1994). Keratinocyte growth factor receptor ligands induce transforming growth factor α expression and activate the epidermal growth factor receptor signaling pathway in cultured epidermal keratinocytes. Cell Growth & Diff. 5, 1283–1292.

Dominey, A. M., Wang, X–J., King, L. E., Jr., Nanney, L. B., Gagne, T. A., Sellheyer, K., Bundman, D. S., Longley, M. A., Rothnagel, J. A., Greenhalgh, D. A., & Roop, D. R. (1993). Targeted overexpression of transforming growth factor α in the epidermis of transgenic mice elicits hyperplasia, hyperkeratosis, and spontaneous, squamous papillomas. Cell Growth Diff. 4, 1071–1082.

Driman, D. K., Kobrin, M. S., Kudlow, J. E., & Asa, S. L. (1992). Transforming growth factor α in normal and neoplastic human endocrine tissues. Hum. Pathol. 23, 1360–1365.

Ebner, R. & Derynck, R. (1991). Epidermal growth factor and transforming growth factor α: Differential intracellular routing and processing of ligand-receptor complexes. Cell Regul. 2, 599–612.

Elder, J. T., Fisher, G. J., Lindquist, P. B., Bennett, G. L., Pittelkow, M. R., Coffey, R. J., Jr., Ellingsworth, L., Derynck, R., & Voorhees, J. J. (1989). Overexpression of transforming growth factor α in psoriatic epidermis. Science 243, 811–814.

Evarts, R. P., Hue, Z., Fujio, K., Marsden, E. R., & Thorgeirsson, S. S. (1993). Activation of hepatic stem cell compartment in the rat: Role of transforming growth factor α, hepatocyte growth factor, and acidic fibroblast growth factor in early proliferation. Cell Growth Diff. 4, 555–561.

Evarts, R. P., Nakatsukasa, H., Marsden, E. R., Hue, Z., & Thorgeirsson, S. S. (1992). Expression of transforming growth factor α in regenerating liver and during hepatic differentiation. Mol. Carcinog. 5, 25–31.

Farrall, M., Buetow, K. H., & Murray, J. C. (1993). Resolving an apparent paradox concerning the role of TGFα in CL/P. Am. J. Hum. Genet. 52, 434–436.

Feild, J. A., Reid, R. H., Rieman, D. J., Page–Kline, T., Sathe, G., Greig, R. G., & Anzano, M. A. (1992). Structure-function analysis of human transforming growth factor α by site-directed mutagenesis. Biochem. J. 283, 91–98.

Filmus, J., Shi, W., & Spencer, T. (1993). Role of transforming growth factor α (TGF-α) in the transformation of *ras*-transfected rat intestinal epithelial cells. Oncogene 8, 1017–1022.

Finzi, E., Fleming, T., & Pierce, J. H. (1990). Retroviral expression of transforming growth factor α does not transform fibroblasts or keratinocytes. Soc. Invest. Dermatol. 95, 382–387.

Finzi, E., Fleming, T., Segatto, O., Pennington, C. Y., Bringman, T. S., Derynck, R., & Aaronson, S. A. (1987). The human transforming growth factor type α coding sequence is not a direct-acting oncogene when overexpressed in NIH 3T3 cells. Proc. Natl. Acad. Sci. USA 84, 3733–3737.

Flanagan, J. G., Chen, D. C., & Leder, P. (1991). Transmembrane form of the kit ligand growth factor is determined by alternative splicing and is missing in the Sl^d mutant. Cell 64, 1025–1035.

Fowler, K. J., Mann, G. B., & Dunn, A. R. (1993). Linkage of the murine-transforming growth factor α gene with *Igk*, *Ly-2*, and *Fabp 1* on chromosome 6. Genomics 16, 782–784.

Fowler, K. J., Walker, F., Alexander, W., Hibbs, M. L., Nice, E. C., Bohmer, R. M., Mann, G. B., Thumwood, C., Maglitto, R., Danks, J. A., Chetty, R., Burgess, A. W., & Dunn, A. R. (1995). A mutation in the epidermal growth factor receptor in waved-2 mice has a profound effect on receptor biochemistry that results in impaired lactation. Proc. Natl. Acad. Sci. USA 92, 1465–1469.

Gaido, K. W., Maness, S. C., Leonard, L. S., & Greenlee, W. F. (1992). 2,3,7,8-Tetrachlorodibenzo-p-dioxin-dependent regulation of transforming growth factors α and β 2 expression in a human keratinocyte cell line involves both transcriptional and posttranscriptional control. J. Biol. Chem 267, 24591–24595.

Garcia, J. C., Gehm, B. D., & Rosner, M. R. (1989). An evolutionarily conserved enzyme degrades transforming growth factor α as well as insulin. J. Cell Biol. 109, 1301–1307.

Gentry, L. E., Twardzik, D. R., Lim, G. J., Ranchalis, J. E., & Lee, D. C. (1987). Expression and characterization of transforming growth factor α precursor protein in transfected mammalian cells. Mol. Cell. Biol. 7, 1585–1591.

Glinsmann–Gibson, B. J. & Korc, M. (1991). Regulation of transforming growth factor α mRNA expression in T_3M_4 human pancreatic carcinoma cells. Pancreas 6, 142–149.

Godwin, A.K. & Lieberman, M. W. (1990). Early and late responses to induction of *ras*T24 expression in Rat-1 cells. Oncogene 5, 1231–1241.

Gomella, L. G., Sargent, E. R., Wade, T. P., Anglard, P., Linehan, W. M., & Kasid, A. (1989). Expression of transforming growth factor α in normal human adult kidney and enhanced expression of transforming growth factors α and β1 in renal cell carcinoma. Cancer Res. 49, 6972–6975.

Gong, Y., Anzai, Y., Murphy, L. C., Ballejo, G., Holinka, C. F., Gurpide, E., & Murphy, L. J. (1991). Transforming growth factor gene expression in human endometrial adenocarcinoma cells: Regulation by progestins. Cancer Res. 51, 5476–5481.

Gottlieb, A. B., Chang, C. K., Posnett, D. N., Fanelli, B., & Tam, J. P. (1988). Detection of transforming growth factor α in normal, malignant, and hyperproliferative human keratinocytes. J. Exp. Med. 167, 670–675.

Gray, A., Dull, T. J., & Ullrich, A. (1983). Nucleotide sequence of epidermal growth factor cDNA predicts a 128,000-molecular weight protein precursor. Nature 303, 722–725.

Green, M. R., Phil, D., & Couchman, J. R. (1984). Distribution of epidermal growth factor receptors in rat tissues during embryonic skin development, hair formation, and the adult hair growth cycle. J. Invest. Dermatol. 83, 118–123.

Guy, P. M., Platko, J. V., Cantley, L. C., Cerione, R. A., & Carraway, K. L., III. (1994). Insect cell-expressed $p180^{erbB3}$ possesses an impaired tyrosine kinase activity. Proc. Natl. Acad. Sci. USA 91, 8132–8136.

Halter, S. A., Dempsey, P., Matsui, Y., Stokes, M. K., Graves–Deal, R., Hogan, B. L. M., & Coffey, R. J. (1992). Distinctive patterns of hyperplasia in transgenic mice with mouse mammary tumor virus-transforming growth factor α. Am. J. Pathol. 140, 1131–1146.

Hamburger, A.W. & Pinnamaneni, G. (1993). Interferon-induced enhancement of transforming growth factor α expression in a human breast cancer cell line. Proc. Soc. Exp. Biol. Med. 202, 64–68.

Han, V. K. M., Hunter, E. S., III, Pratt, R. M., Zendegui, J. G., & Lee, D. C. (1987). Expression of rat-transforming growth factor alpha mRNA during development occurs predominantly in the maternal decidua. Mol. Cell. Biol. 7, 2335–2343.

Hanauske, A. R., Buchok, J., Scheithauer, W., & VonHoff, D. D. (1987). Human colon cancer cell lines secrete TGFα-like activity. Br. J. Cancer 55, 57–59.

Handford, P. A., Mayhew, M., Baron, M., Winship, P. R., Campbell, I. D., & Brownlee, G. G. (1991). Key residues involved in calcium-binding motifs in EGF-like domains. Nature 351, 164–167.

Harano, T. & Mizuno, K. (1994). Phorbol ester-induced activation of a membrane-bound candidate protransforming growth factor α processing enzyme. J. Biol. Chem. 269, 20305–20311.

Heimbrook, D. C., Stirdivant, S. M., Ahern, J. D., Balishin, N. L., Patrick, D. R., Edwards, G. M., Defeo–Jones, D., Fitzgerald, D. J., Pastan, I., & Oliff, A. (1990). Transforming growth factor α-*Pseudomonas* exotoxin fusion protein prolongs survival of nude mice bearing tumor xenografts. Proc. Natl. Acad. Sci. USA 87, 4697–4701.

Higashiyama, S., Abraham, J. A., Miller, J., Fiddes, J. C., & Klagsbrun, M. (1991). A heparin–binding growth factor secreted by macrophage-like cells that is related to EGF. Science 251, 936–939.

Hill, R. J. & Sternberg, P. W. (1992). The gene *lin-3* encodes in inductive signal for vulval development in *C. elegans*. Nature 358, 470–476.

Holmes, W. E., Sliwkowski, M. X., Akita, R. W., Henzel, W. J., Lee, J., Park, J. W., Yansura, D., Abadi, N., Raab, H., Lewis, G. D., Shepard, H. M., Kuang, W–J., Wood, W. I., Goeddel, D. V., & Vandlen, R. L. (1992). Identification of heregulin, a specific activator of p185^{erbB2}. Science 256, 1205–1210.

Huang, L. H., Cheng, H., Pardi, A., Tam, J. P., & Sweeney, W. V. (1991). Sequence-specific H NMR assignments, secondary structure, and location of the calcium-binding site in the first epidermal growth factor-like domain of blood coagulation factor IX. Biochemistry 30, 7402–7409.

Humphreys–Beher, M. G., Macauley, S. P., Chegini, N., van Setten, G., Purushotham, K., Stewart, C., Wheeler, T. T., & Schultz, G. S. (1994). Characterization of the synthesis and secretion of transforming growth factor α from salivary glands and saliva. Endocrinology 134, 963–970.

Ibbotson, K. J., Twardzik, D. R., D'Souza, S. M., Hargreaves, W. R., Todaro, G. J., & Mundy, G. R. (1985). Stimulation of bone resorption *in vitro* by synthetic transforming growth factor α. Science 228, 1007–1009.

Ignotz, R. A., Kelly, B., Davis, R. J., & Massagué, J. (1986). Biologically active precursor for transforming growth factor type α released by retrovirally transformed cells. Proc. Natl. Acad. Sci. USA 83, 6307–6311.

Imler, J. L., Schatz, C., Wasylyk, C., Chatton, B., & Wasylyk, B. (1988). A Harvey-*ras* responsive transcription element is also responsive to a tumor-promoter and to serum. Nature 332, 275–278.

Jakobivits, E. B., Schlokat, U., Vannice, J. L., Derynck, R., & Levinson, A. D. (1988). The human transforming growth factor α promoter directs transcription initiation from a single site in the absence of a TATA sequence. Mol. Cell. Biol. 8, 5549–5554.

Jakowlew, S. B., Kondaiah, P., Dillard, P. J., Sporn, M. B., & Roberts, A. B. (1988). A novel low molecular weight ribonucleic acid (RNA) related to transforming growth factor α messenger RNA. Mol. Endocrinol. 2, 1056–1063.

Jhappan, C., Stahle, C., Harkins, R. N., Fausto, N., Smith, G. H., & Merlino, G. T. (1990). TGFα overexpression in transgenic mice induces liver neoplasia and abnormal development of the mammary gland and pancreas. Cell 61, 1137–1146.

Kalthoff, H., Roeder, C., Brockhaus, M., Thiele, H. G., & Schmiegel, W. (1993). Tumor necrosis factor (TNF) upregulates the expression of p75 but not p55 TNF receptors, and both receptors mediate, independently of each other, upregulation of transforming growth factor α and epidermal growth factor receptor mRNA. J. Biol. Chem. 268, 2762–2766.

Kaplan, P. L., Anderson, M., & Ozanne, B. (1982). Transforming growth factor(s) production enables cells to grow in the absence of serum: An autocrine system. Proc. Natl. Acad. Sci. 79, 485–489.

Keeler, C. E. (1935). A second rexoid coat character in the house mouse. J. Hered. 26, 189–191.

Kimura, H., Fischer, W. H., & Schubert, D. (1990). Structure, expression, and function of a Schwannoma-derived growth factor. Nature 348, 257–260.

Kingsley, C. & Winoto, A. (1992). Cloning of GT box-binding proteins: A novel Sp1 multigene family regulating T-cell receptor gene expression. Mol. Cell. Biol. 12, 4251–4261.

Kitadai, Y., Yamazaki, H., Yasui, W., Kyo, E., Yokozaki, H., Kajiyama, G., Johnson, A. C., Pastan, I., & Tahara, E. (1993). GC factor represses transcription of several growth factor/receptor genes and causes growth inhibition of human gastric carcinoma cell lines. Cell Growth Diff. 4, 291–296.

Klein, S. B., Fisher, G. J., Jensen, T. C., Mendelsohn, J., Voorhees, J. J., & Elder, J. T. (1992). Regulation of TGF-α expression in human keratinocytes: PKC-dependent and -independent pathways. J. Cell. Physiol. 151, 326–336.

Kline, T. P., Brown, F. K., Brown, S. C., Jeffs, P. W., Kopple, K. D., & Mueller, L. (1990). Solution structures of human transforming growth factor α derived from ^{1}H NMR data. Biochemistry 29, 7805–7813.

Kobrin, M. S., Asa, S. L., Samsoondar, J., & Kudlow, J. E. (1987). α-Transforming growth factor in the bovine anterior pituitary gland: Secretion by dispersed cells and immunohistochemical localization. Endocrinology 121, 1412–1416.

Koch, C. A., Anderson, D., Moran, M. F., Ellis, C., & Pawson, T. (1991). SH2 and SH3 domains: Elements that control interactions of cytoplasmic signaling proteins. Science 252, 668–674.

Kohda, D., Shimada, I., Miyake, T., Fuwa, T., & Inagaki, F. (1989). Polypeptide chain fold of human transforming growth factor α analogous to those of mouse and human epidermal growth factors as studied by two-dimensional ^{1}H NMR. Biochemistry 28, 953–958.

Korc, M. & Finman, J. E. (1989). Attenuated processing of epidermal growth factor in the face of marked degradation of transforming growth factor α. J. Biol. Chem. 264, 14990–14999.

Kornblum, H. I., Chugani, H. T., Tatsukawa, K., & Gall, C. M. (1994). Cerebral hemidecortication alters expression of transforming growth factor α mRNA in the neostriatum of developing rats. Mol. Brain Res. 21, 107–114.

Kudlow, J. E., Kobrin, M. S., Purchio, A. F., Twardzik, D. R., Hernandez, E. R., Asa, S. L., & Adashi, E. Y. (1987). Ovarian transforming growth factor α gene expression: Immunohistochemical localization to the theca-interstitial cells. Endocrinology 121, 1577–1579.

Kudlow, J. E., Leung, A. W. C., Kobrin, M. S., Paterson, A. J., & Asa, S. L. (1989). Transforming growth factor α in the mammalian brain: Immunohistochemical detection in neurons and characterization of its mRNA. J. Biol. Chem. 264, 3880–3883.

Laird, A. D., Brown, P. I., & Fausto, N. (1994). Inhibition of tumor growth in liver epithelial cells transfected with a transforming growth factor α antisense gene. Cancer Res. 54, 4224–4232.

Lax, I., Johnson, A., Howk, R., Sap, J., Bellot, F., Winkler, M., Ullrich, A., Vennstrom, B., Schlessinger, J., & Givol, D. (1988). Chicken epidermal growth factor (EGF) receptor: cDNA cloning, expression in mouse cells, and differential binding of EGF and transforming growth factor α. Mol. Cell. Biol. 8, 1970–1978.

Lazar, E., Vicenzi, E., Van Obberghen–Schillin, E., Wolff, B., Dalton, S., Watanabe, S., & Sporn, M. B. (1989). Transforming growth factor α: An aromatic side chain at position 38 is essential for biological activity. Mol. Cell. Biol. 9, 860–864.

Lazar, E., Watanabe, S., Dalton, S., & Sporn, M. B. (1988). Transforming growth factor α: Mutation of aspartic acid 47 and leucine 48 results in different biological activities. Mol. Cell. Biol. 8, 1247–1252.

Lee, D. C., Rochford, R., Todaro, G. J., & Villarreal, L. P. (1985a). Developmental expression of rat transforming growth factor α mRNA. Mol. Cell. Biol. 5, 3644–3646.

Lee, D. C., Rose, T. M., Webb, N. R., & Todaro, G. J. (1985b). Cloning and sequence analysis of a cDNA for rat transforming growth factor α. Nature 313, 489–491.

Lee, G.–H., Merlino, G., & Fausto, N. (1992). Development of liver tumors in transforming growth factor α transgenic mice. Cancer Res. 52, 5162–5170.

Lee, L. W., Raymond, V. W., Tsao, M.–S., Lee, D. C., Earp, H. S., & Grisham, J. W. (1991). Clonal cosegregation of tumorigenicity with overexpression of c-*myc* and transforming growth factor α genes in chemically transformed rat liver epithelial cells. Cancer Res. 51, 5238–5244.

Lee, S. W., Morhenn, V. B., Ilnicka, M., Eugui, E. M., & Allison, A. C. (1991). Autocrine stimulation of interleukin-1 α and transforming growth factor α production in human keratinocytes and its antagonism by glucocorticoids. J. Invest. Dermatol. 97, 106–110.

Linsley, P. S., Hargreaves, W. R., Twardzik, D. R., & Todaro, G. J. (1985). Detection of larger polypeptides structurally and functionally related to type I transforming growth factor. Proc. Natl. Acad. Sci. USA 82, 356–360.

Liscia, D. S., Merlo, G., Ciardiello, F., Kim, N., Smith, G. H., Callahan, R., & Salomon, D. S. (1990). Transforming growth factor α messenger RNA localization in the developing adult rat and human mammary gland by *in situ* hybridization. Dev. Biol. 140, 123–131.

Liu, C. & Tsao, M.–S. (1993). *In vitro* and *in vivo* expressions of transforming growth factor α and tyrosine kinase receptors in human non-small-cell lung carcinomas. Am. J. Pathol. 142, 1155–1160.

Liu, C., Tsao, M.–S., & Grisham, J. W. (1988). Transforming growth factors produced by normal and neoplastically transformed rat liver epithelia cells in culture. Cancer Res. 48, 850–855.

Liu, C., Woo, A., & Tsao M.-S. (1990). Expression of transforming growth factor α in primary human colon and lung carcinomas. Br. J. Cancer 62, 425–429.

Liu, S. C., Sanfilippo, B., Perroteau, I., Derynck, R., Salomon, D. S., & Kidwell, W. R. (1987). Expression of transforming growth factor α (TGFα) in differentiated rat mammary tumors: Estrogen induction of TGFα production. Mol. Endocrinol. 1, 683–692.

Lobb, D. K., Kobrin, M. S., Kudlow, J. E., & Dorrington, J. H. (1989). Transforming growth factor α in the adult bovine ovary: Identification in growing ovarian follicles. Biol. Reprod. 40, 1087–1093.

Luetteke, N. C., Lee, D. C., Palmiter, R. D., Brinster, R. L., & Sandgren, E. P. (1993a). Regulation of fat and muscle development by TGFα in transgenic mice and cultured cells. Cell Growth & Diff. 4, 203–213.

Luetteke, N. C., Michalopoulos, G. K., Teixido, J., Gilmore, R., Massagué, J., & Lee, D. C. (1988). Characterization of high molecular weight transforming growth factor α produced by rat hepatocellular carcinoma cells. Biochemistry 27, 6487–6494.

Luetteke, N. C., Phillips, H. K., Qiu, T. H., Copeland, N. G., Earp, H. S., Jenkins, N. A., & Lee, D. C. (1994). The mouse *Waved-2* phenotype results from a point mutation in the EGF receptor tyrosine kinase. Genes Dev. 8, 399–413.

Luetteke, N. C., Qiu, T. H., Oliver, P., Smithies, O., & Lee, D. C. (1993b). TGFα deficiency results in hair follicle and eye abnormalities in targeted and *waved-1* mice. Cell 73, 263–278.

Lynch, M. J., Pelosi, L., Carboni, J. M., Merwin, J., Coleman, K., Wang, R. R. C., Lin, P–F. M., Henry, D. L., & Brattain, M. G. (1993). Transforming growth factor β1 induces transforming growth factor α promoter activity and transforming growth factor α secretion in the human colon adenocarcinoma cell line FET1. Cancer Res. 53, 4041–4047.

Ma, Y. J., Junier, M., Costa, M. E., & Ojeda, S. R. (1992). Transforming growth factor α (TGFα) gene expression in the hypothalamus is developmentally regulated and linked to sexual maturation. Neuron 9, 657–670.

Madtes, D. K., Raines, E. W., Sakariassen, K. S., Assoian, R. K., Sporn, M. B., Bell, G. I., & Ross, R. (1988). Induction of transforming growth factor α in activated human alveolar macrophages. Cell 53, 285–293.

Majello, B., De Luca, P., Hagen, G., Suske, G., & Lania, L. (1994). Different members of the Sp1 multigene family exert opposite transcriptional regulation of the long terminal repeat of HIV-1. Nucleic Acids Res. 22, 4914–4921.

Mann, G. B., Fowler, K. J., Gabriel, A., Nice, E. C., Williams, R. L., & Dunn, A. R. (1993). Mice with a null mutation of the TGFα gene have abnormal skin architecture, wavy hair, and curly whiskers and often develop corneal inflammation. Cell 73, 249–261.

Marquardt, H., Hunkapiller, M. W., Hood, L. E., & Todaro, G. J. (1984). Rat-transforming growth factor type 1: Structure and relation to epidermal growth factor. Science 223, 1079–1082.

Marquardt, H., Hunkapiller, M. W., Hood, L. E., Twardzik, D. R., De Larco, J. E., Stephenson, J. R., & Todaro, G. J. (1983). Transforming growth factors produced by retrovirus-transformed rodent fibroblasts and human melanoma cells: Amino acid sequence homology with epidermal growth factor. Proc. Natl. Acad. Sci. USA 80, 4684–4688.

Massagué, J. (1983). Epidermal growth factor-like transforming growth factor. II. Interaction with epidermal growth factor receptors in human placenta membranes and A431 cells. J. Biol. Chem. 258, 13614–13620.

Massagué, J., Czech, M. P., Iwata, K., DeLarco, J. E., & Todaro, G. J. (1982). Affinity labeling of a transforming growth factor receptor that does not interact with epidermal growth factor. Proc. Natl. Acad. Sci. USA 79, 6822–6826.

Matsui, Y., Halter, S. A., Holt, J. T., Hogan, B. L. M., & Coffey, R. J. (1990). Development of mammary hyperplasia and neoplasia in MMTV-TGFα transgenic mice. Cell 61, 1147–1155.

McClain, E. A., Paterson, A. J., Roos, M. D., Wei, X., & Kudlow, J. E. (1992). Glucose and glucosamine regulate growth factor gene expression in vascular smooth muscle cells. Proc. Natl. Acad. Sci. USA 89, 8150–8154.

McGeady, M. L., Kerby, S., Shankar, V., Ciardiello, F., Salomon, D., & Seidman, M. (1989). Infection with a TGFα retroviral vector transforms normal mouse mammary epithelial cells but not normal rat fibroblasts. Oncogene 4, 1375–1382.

Mead, J. E. & Fausto, N. (1989). Transforming growth factor α may be a physiological regulator of liver regeneration by means of an autocrine mechanism. Proc. Natl. Acad. Sci. USA 86, 1558–1562.

Mesri, E. A., Kreitman, R. J., Fu, Y.-M., Epstein, S. E., & Pastan, I. (1993). Heparin-binding transforming growth factor α-*Pseudomonas* exotoxin A. J. Biol. Chem. 268, 4853–4862.

Miettinen, P. J. (1993). Transforming growth factor α and epidermal growth factor expression in human fetal gastrointestinal tract. Pediatr. Res. 33, 481–486.

Modjtahedi, N., Haddada, H., Lamonerie, T., Lazar, E., Lavialle, C., & Brison, O. (1992). TGFα production correlates with tumorigenicity in clones of the SW613-S human colon carcinoma cell line. Int. J. Cancer 52, 483–490.

Montelione, G. T., Winkler, M. E., Burton, L. E., Rinderknecht, E., Sporn, M. B., & Wagner, G. (1989). Sequence-specific ^{1}H NMR assignments and identification of two small antiparallel β-sheets in the solution structure of recombinant human transforming growth factor α. Proc. Natl. Acad. Sci. USA 86, 1519–1523.

Morishige, K., Kurachi, H., Amemiya, K., Fujita, Y., Yamamoto, T., Miyake, A., & Tanizawa, O. (1991a). Evidence for the involvement of transforming growth factor α and epidermal growth factor receptor autocrine growth mechanism in primary human ovarian cancers *in vitro*. Cancer Res. 51, 5322–5328.

Mroczkowski, B., Reich, M., Chen, K., Bell, G. I., & Cohen, S. (1989). Recombinant human epidermal growth factor precursor is a glycosylated membrane protein with biological activity. Mol. Cell. Biol. 9, 2771–2778.

Mueller, S. G., Kobrin, M. S., Paterson, A. J., & Kudlow, J. E. (1989). Transforming growth factor α expression in the anterior pituitary gland: Regulation by epidermal growth factor and phorbol ester in dispersed cells. Mol. Endocrinol. 3, 976–983.

Mueller, S. G. & Kudlow, J. E. (1991). Transforming growth factor β (TGFβ) inhibits TGFα expression in bovine anterior pituitary-derived cells. Mol. Endocrinol. 5, 1439–1446.

Mueller, S. G., Paterson, A. J., & Kudlow, J. E. (1990). Transforming growth factor α in arterioles: Cell surface processing of its precursor by elastases. Mol. Cell. Biol. 10, 4596–4602.

Murakami, H., Sanderson, N. D., Nagy, P., Marino, P. A., Merlino, G., & Thorgeirsson, S. S. (1993). Transgenic mouse model for synergistic effects of nuclear oncogenes and growth factors in tumorigenesis: Interaction of c-*myc* and transforming growth factor α in hepatic oncogenesis. Cancer Res. 53, 1719–1723.

Murphy, L. C. & Dotzlaw, H. (1989). Regulation of transforming growth factor α and transforming growth factor β messenger ribonucleic acid abundance in T-47D human breast cancer cells. Mol. Endocrinol. 3, 611–617.

Murphy, L. C., Dotzlaw, H., Wong, M. S., Miller, T., & Murphy, L. J. (1991). Mechanisms involved in the evolution of progestin resistance in human breast cancer cells. Cancer Res. 51, 2051–2057.

Musgrove, E. A., Lee, C. S. L., & Sutherland, R. L. (1991). Progestins both stimulate and inhibit breast cancer cell cycle progression while increasing expression of transforming growth factor α, epidermal growth factor receptor, c-*fos*, and c-*myc* genes. Mol. Cell. Biol. 11, 5032–5043.

Mydlo, J. H., Michaeli, J., Cordon-Cardo, C., Goldenberg, A. S., Heston, W. D. W., & Fair, W. R. (1989). Expression of transforming growth factor α and epidermal growth factor messenger RNA in neoplastic and nonneoplastic human kidney tissue. Cancer Res. 49, 3407–3411.

Nabell. L. M., Rampyari, H. R., Sayeski, P. P., Paterson, A. J., & Kudlow, J. E. (1994). Human immunodeficiency virus 1 tat stimulates transcription of the transforming growth factor α gene in an epidermal growth factor-dependent manner. Cell Growth Differ. 5, 87–93.

Naglich, J. G., Metherall, J. E., Russell, D. W., & Eidels, L. (1992). Expression cloning of a diptheria toxin receptor: Identity with a heparin-binding EGF-like growth factor precursor. Cell 69, 1051–1061.

Nelson, K. G., Takahashi, T., Lee, D. C., Luetteke, N. C., Bosserts, N. L., Ross, K., Eitzman, B. E., & McLachlan, J. A. (1992). Transforming growth factor α is a potential mediator of estrogen action in the mouse uterus. Endocrinology 131, 1657–1664.

Nestor, J. J., Newman, S. R., De Lustro, B., Todaro, G. J., & Schreiber, A. B. (1985). A synthetic fragment of rat-transforming growth factor α with receptor binding and antigenic properties. Biochem. Biophys. Res. Commun. 129, 226–232.

Neuman–Silberberg, F. S. & Schüpbach, T. (1993). The *Drosophila* dorsoventral patterning gene *gurken* produces a dorsally localized RNA and encodes a TGFα-like protein. Cell 75, 165–174.

Ohmura, E., Okada, M., Onoda, N., Kamiya, Y., Murakami, H., Tsushima, T., & Shizume, K. (1990). Insulin-like growth factor I and transforming growth factor α as autocrine growth factors in human pancreatic cancer cell growth. Cancer Res. 50, 103–107.

Pai, L. H., Galley, M. G., Fitzgerald, D. J., & Pastan, I. (1991). Antitumor activity of a transforming growth factor *α-pseudomonas* exotoxin fusion protein (TGFα–PE40). Cancer Res. 51, 2808–2812.

Pandiella, A., Boysenberry, M. W., Huang, E. J., Besmer, P., & Massagué, J. (1992). Cleavage of membrane-anchored growth factors involves distinct protease activities regulated through common mechanisms. J. Biol. Chem. 267, 24028–24033.

Pandiella, A. & Massagué, J. (1991a). Multiple signals activate cleavage of the membrane transforming growth factor α precursor. J. Biol. Chem. 266, 5769–5773.

Pandiella, A. & Massagué, J. (1991b). Cleavage of the membrane precursor for transforming growth factor α is a regulated process. Proc. Natl. Acad. Sci. USA 88, 1726–1730.

Pathak, B. G., Gilbert, D. J., Harrison, C. A., Luetteke, N. C., Chen, X., Klagsbrun, M., Plowman, G. D., Copeland, N. G., Jenkins, N. A., & Lee, D. C. (1995). Murine chromosomal location of three EGF receptor ligands: Amphiregulin (*Areg*), betacellulin (*Btc*) and heparin-binding EGF (*Hegfl*). Genomics, in press.

Peles, E., Bacus, S. S., Koski, R. A., Lu, H. S., Wen, D., Ogden, S. G., Levy, R. B., & Yarden, Y. (1992). Isolation of the Neu/HER-2 stimulatory ligand: A 44-kd glycoprotein that induces differentiation of mammary tumor cells. Cell 69, 205–216.

Pfeffer, S. & Ullrich, A. (1985). Epidermal growth factor: Is the precursor a receptor? Nature 313, 184.

Pittelkow, M. R., Cook, P. W., Shipley, G. D., Derynck, R., & Coffey, R. J., Jr. (1993). Autonomous growth of human keratinocytes requires epidermal growth factor receptor occupancy. Cell Growth Diff. 4, 513–521.

Pittelkow, M. R., Lindquist, P. B., Abraham, R. T., Graves–Deal, R., Derynck, R., & Coffey, R. J., Jr. (1989). Induction of transforming growth factor α expression in human keratinocytes by phorbol esters. J. Biol. Chem. 264, 5164–5171.

Plowman, G. D., Culouscou, J.–M., Whitney, G. S., Green, J. M., Carlton, G. W., Foy, L., Neubauer, M. G., & Shoyab, M. (1993). Ligand-specific activation of HER4/p180^{erbB4}, a fourth member of the epidermal growth factor receptor family. Proc. Natl. Acad. Sci. USA 90, 1746–1750.

Polk, W. H., Soroka, C. J., Goldenring, J. R., Modlin, I. M., Bundz, S., Dempsey, P. J., & Coffey, R. J. (1991). Immunolocalization of transforming growth factor α and its receptor to the parietal cell in normal gastric mucosa. Gastroenterology 100, A660.

Price, J. V., Clifford, R. J., & Schupbach, T. (1989). The maternal ventralizing locus *torpedo* is allelic to *faint little ball*, an embryonic lethal, and encodes the *Drosophila* EGF receptor homolog. Cell 56, 1085–1092.

Raja, R. H., Paterson, A. J., Shin, T. H., & Kudlow, J. E. (1991). Transcriptional regulation of the human transforming growth factor α gene. Mol. Endocrinol. 5, 514–520.

Rappolee, D. A., Brenner, C. A., Schultz, R., Mark, D., & Werb, Z. (1988a). Developmental expression of PDGF, TGFα, and TGFβ genes in preimplantation mouse embryos. Science 241, 1823–1825.

Rappolee, D. A., Mark, D., Banda, M. J., & Werb, Z. (1988b). Wound macrophages express TGFα and other growth factors *in vivo*: Analysis by mRNA phenotyping. Science 241, 708–712.

Raz, E. & Shilo, B.-A. (1992). Dissection of the faint little ball (flb) phenotype: Determination of the development of the *Drosophila* central nervous system by early interactions in the ectoderm. Development 114, 113–123.

Raymond, V. W., Lee, D. C., Grisham, J. W., & Earp, H. S. (1989). Regulation of transforming growth factor α messenger RNA expression in a chemically transformed rat hepatic epithelial cell line by phorbol ester and hormones. Cancer Res. 49, 3608–3612.

Reddy, K. B., Yee, D., Hilsenbeck, S. G., Coffey, R. J., & Osborne, C. K. (1994). Inhibition of estrogen-induced breast cancer cell proliferation by reduction in autocrine transforming growth factor α expression. Cell Growth Diff. 1275–1282.

Roberts, A. B., Anzano, M. A., Lamb, L. C., Smith, J. M., & Sporn, M. B. (1981). New class of transforming growth factors potentiated by epidermal growth factor: Isolation from non-neoplastic tissues. Proc. Natl. Acad. Sci. USA 78, 5339–5343.

Roberts, A. B., Lamb, L. C., Newton, D. L., Sporn, M. B., DeLarco, J. E., & Todaro, G. J. (1980). Transforming growth factors: Isolation of polypeptides from virally and chemically transformed cells by acid/ethanol extraction. Proc. Natl. Acad. Sci. USA 77, 3494–3498.

Roberts, A. B. & Sporn, M. B. (1990). The transforming growth factor βs. In: Peptide Growth Factors and their Receptors, Handbook of Experimental Pharmacology 95/11 (Sporn, M. B. & Roberts, A. B., eds.), Springer–Verlag, Berlin, pp. 419–472.

Rosenthal, A., Lindquist, P. B., Bringman, T. S., Goeddel, D. V., & Derynck, R. (1986). Expression in rat fibroblasts of a human transforming growth factor α cDNA results in transformation. Cell 46, 301–309.

Rutledge, B. J., Zhang, K., Bier, E., Jan, Y. N., & Perrimon, N. (1992). The *Drosophila spitz* gene encodes a putative EGF-like growth factor involved in dorsal-ventral axis formation and neurogenesis. Genes Dev. 6, 1503–1517.

Saeki, T., Cristiano, A., Lynch, M. J., Brattain, M., Kim, N., Normanno, N., Kenney, N., Ciardiello, F., & Salomon, D. S. (1991). Regulation by estrogen through the 5′-flanking region of the transforming growth factor α gene. Mol. Endocrinol. 5, 1955–1963.

Salomon, D. S., Perroteau, I., Kidwell, W. R., Tam, J., & Derynck, R. (1987). Loss of growth responsiveness to epidermal growth factor and enhanced production of α-transforming growth factors in *ras*-transformed mouse mammary epithelial cells. J. Cell. Physiol. 130, 397–409.

Sandgren, E. P., Luetteke, N. C., Palmiter, R. D., Brinster, R. L., & Lee, D. C. (1990). Overexpression of TGFα in transgenic mice: Induction of epithelial hyperplasia, pancreatic metaplasia, and carcinoma of the breast. Cell 61, 1121–1135.

Sandgren, E. P., Luetteke, N. C., Qiu, T. H., Palmiter, R. D., Brinster, R. L., & Lee, D. C. (1993). Transforming growth factor α dramatically enhances oncogene-induced carcinogenesis in transgenic mouse pancreas and liver. Mol. Cell. Biol. 13, 320–330.

Sandgren, E. P., Schroeder, J., Qiu, T. H., Palmiter, R. D., Brinster, R. L., & Lee, D. C. Inhibition of mammary gland involution is associated with TGFα- but not c-myc-induced tumorigenesis in transgenic mice, submitted.

Sassani, R., Bartlett, S. P., Feng, H., Goldner–Sauve, A., Haq, A. K., Buetow, K. H., & Gasser, D. L. (1993). Association between alleles of the transforming growth factor α locus and the occurrence of cleft lip. Am. J. Med. Genet. 45, 565–569.

Schejter, E. D. & Shilo, B.–Z. (1989). The *Drosophila* EGF receptor homolog (DER) gene is allelic to *faint little ball*, a locus essential for embryonic development. Cell 56, 1093–1104.

Schmiegel, W., Roeder, C., Schmielau, J., Rodeck, U., & Kalthoff, H. (1993). Tumor necrosis factor α induces the expression of transforming growth factor α and the epidermal growth factor receptor in human pancreatic cancer cells. Proc. Natl. Acad. Sci. USA 90, 863–867.

Schreiber, A. B., Winkler, M. E., & Derynck, R. (1986). Transforming growth factor α: A more potent angiogenic mediator than epidermal growth factor. Science 232, 1250–1253.

Schultz, G. S., Chegini, N., Grant, M., Khaw, P., & MacKay, S. (1992). Effects of growth factors on corneal wound healing. Acta Ophthalmol. 70, 60–66.

Schultz, G. S., White, M., Mitchell, R., Brown, G., Lynch, J., Twardzik, D. R., & Todaro, G. J. (1987). Epithelial wound healing enhanced by transforming growth factor α and vaccinia growth factor. Science 235, 350–352.

Schuuring, E., Verhoeven, E., van Tinteren, H., Peterse, J. L., Nunnink, B., Thunnissen, F. B., Devilee, P., Cornelisse, C. J., van de Vijver, M. J., Mooi, W. J., & Michalides, R. J. A. M. (1992). Amplification of genes within the chromosome 11q13 region is indicative of poor prognosis in patients with operable breast cancer. Cancer Res. 52, 5229–5234.

Scott, J., Urdea, M., Quiroga, M., Sanchez–Pescador, R., Fong, N., Selby, M., Rutter, W. J., & Bell, G. I. (1983). Structure of a mouse submaxillary gland messenger RNA encoding epidermal growth factor and seven related proteins. Science 221, 236–240.

Selander–Sunnerhagen, M., Ullner, M., Persson, E., Teleman, O., Stenflo, J., & Drakenberg, T. (1992). How an epidermal growth factor (EGF)-like domain binds calcium. J. Biol. Chem. 267, 19642–19649.

Seroogy, K. B., Han, V. K. M., & Lee, D. C. (1991). Regional expression of transforming growth factor α mRNA in the rat central nervous system. Neurosci. Lett. 125, 241–245.

Seroogy, K. B., Lundgren, K. H., Lee, D. C., Guthrie, K. M., & Gall, C. M. (1993). Cellular localization of transforming growth factor α mRNA in rat forebrain. J. Neurochem. 60, 1777–1782.

Seroogy, K. B., Numan, S., Gall, C. M., Lee, D. C., & Kornblum, H. I. (1994). Expression of EGF receptor mRNA in rat nigrostriatal system. NeuroReport 6, 105–108.

Seroogy, K. B., Gall, C. M., Lee, D. C., & Kornblum, H. I. (1995). Proliferative zones of postnatal rat brain express epidermal growth factor receptor mRNA. Brain Res. 670, 157–164.

Shankar, V., Ciardiello, F., Kim, N., Derynck, R., Liscia, D. S., Merlo, G., Langton, B. C., Sheer, D., Callahan, R., Bassin, R. H., Lippman, M. E., Hynes, N., & Salomon, D. S. (1989). Transformation of an established mouse mammary epithelial cell line following transfection with a human transforming growth factor α cDNA. Mol. Carcinog. 2, 1–11.

Sherwin, S. A., Twardzik, D. R., Bohn, W. H., Cockley, K. D., & Todaro, G. J. (1983). High molecular weight transforming growth factor activity in the urine of patients with disseminated cancer. Cancer Res. 43, 403–407.

Shin, T. H., Paterson, A. J., Grant, J. H., III, Meluch, A. A., & Kudlow, J. E. (1992). 5-Azacytidine treatment of HA-A melanoma cells induces Sp1 activity and concomitant transforming growth factor α expression. Mol. Cell. Biol. 12, 3998–4006.

Shing, Y., Christofori, G., Hanahan, D., Ono, Y., Sasada, R., Igarashi, K., & Folkman, J. (1993). Betacellulin: A mitogen from pancreatic β-cell tumors. Science 259, 1604–1607.

Shoyab, M., McDonald, V. L., Bradley, J. G., & Todaro, G. J. (1988). Amphiregulin: A bifunctional growth-modulating glycoprotein produced by the phorbol 12-myristate 13-acetate-treated human breast adenocarcinoma cell line MCF-7. Proc. Natl. Acad. Sci. USA 85, 6528–6532.

Shoyab, M., Plowman, G. D., McDonald, V. L., Bradley, J. G., & Todaro, G. J. (1989). Structure and function of human amphiregulin: A member of the epidermal growth factor family. Science 243, 1074–1076.

Shum, L., Reeves, S. A., Kuo, A. C., Fromer, E. S., & Derynck, R. (1994). Association of the transmembrane TGFα precursor with a protein kinase complex. J. Cell Biol. 125, 903–916.

Siegfried, J. M. (1987). Detection of human lung epithelial cell growth factors produced by a lung carcinoma cell line: Use in culture of primary solid lung tumors. Cancer Res. 47, 2903–2910.

Skinner, M. K., Takacs, K., & Coffey, R. J. (1989). Transforming growth factor α gene expression and action in the seminiferous tubule: Peritubular cell–sertoli cell interactions. Endocrinology 124, 845–854.

Smith, J. J., Derynck, R., & Korc, M. (1987). Production of transforming growth factor α in human pancreatic cancer cells: Evidence for a superagonist autocrine cycle. Proc. Natl. Acad. Sci. USA 84, 7567–7570.

Snedeker, S. M., Brown, C. F., & Augustine, R. P. (1991). Expression and functional properties of transforming growth factor α and epidermal growth factor during mouse mammary gland ductal morphogenesis. Proc. Natl. Acad. Sci. USA 88, 8167–8171.

Soltoff, S. P., Carraway, K. L., III, Prigent, S. A., Gullick, W. G., & Cantley, L. C. (1994). ErbB3 is involved in activation of phosphatidylinositol-3-kinase by epidermal growth factor. Mol. Cell. Biol. 14, 3550–3558.

Sporn, M. B. & Todaro, G. J. (1980). Autocrine secretion and malignant transformation of cells. N. Engl. J. Med. 303, 878–880.

Stoll, C., Qian, J. F., Feingold, J., Sauvage, P., & May, E. (1993). Genetic variation in transforming growth factor α: Possible association of BamHI polymorphism with bilateral sporadic cleft lip and palate. Hum. Genet. 92, 81–82.

Strandjord, T. P., Clark, J. G., & Madtes, D. K. (1994). Expression of TGF-α, EGF, and EGF receptor in fetal rat lung. Am. J. Physiol. 11, L384–L389.

Strange, R., Li, R., Saurer, S., Burkhardt, A., & Friis, R. R. (1992). Apoptotic cell death and tissue remodeling during mouse mammary gland involution. Development 115, 49–58.

Stromberg, K., Hudgins, W. R., & Orth, D. N. (1987). Urinary TGFs in neoplasia: Immunoreactive TGFα in the urine of patients with disseminated breast carcinoma. Biochem. Biophys. Res. Commun. 144, 1059–1068.

Stroobant, P., Rice, A. P., Gullick, W. J., Cheng, D. J., Kerr, I. M., & Waterfield, M. D. (1985). Purification and characterization of vaccinia virus growth factor. Cell 42, 383–393.

Takagi, H., Sharp, R., Hammermeister, C., Goodrow, T., Bradley, M. O., Fausto, N., & Merlino, G. (1992). Molecular and genetic analysis of liver oncogenesis in transforming growth factor α transgenic mice. Cancer Res. 52, 5171–5177.

Takagi, H., Sharp, R., Takayama, H., Anver, M. R., Ward, J. M., & Merlino, G. (1993). Collaboration between growth factors and diverse chemical carcinogens in hepatocarcinogenesis of transforming growth factor α transgenic mice. Cancer Res. 53, 4329–4336.

Tam, J. P. (1985). Physiological effects of transforming growth factor in the newborn mouse. Science 229, 673–675.

Tamada, H., Das, S. K., Andrews, G. K., & Dey, S. K. (1991). Cell-type-specific expression of transforming growth factor α in the mouse uterus during the peri-implantation period. Biol. Reprod. 45, 365–372.

Tanaka, S., Imanishi, K., Yoshihara, M., Haruma, K., Sumii, K., Kajiyama, G., & Akamatsu, S. (1991). Immunoreactive transforming growth factor α is commonly present in colorectal neoplasia. Am. J. Pathol. 139, 123–129.

Tateishi, M., Ishida, T., Mitsudomi, T., Kaneko, S., & Sugimachi, K. (1990). Immunohistochemical evidence of autocrine growth factors in adenocarcinoma of the human lung. Cancer Res. 50, 7077–7080.

Teerds, K. J., Rommerts, F. F. G., & Dorrington, J. H. (1990). Immunohistochemical detection of transforming growth factor α in Leydig cells during the development of the rat testis. Mol. Cell. Endocrinol. 69, R1–R6.

Teixidó, J., Gilmore, R., Lee, D. C., & Massagué, J. (1987). Integral membrane glycoprotein properties of the prohormone protransforming growth factor α. Nature 326, 883–885.

Teixidó, J. & Massagué, J. (1988). Structural properties of a soluble bioactive precursor for transforming growth factor α. J. Biol. Chem. 263, 3924–3929.

Teixidó, J., Wong, S. T., Lee, D. C., & Massagué, J. (1990). Generation of transforming growth factor α from the cell surface by an O-glycosylation-independent multistep process. J. Biol. Chem. 265, 6410–6415.

Thomas, D. M., Nasim, M. M., Gullick, W. J., & Alison, M. R. (1992). Immunoreactivity of transforming growth factor α in the normal adult gastrointestinal tract. Gut 33, 628–631.

Todaro, G. J., De Larco, J. E., & Cohen, S. (1976). Transformation by murine and feline sarcoma viruses specifically blocks binding of epidermal growth factor to cells. Nature 264, 26–31.

Todaro, G. J., De Larco, J. E., Nissley, S. P., & Rechler, M. M. (1997). MSA and EGF receptors on sarcoma virus transformed cells and human fibrosarcoma cells in culture. Nature 267, 526–528.

Todaro, G. J., Fryling, C., & De Larco, J. E. (1980). Transforming growth factors produced by certain human tumor cells: Polypeptides that interact with epidermal growth factor receptors. Proc. Natl. Acad. Sci. USA 77, 5258–5262.

Todaro, G. J., Lee, D. C., Webb, N. R., Rose, T. M., & Brown, J. P. (1985). Rat type-α transforming growth factor: Structure and possible function as a membrane receptor. Cancer Cells 3, 51–58.

Toyoda, H., Komurasaki, D., Uchida, Y., Takayama, T., Isobe, T., Okuyama, T., & Honada, K. (1995). Epiregulin: A novel epidermal growth factor with mitogenic activity for rat primary hepatocytes. J. Biol. Chem. 270, 7495–7500.

Tricoli, J. V., Nakai, H., Byers, M. G., Bell, G. I., & Shows, T. B. (1986). The gene for human transforming growth factor α is on the short arm of chromosome 2. Cytogenet. Cell Genet. 42, 94–98.

Twardzik, D. R. (1985). Differential expression of transforming growth factor α during prenatal development of the mouse. Cancer Res. 45, 5413–5416.

Twardzik, D. R., Ranchalis, J. E., & Todaro, G. J. (1982a). Mouse embryonic-transforming growth factors related to those isolated from tumor cells. Cancer Res. 42, 590–593.

Twardzik, D. R., Todaro, G. J., Marquardt, H., Reynolds, F. H., Jr., & Stephenson, J. R. (1982c). Transformation induced by Abelson murine leukemia virus involves production of a polypeptide growth factor. Science 216, 894–897.

Twardzik, D. R., Todaro, G. J., Reynolds, F. H., Jr., & Stephenson, J. R. (1983). Similar transforming growth factors (TGFs) produced by cells transformed by different isolates of feline sarcoma virus. Virology 124, 201–207.

Untawale, S., Zorbas, M. A., Hodgson, C. P., Coffey, R. J., Gallick, G. E., North, S. M., Wildrick, D. M., Olive, M., Thick, M., Yeoman, L. C., & Boman, B. M. (1993). Transforming growth factor α production and autoinduction in a colorectal carcinoma cell line (DiFi) with an amplified epidermal growth factor receptor gene. Cancer Res. 53, 1630–1636.

Upton, C., Macen, J. L., & McFadden, G. (1987). Mapping and sequencing of a gene from myxoma virus that is related to those encoding epidermal growth factor and transforming growth factor α. J. Virol. 61, 1271–1275.

Vassar, R. & Fuchs, E. (1991). Transgenic mice provide new insights into the role of TGFα during epidermal development and differentiation. Genes Dev. 5, 714–727.

Vassar, R., Hutton, M. E., & Fuchs, E. (1992). Transgenic overexpression of transforming growth factor α bypasses the need for c-Ha-*ras* mutations in mouse skin tumorigenesis. Mol. Cell. Biol. 12, 4643–4653.

Vonderhaar, B. K. (1987). Local effects of EGF, TGFα, and EGF-like growth factors on lobuloalveolar development of the mouse mammary gland *in vivo*. J. Cell. Physiol. 132, 581–584.

Watanabe, S., Lazar, E., & Sporn, M. B. (1987). Transformation of normal rat kidney (NRK) cells by an infectious retrovirus carrying a synthetic rat type α transforming growth factor gene. Proc. Natl. Acad. Sci. USA 84, 1258–1262.

Wen, D., Peles, E., Cuppies, R., Suggs, S. V., Bacus, S. S., Luo, Y., Trail, G., Hu, S., Silbiger, S. M., Levy, R. B., Koski, R. A., Lu, H. S., & Yarden, Y. (1992). Neu differentiation factor: A transmembrane glycoprotein containing an EGF domain and an immunoglobulin homology unit. Cell 69, 559–572.

Wilcox, J. N., & Derynck, R. (1988a). Developmental expression of transforming growth factors α and β in mouse fetus. Mol. Cell. Biol. 8, 3415–3422.

Wilcox, J. N. & Derynck, R. (1988b). Localization of cells synthesizing transforming growth factor α mRNA in the mouse brain. J. Neurosci. 8, 1901–1904.

Wong, S. T., Winchell, L. F., McCune, B. K., Earp, H. S., Teixido, J., Massagué, J., Herman, B., & Lee, D. C. (1989). The TGFα precursor expressed on the cell surface binds to the EGF receptor on adjacent cells, leading to signal transduction. Cell 56, 495–506.

Yang, S–G., Winkler, M. E., & Hollenberg, M. D. (1990). Contribution of the C-terminal dipeptide of transforming growth factor α to its activity: Biochemical and pharmacologic profiles. Eur. J. Pharm. 188, 289–300.

Yeh, Y–C., Tsai, J–F., Chuang, L–Y., Yeh, H–W., Tsai, J–H., Florine, D. L., & Tam, J. P. (1987). Elevation of transforming growth factor α and its relationship to the epidermal growth factor and α-fetoprotein levels in patients with hepatocellular carcinoma. Cancer Res. 47, 896–901.

Zipfel, P. A., Ziober, B. L., Morris, S. L., & Mulder, K. M. (1993). Upregulation of transforming growth factor α expression by transforming growth factor β1, epidermal growth factor, and *N*,*N*-dimethylformamide in human colon carcinoma cells. Cell Growth & Differentiation 4, 637–645.

TRANSFORMING GROWTH FACTOR βS: BIOCHEMISTRY AND BIOLOGICAL ACTIVITIES *IN VITRO* AND *IN VIVO*

Wei Cui and Rosemary J. Akhurst

Abstract 320
I. Introduction 320
II. Biochemistry and Molecular Biology of TGFβ 321
A. Molecular Structure of TGFβ 321
B. Transcriptional and Posttranscriptional Control of TGFβ Gene Expression . . 323
C. Activation of Latent TGFβ 325
III. Effects of TGFβ *In Vitro* 326
A. Control of Cell Proliferation 326
B. Modulation of Extracellular Matrix 327
C. Control of Cellular Differentiation 328
D. Effects on the Immune System 329
E. Induction of Apoptosis 329
IV. Differential Functions of TGFβ Isoforms 330
V. Biological Activities of TGFβ *In Vivo* 331
A. Ectopic Expression of TGFβ1 in Transgenic Mice 331

Growth Factors and Cytokines in Health and Disease
Volume 1B, pages 319–356.

ISBN: 0-7623-0091-4

B. TGFβ Gene "Knock-Out" Mice 334
C. TGFβ in Wound Healing and Tissue Repair 338
D. Role of TGFβ in Multistage Epithelial Carcinogenesis 340
VII. Conclusions 343
Abbreviations 343
References 344

ABSTRACT

TGFβs are a subset of three genetically distinct and closely related growth factors, belonging to a larger superfamily of related molecules. They share most biological activities, and their actions are potentiated by the same set of cell-surface receptors (see Massague's chapter). Production of TGFβs is regulated at many levels, including transcriptional, posttranscriptional and translational control. Furthermore, these molecules are synthesized as latent forms, which require activation at the cell surface, thus, introducing a further regulatory point for control of biological activity. Reports and reviews on the *in vitro* activities of TGFβs have been very extensive. In general, they are potent negative growth regulators, they can influence cellular differentiation in a positive or negative fashion, and they are major modulators of extracellular matrix composition and cell/substratum interactions. *In vivo*, they play a role in a number of biological processes common to embryonic and fetal development, tissue repair, and the immune response. Aberrations in the production and response to TGFβ can, therefore, lead to the outgrowth of neoplastic lesions.

I. INTRODUCTION

It has been just over a decade since the first reports which documented the peptide, transforming growth factor beta (TGFβ) as a distinct molecule (Assoian et al., 1983; Frolik et al., 1983; Roberts et al., 1983). Two years later, the first human cDNA encoding TGFβ1 was molecularly cloned (Derynck et al., 1985). Since then, a remarkable understanding of the structure and function of this family of peptides has been achieved. The TGFβs are now known to be the archetype of a larger superfamily of related molecules with more than thirty individual members (Kingsley, 1994). Five different genes for TGFβ have been isolated to date, though only three isoforms, namely TGFβ1, β2, and β3, have been found in mammals. TGFβ4, found only in the chick, is probably the avian homologue of TGFβ1 (Burt and Paton, 1992). TGFβ5 (Kondaiah et al., 1990), which is only found in the tetraploid organism, *Xenopus*, is possibly a polymorphic variant of one of the TGFβs.

Most studies on TGFβ have attempted to understand its biological function and mechanism of action on cells. Immense progress has been made in our conceptual knowledge of TGFβ action, especially with the development of transgenic techniques, which has allowed its *in vivo* functions to be addressed. TGFβ is accepted as the prototypical multifunctional growth factor, controlling cell proliferation and differentiation, modulating the synthesis of extracellular matrix (ECM) and the biological activities of other growth factors. TGFβ appears to play a vital role in tissue homeostasis and in the immune response, and aberrations in cellular production and sensitivity to TGFβs probably contribute to multistage carcinogenesis.

This chapter reviews the recent advances in knowledge about TGFβ structure and biological function, emphasizing its biological effects *in vivo* and its role in multistage carcinogenesis of the skin.

II. BIOCHEMISTRY AND MOLECULAR BIOLOGY OF TGFβ

A. Molecular Structure of TGFβ

TGFβ was originally purified as a homodimeric protein of 25kDa from human placenta (Frolik et al., 1983), platelets (Assoian et al., 1983), and from bovine kidney (Roberts et al., 1983). This form is now known as TGFβ1. Further biochemical purification and molecular cloning identified three genetically distinct isoforms in mammals, designated TGFβ1, β2, and β3, respectively. The three genes encode precursors of 390 (TGFβ1) and 412 (TGFβ2 and β3) amino acids and have been genomically mapped to human chromosomes 19, 1 and 14, respectively (Fujii et al., 1986; Barton et al., 1988; ten Dijke et al., 1988). The three genes have a similar seven-exon gene structure and exhibit high DNA sequence conservation, suggesting that they are derived from a common ancestor. The TGFβs are synthesized as large precursor molecules (pre-pro-TGFβ) (Figure 1A), each with an N-terminal, short-signal peptide. The carboxy-terminal bioactive domain contains 112 amino acids and is generated by proteolytic cleavage from the longer N-terminal fragment. The latter is termed the latency-associated peptide (βLAP), and remains noncovalently associated with the mature dimer, keeping TGFβ in a latent state.

The small bioactive peptide is highly conserved between isoforms (70–80% amino acid homology) and is virtually identical in individual isoforms from different species. The three-dimensional structure of the mature bioactive dimer of TGFβ2 has been determined by X-ray crystallography at high resolution (Daopin et al., 1992; Schlunegger and Grutter, 1992). This revealed that the molecule has an unusual elongated nonglobular conformation (Figure 1B). Eight of the nine cysteine residues in each monomer chain are involved in tightly associated intrachain disulfide bonds, and there is only a single disulfide bond cross-linking the two monomers, suggesting that hydrophobic interactions are important in stabilizing the dimer. Two of the intrachain disulfide bonds are involved in an unusually tight eight-member ring structure at the core of the molecule, and the third intrachain disulfide bond points directly through the ring. This ring structure is also conserved in other TGFβ superfamily members (Kingsley, 1994), and a similar structural motif is seen in other unrelated growth factors (McDonald and Hendrickson, 1993). The fourth intramolecular disulfide bond, cross-linking the amino termini of the mature dimer, is specific to the TGFβs alone. Nuclear magnetic

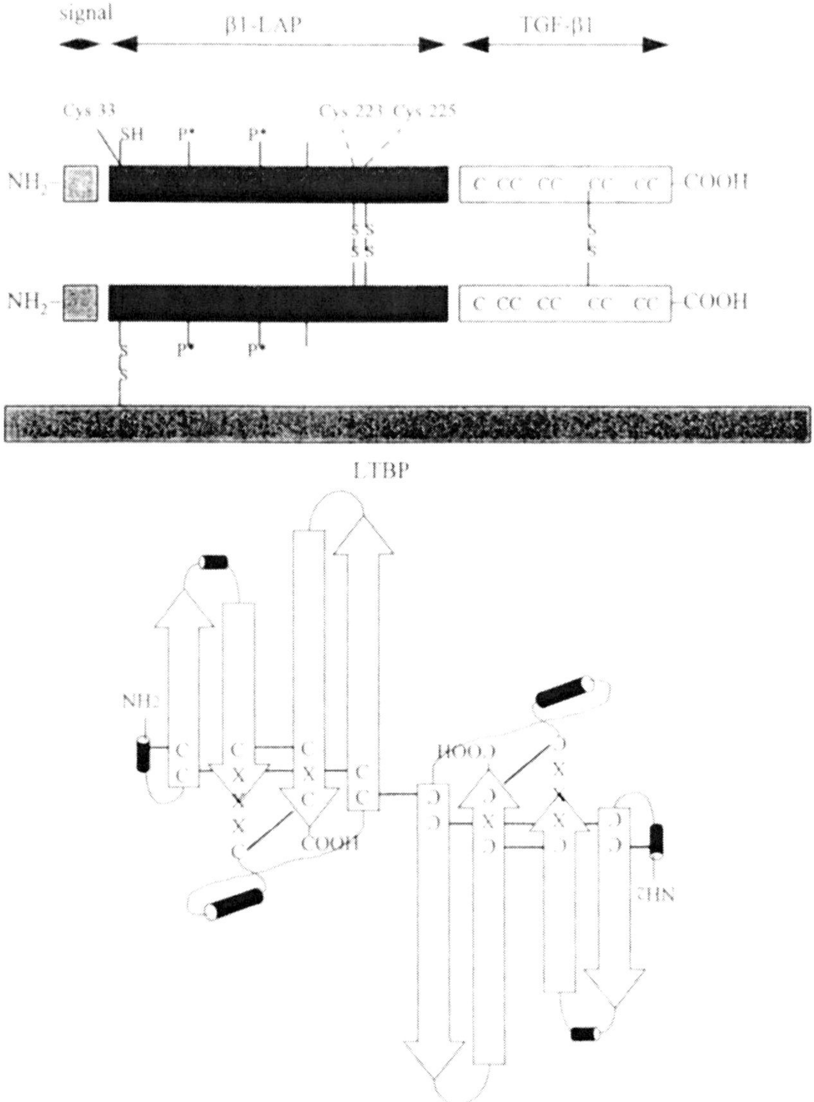

Figure 1. Structure of TGFβs. (A) Latent TGFβ1 complex. TGFβ1 protein is synthesized as a larger precursor which contains an amino-terminal signal sequence (29 aa) followed by a β1-LAP domain (249 aa) and the carboxy-terminal mature TGFβ1 (112 aa). The three conserved cysteine residues in the βLAP are indicated (c) as are the nine cysteine residues in the mature dimer. One molecule of LTBP is bound to β-LAP dimer by disulfide bond(s). In the β1-LAP domain, two of the N-linked carbohydrates may contain man-6-P (shown as P*), which may contribute to activation of TGFβ1. (B) Crystal structure of the mature form of TGFβ2. Eight cysteine residues in each monomer form four intrachain disulfide bonds, and only one disulfide bond links the two monomers. Two intrachain disulfide bonds are involved in a c-x-x-x-c-c-x-c eight member ring.

resonance (NMR) analysis of TGFβ1 in solution has confirmed many aspects of the TGFβ2 crystallography data (Archer et al., 1993).

Although the structure of mature TGFβs is highly conserved among the three isoforms, the sequences of the βLAPs are markedly dissimilar. The βLAPs do, however, share some important structural features, such as three cysteine residues. Two of these (Cys 223 and 225) are involved in interchain disulfide bonding of the β1-LAP dimer. Site-directed mutagenesis of these cysteines to serine results in the inability of β1-LAP to dimerize, thus leading to production of TGFβ1 in a predominantly bioactive form (Brunner et al., 1989). The βLAPs also possess several isoform-specific, N-linked glycosylation sites, where mannose-6-phosphate (man-6-P) residues are added posttranslationally.

The βLAP region is probably important for the protein folding, secretion, and activation (Gray and Mason, 1990; Sha et al., 1991; Lopez et al., 1992). It has been postulated that divergence in the βLAP region of each TGFβ isoform specifies interisoform functional distinctions (Denhez et al., 1990; Roberts and Sporn, 1990). β1LAP and β3LAP both have cell-adhesion, receptor-recognition sequences (RGD sequences), whereas β2LAP does not. However, functional differences between the isoforms are easily discernible using recombinant bioactive TGFβs, which do not contain the LAP region. Thus, although βLAP might not directly contribute to functional distinctions between TGFβ2 and the other two isoforms (Qian et al., 1992), it might indirectly lead to functional distinctions, by influencing the affinity of the latent complex for binding and localizing within the ECM. All three βLAPs are capable of binding interchangeably to each of the mature isoform homodimers (Gentry and Nash, 1990; Miller et al., 1992), though whether this leads to functional distinctions has not been addressed.

B. Transcriptional and Posttranscriptional Control of TGFβ Gene Expression

The promoters of all three human TGFβ isoforms have been cloned and characterized (Kim et al., 1989; Lafyatis et al., 1990; Noma et al., 1991; O'Reilly et al., 1992; Arrick et al., 1994). Each displays distinct regulatory elements (Figure 2), suggesting that expression of each isoform can be independently regulated. In contrast to dissimilarity of promoter sequences between isoforms, there is considerable conservation of gene promoter homology between species (Geiser et al., 1991). This would be consistent with the diverse expression patterns of the three isoforms, as well as the conserved expression patterns of the three isoforms between human and mouse (Gatherer et al., 1990; Millan et al., 1991).

Several studies have demonstrated that posttranscriptional and translational regulation of TGFβ are important in controlling biosynthesis of TGFβ (Kehrl et al., 1986a; Assoian et al., 1987; Glick et al., 1989; Coletta et al., 1990). Release of bioactive TGFβ1 from activated monocytes is certainly regulated posttranslationally (Assoian et al., 1987). In addition, the increased secretion of TGFβ2 in response to retinoic acid treatment in a number of biological systems is predominantly

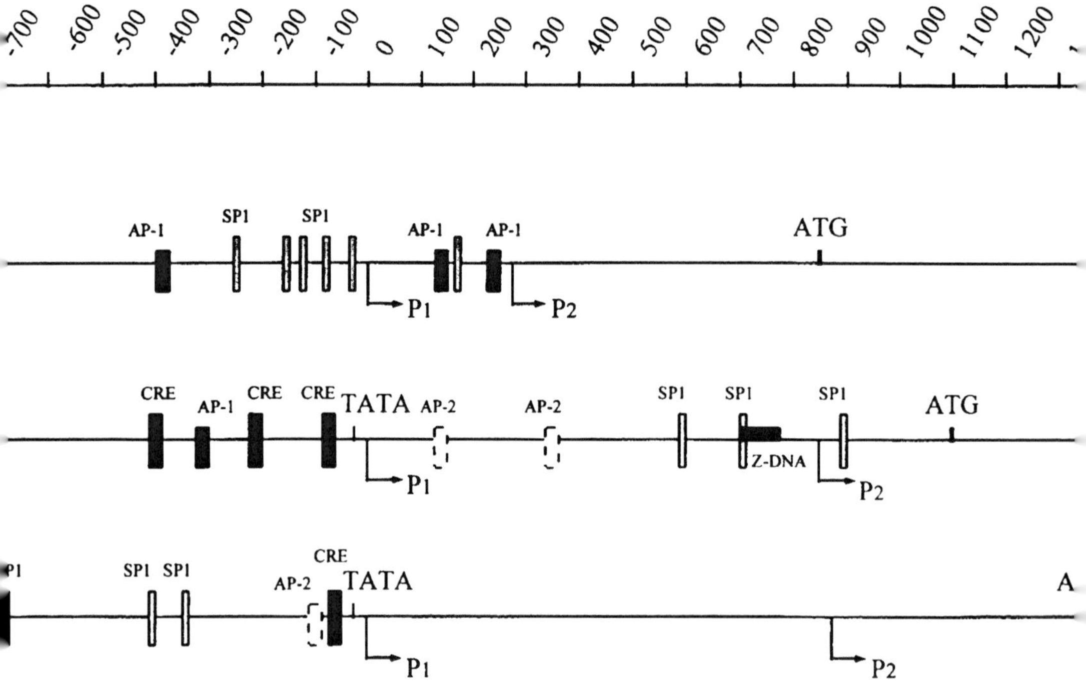

Divergent promoters of three human TGFβ genes. All three TGFβ gene promoters have long 5′-noncod
wo major transcription start sites. However, TGFβ1 gene promoter has no TATA box, whereas TGFβ2 and β3
sus TATA box. Moreover, TGFβ1 promoter has several AP-1-like binding sequences and is regulated by
ional factors. In contrast, TGFβ3 contains no AP-1-like sequence but has a consensus cAMP-responsive elem
ting the promoter activity by forskolin. TGFβ2 promoter contains several CRE consensus sequences, AP-1 a
-nucleotide region characteristic of Z-DNA helices. The precise role of these elements in regulating TGFβ
o be determined.

regulated translationally and posttranslationally (Glick et al., 1989; Abbott and Birnhaum, 1990).

Recent studies have suggested that noncoding sequences at the 5′ end of TGFβ mRNA may exert cell-specific inhibitory effects on translation (Arrick et al., 1991; Robinson et al., 1991; Kim et al., 1992; Romeo et al., 1993; Arrick et al., 1994). All three TGFβ mRNA transcripts have been shown to contain long, often GC-rich 5′ untranslated regions (UTRs), in contrast to the vast majority of eukaryotic mRNAs (Kozak, 1987). By transfection of cells with chimeric plasmid constructs and deletion mutagenesis studies *in vitro*, a stable stem-and-loop structure within the 5′ UTR of TGFβ1 mRNA is involved in reducing translational efficiency in a tissue-specific manner (Kim et al., 1992; Arrick et al., 1994). Moreover, using similar techniques, the 3′ untranslated region of TGFβ1 mRNA, which also contains a GC-rich domain, has a bifunctional effect on overall protein expression (Scotto and Assoian, 1993). However, this does not appear to be the case for TGFβ2 and β3, because neither 3′ UTRs have this GC-rich domain.

C. Activation of Latent TGFβ

Unlike other peptide growth factors, all three TGFβs are synthesized and secreted almost exclusively as biologically inactive complexes (Derynck et al., 1985; Brown et al., 1990). These latent complexes can not bind to cell-surface receptors (Pircher et al., 1986) and are, therefore, inactive biologically. Thus, activation of latent TGFβ is a possible control point for regulation of its bioactivity *in vivo*.

By transfection of TGFβ1 cDNAs into mammalian cells, recombinant TGFβ1 was found synthesized as a "small latent complex" (Gentry et al., 1987; Madisen et al., 1990; Brown et al., 1990) composed of the dimeric βLAP in association with potentially bioactive TGFβ. However, *in vivo*, in human platelets and various rat tissues, endogenous TGFβ1 is secreted as a "large latent complex" (Miyazono et al., 1988; Tsuji et al., 1990) comprising an additional component, named the latent TGFβ-binding protein (LTBP), which is covalently linked to βLAP (Figure1A). LTBP is a glycoprotein of 125–160kDa, containing several epithelial growth factor-like (EGF-like) repeats (Kanzaki et al., 1990). It is structurally related to fibrillin, the ECM protein which is genetically defective in Marfan syndrome individuals (Maslen et al., 1991). The function of LTBP is not fully understood, but it is clearly unnecessary for conferring latency (Kanzaki et al., 1990). It may be important for accurate and efficient protein folding and secretion from the cell (Miyazono et al., 1991). Recent data suggest that it may also play an important role in efficient activation of latent TGFβ at the cell surface (Flaumenhaft et al., 1993).

The latent TGFβ complexes can be activated *in vitro* by acidification, alkalinization, heating, or the use of chaotropic agents, such as SDS and urea, though the mechanism of activation *in vivo* is still a subject of much study. Miyazono and Heldin (1989) suggested that glycosylation might be important because enzymatic removal of carbohydrate residues resulted in inactivation. It was demonstrated that binding of βLAP to man-6-P /IGF-II receptor is required for activation *in vivo*

(Kovacina et al., 1989; Dennis and Rifkin, 1991). This event, however, cannot be essential because man-6-P was not observed in β1-LAP of a human erythroleukemia cell line (Miyazono et al., 1993). Furthermore, other processes must also be invoked because many cells, which express man-6-P receptors, fail to activate latent TGFβ (Massague et al., 1992). Activation is thought to occur at the cell surface, requiring cell–cell contact in certain cell types (Sato and Rifkin, 1989). For example, activation of TGFβ occurred only in cocultures of endothelial cells together with smooth muscle cells, but failed in cultures of each cell type alone (Antonelli–Orlidge et al., 1989; Sato and Rifkin, 1989). Plasmin was implicated in this activation, because the process was blocked by adding plasmin inhibitor or antiplasmin antibodies (Sato and Rifkin, 1989, Lyons et al., 1990). Physiological activation of latent TGFβ by plasmin is thought to occur by proteolytic nicking of the amino terminus of βLAP, leading to release of the bioactive dimer (Lyons et al., 1990). It has also been demonstrated that urokinase-type plasminogen activator (uPA) is required for TGFβ activation in endothelial cells (Flaumenhaft et al., 1992). Finally, both LTBP (Flaumenhaft et al., 1993) and transglutaminase type II (Kojima et al., 1993) have been found to be necessary for TGFβ activation of endothelial cells. The finding of the requirement for transglutaminase was surprising, but might be related to its activity in covalently cross-linking proteins to the cell surface, possibly immobilizing plasmin/uPA or concentrating the latent complex at the cell surface by cross-linking of LTBP molecules (Kojima et al., 1993).

III. EFFECTS OF TGFβ *IN VITRO*

The biological effects of the TGFβs *in vitro* are remarkably diverse. They can stimulate or inhibit cell proliferation, downregulate or upregulate cell differentiation, or alter the cellular phenotype. Their precise action depends on the cell type, the proliferative or differentiative state of that cell, and the culture conditions.

A. Control of Cell Proliferation

TGFβs can affect the proliferation of cells from almost every lineage. Although TGFβ was originally identified by its ability to induce a transformed phenotype in normal rat kidney fibroblast cells (Moses et al., 1981; Roberts et al., 1981), intensive studies in the last decade have demonstrated that the predominant effect of TGFβ on cell proliferation is inhibitory. Its ability to synergize with TGFα in a transformation assay (Anzano et al., 1982) was probably due to its effects in modulating ECM deposition. All three TGFβ isoforms possess reversible growth inhibitory activities on a wide variety of cell types, including epithelial, endothelial, hematopoietic, fibroblastic, lymphoid, and neuronal cells (reviewed by Roberts and Sporn, 1990; Massague, 1990).

In contrast to its predominant growth inhibitory effects, reports of TGFβ acting as a growth promoter are relatively few. These growth stimulatory effects are mainly indirect, mediated either through ECM proteins or via other growth factors.

There is no clear evidence that TGFβ receptors can directly elicit mitogenic signals (Massague et al., 1992). For instance, the growth promoting effect of TGFβ on NRK fibroblasts in soft agar was actually the result of enhanced expression of the ECM proteins, fibronectin, type I collagen and proteoglycans, and increased integrin-mediated adhesion to these molecules (Ignotz and Massague, 1986). Likewise, TGFβ1 induced mitogenesis in Balb/c3T3 cells by stimulating the production of basic fibroblast growth factor-binding proteoglycans (Nugent and Edelman, 1992). The proliferative response to TGFβ in aortic smooth muscle cells was due to induction of platelet-derived growth factor (PDGF) synthesis, demonstrated by the fact that this response was ablated when neutralizing antibody to PDGF was added into the medium (Battegay et al., 1990). Finally, in neuronal cell culture, TGFβ protected neurons from hypoxia-induced degeneration and death leading to increased survival (Lindholm et al., 1992; Prehn et al., 1993), by upregulating nerve growth factor (NGF) production (Buchman et al., 1994).

B. Modulation of Extracellular Matrix

One of the most striking bioactivities of TGFβ is modulation of ECM deposition, composition, and cell-substratum interactions. The ECM is not only a support for tissue structure, but more importantly, a complex and dynamic molecular substratum which can modify cellular phenotype (Madri et al., 1983; Li et al., 1987). The ECM can affect cell proliferation, differentiation, morphology, adhesion, migration, and interaction between cells. Thus, TGFβ may affect cell proliferation and differentiation by modulation of ECM. These indirect effects are probably central to normal developmental processes and relevant to wound healing and carcinogenesis.

In general, TGFβ promotes ECM formation and can enhance the response of cells to ECM via alterations in integrin expression. It stimulates the deposition of several ECM components, including fibronectins, interstitial collagens, thrombospondin, tenascin, laminin, and chondroitin/dermatan sulfate proteoglycans (reviewed by Roberts and Sporn, 1990). The enhancement of ECM protein synthesis by TGFβ results from the increased transcription of the corresponding genes and, in some instances, stabilization of their mRNAs.

In addition, TGFβ potentiates ECM production by its ability to inhibit matrix protein degradation, which is achieved through its action on genes encoding proteases and proteases inhibitors. TGFβ decreases synthesis and secretion of several proteases, including tissue plasminogen activator, uPA, thiol protease, collagenase, and transin, but increases expression of the protease inhibitors, plasminogen activator inhibitor type I (PAI-I), urokinase, and the tissue inhibitor of metalloproteinases (TIMP) (Roberts and Sporn, 1990).

TGFβ also regulates cell-substratum interactions by altering expression of integrins, a major class of cell-adhesion receptor. TGFβ selectively increases the expression of integrins which bind fibronectin, collagen, and other ECM proteins (Ignotz et al., 1989, Heino and Massague, 1989).

C. Control of Cellular Differentiation

TGFβ acts as a bifunctional modulator of cell differentiation independently of its effects on proliferation. The various biological effects of TGFβ on differentiation are heavily dependent upon the differentiative state of the target cells and their environment. TGFβ has been reported to be a potent inhibitor of adipocyte and chondrocyte differentiation (Torti et al., 1989; Kato et al., 1988). Nevertheless, it promotes osteoblast and osteoclast differentiation (Noda and Rodan, 1987; Fiorelli et al., 1994). When primary myoblasts and established cell lines of both smooth and skeletal muscle were grown in low-mitogen, differentiation-inducing conditions, terminal differentiation was prevented by the presence of TGFβ (Olson et al., 1986, Massague et al., 1986; Florini et al., 1986). In contrast, when skeletal myoblasts were cultured in mitogen-rich conditions, TGFβ induced their terminal differentiation (Zentrella and Massague, 1992). The inhibitory effect of TGFβ on myoblast differentiation was subsequently shown to be due to indirect effects of TGFβ on the ECM (Heino and Massague, 1990), whereas the differentiation stimulatory effects are more likely to be direct (Zentrella and Massague, 1992). TGFβ also induces differentiation of the erythoid cell line K562 (Chen et al., 1989) and stimulates granulocyte/macrophage colony-stimulating factor (GM-CSF)-induced granulopoiesis (Keller et al., 1991, 1994).

TGFβ induces differentiation of human bronchial epithelial cells and certain colon carcinoma cells (Masui et al., 1986; Chakrabarty et al., 1989), but its reported effects on keratinocyte differentiation have been controversial (Reiss and Sartorelli, 1987; Coffey et al., 1988; Mansbridge and Hanawalt, 1988). TGFβ alone inhibited differentiation of human foreskin-derived keratinocytes in a dose-dependent manner but, in combination with epidermal growth factor, it enhanced their differentiation (Reiss and Sartorelli, 1987). TGFβ modulates keratinocyte phenotype in culture, by altering the profile of keratin genes expressed (Reiss and Sartorelli, 1987; Mansbridge and Hanawalt, 1988). In contrast, TGFβ had no differentiative effect on the murine keratinocyte cell line, Balb/Mk (Coffey et al., 1988). Finally, Matsumoto et al. (1990) demonstrated that TGFβ enhanced differentiation of human keratinocytes under high Ca^{2+} conditions, but inhibited their differentiation under low Ca^{2+} conditions. It is quite clear that the specific effects of TGFβ on cells in culture depend on the exact culture conditions. Furthermore, the fact that simply culturing keratinocytes can alter the expression of TGFβ receptors (Cui et al., 1995) and that complete keratinocyte differentiation is difficult to achieve *in vitro*, makes it difficult to draw any precise conclusions regarding the differentiative activity of TGFβ on keratinocytes from studies performed *in vitro*.

Contrasting effects of TGFβ on differentiation have also been seen in endothelial cells. TGFβ1 is a potent endothelial growth inhibitor in monolayer cultures (Heimark et al., 1986; Muller et al., 1987). However, when microvasular endothelial cells were seeded into a three-dimensional culture system, TGFβ1 did not affect cell proliferation but could induce the differentiation of capillary tube-like structures (Madri et al., 1988), with a concomitant increase in expression of platelet endothelial cell-adhesion molecule (PECAM), fibronectin, and certain integrins (Merwin et al., 1990; Basson et al., 1992; Madri et al., 1992). These differential biological responses might also be due to the differential expression of TGFβ receptors under different culture conditions (S. Sankar and J. Madri, personal communication).

D. Effects on the Immune System

TGFβ is a potent modulator of the immune system and inflammatory responses, generally functioning as an immunosuppressor (Wahl, 1992). TGFβ can inhibit proliferation of thymocytes (Ristow, 1986), T lymphocytes (Kehrl et al., 1986b) and large granular lymphocytes (Ortaldo et al., 1991), and can suppress B cell proliferation and immunoglobulin (IgG and IgM) secretion (Kehrl et al., 1986a; 1991). It also restrains the activity of natural killer cells (Rook et al., 1986) as well as lymphokine-activated killer cells (Espevik et al., 1988). Furthermore, it suppresses the production and activity of some cytokines, such as interleukin-1 (IL-1), IL-2, and IL-3 (Wahl et al., 1988; Kehrl et al., 1986a, b; Ohta et al., 1987), interferon-γ (Czarniecki et al., 1988) and tumor necrosis factor α (TNFα).

During inflammation, TGFβ has more complicated effects. At the beginning of the inflammatory response it acts positively to stimulate monocyte migration and macrophage production, acting as a potent chemoattractant (Wahl et al., 1987). After the initiation of an inflammatory response, it then exhibits potent anti-inflammatory effects, such as preventing the adhesiveness of neutrophils and T-lymphocytes to endothelium (Gamble and Vadas 1988; 1991), downregulation of macrophages and antagonism of TNFα function. In addition, TGFβ1 suppresses the synthesis of nitric oxide by blocking the induction of IL-1β, leading to downregulation of inducible nitric oxide synthase (iNOS) (Roberts et al., 1992; Perrella et al., 1994).

E. Induction of Apoptosis

Apoptosis is a genetically encoded, active cell-death program defined by characteristic morphological and biochemical changes (Wyllie, 1985). It can be considered as part of the cell's repertoire in responding to external and/or internal stimuli, in much the same way as mitogenesis or differentiation.

Several studies have provided evidence that TGFβ might act as an inducer of apoptosis in some cell types. Levels of TGFβ1 mRNA were increased during castration-induced regression of the prostate, a process where apoptosis is of major importance (Kyprianou and Isaacs, 1989). In cultures of both primary rat hepatocytes and human Hep 3B hepatoma cells, TGFβ1 appeared to induce cell death with apoptotic features, including DNA fragmentation, condensation of chromatin, cellular rounding and fragmentation, and phagocytosis by neighboring cells (Oberhammer et al., 1991; 1992; Lin and Chou, 1992; Bursch et al., 1993). This apoptotic process is coordinated with inhibition of cell proliferation by TGFβ. Similar phenomena have also been observed in cultures of primary rabbit uterine epithelial cells and endometrial stromal cells treated with TGFβ (Rotello et al., 1991; Moulton, 1994). These effects can be inhibited by addition of TGFβ antibodies to the medium. Recently, Jurgensmeier et al. (1994) cocultured transformed fibroblasts with normal fibroblasts and found that, in synergy with TGFβ, normal fibroblasts were able to eliminate transformed cells by induction of apoptosis.

The induction of apoptosis by TGFβ is possibly an indirect effect. It may result from changes in ECM or the profile of cytokine production induced by TGFβ. For example, although TGFβ induced apoptosis in Hep 3B cells, it could not cause cytocidal activity in another human hepatoma Hep G2 cells, even though Hep G2 cells contain more TGFβ receptors (Lin and Chou, 1991).

IV. DIFFERENTIAL FUNCTIONS OF TGFβ ISOFORMS

Many studies on the biological activities of TGFβ have concentrated on the effects of recombinant TGFβ1 on cells in culture. As discussed above, there are three independent genes encoding three mammalian TGFβ isoforms. *In vitro* studies have shown that, in general, all three isoforms have similar biological effects, but, in a few biological systems, they have demonstrated very distinct biological actions. TGFβ2 is active in a *Xenopus* animal cap assay for formation of mesoderm from ectoderm, whereas TGFβ1 is not (Rosa et al., 1988). In mammalian cells, TGFβ1 is a very potent endothelial growth inhibitor though this is not true for TGFβ2 (Jennings et al., 1988; Merwin et al., 1991). Besides the qualitative differences between isoforms, there are also considerable quantitative differences in their biological response (Graycar et al., 1989; Cheifetz et al., 1992) which could be important to bioaction *in vivo*.

In general, TGFβ1 and β3 act with a very similar spectrum of bioactivities and potencies on cells *in vitro*, whereas TGFβ2 shows a more restricted spectrum of activity and often has reduced potency on certain cell types (Graycar et al., 1989; Cheifetz et al., 1990). This might be related to the fact that TGFβ1 and β3 bind strongly to the TGFβ type II receptor (TβRII), whereas, in the absence of β-glycan, otherwise known as the type III receptor, TGFβ2 shows very weak binding to TβRII (Massague, 1992; also see Massague's chapter). Certain cell types, notably endothelial and hematopoietic cells, do not express β-glycan, but a functional equivalent, endoglin (Massague et al., 1990), which does not bind TGFβ2 (Massague et al., 1990; Cheifetz et al., 1992). These cell types are refractile to TGFβ2 (Ohta et al., 1987; Jennings et al., 1988). It has been demonstrated that molecular determinants within the mature bioactive region of TGFβ2 (amino acids 40–82) specify its distinct biological activities (Qian et al., 1992), presumably because this divergent amino acid region determines differential binding affinities of TGFβ1 and TGFβ2 for the TβRII (Lin et al., 1992, Massague, 1992).

V. BIOLOGICAL ACTIVITIES OF TGFβ *IN VIVO*

Although studies of TGFβ effects on cells *in vitro* have been invaluable to our understanding of the role that these molecules might play in biological processes *in vivo*, one must be cautious not to extrapolate directly to the *in vivo* situation. Biological responses to TGFβ, which are dictated by expression of the various receptors and intracellular signalling systems (see Massague's chapter) depend on cell type, the proliferative and differentiative status of the cell, and on cell–cell contacts and the cell environment. Cell culture systems rarely properly duplicate the *in vivo* environment. In some tissues, TGFβ has consistent effects both *in vitro* and *in vivo*. However, in others, the effects of TGFβ *in vivo* have been somewhat unpredicted from *in vitro* studies. For example, TGFβ inhibits proliferation of endothelial cells in culture (Heimark et al., 1986; Muller et al., 1987; Jennings et al., 1988) but this growth factor is angiogenic *in vivo* (Roberts et al., 1986; Yang and Moses, 1990). Furthermore, mice lacking TGFβ1 do not exhibit endothelial hyperplasia, but some embryos do have an altered endothelial phenotype (Dickson et al., 1995).

In the last 2 to 3 years, the functions of TGFβ *in vivo* and its potential contribution to a wide variety of pathophysiological processes have received increasing attention. There have been four main approaches to the study of the function of TGFβ *in vivo*: (1) descriptive studies on the sites of TGFβ gene and protein expression during embryogenesis and in pathophysiological conditions, (2) administration of TGFβ and its antagonists *in vivo*, (3) studies on transgenic mice which overexpress TGFβ1 in specific tissues, and (4) studies on TGFβ knockout mice in which individual genes (TGFβ1, 2, or 3) have been disrupted by site-directed homologous recombination using embryonal stem (ES) cells. In the following section, we will review some of the recent advances which have been made by using transgenic technology. Descriptive studies of TGFβ localization and bioactivity during embryogenesis have been comprehensively reviewed elsewhere (Akhurst, 1994), and dosing studies with TGFβ and its antagonists are of more relevance to pharmacological studies than to an understanding of the TGFβs *in vivo*.

A. Ectopic Expression of TGFβ1 in Transgenic Mice

With the development of transgenic technology, it is now possible to transfer selected genes directly into the murine germ line, stably altering the genetic constitution of the mice and generating animals that carry a new genetic program. This transgenic approach provides opportunities to study the *in vivo* functions of genes thought to play important roles in controlling cell growth and differentiation. Many attempts to produce transgenic mice which overexpress TGFβ have not succeeded due to absence of transgene expression or embryonic lethality (Sporn and Roberts, 1992). However, several laboratories have generated TGFβ1-expressing transgenic mice, using tissue-specific gene promoters to restrict the expression of TGFβ1 to specialized organ systems.

Keratin Promoter TGFβ1 in Transgenic Mice

In vitro studies have shown that TGFβ has a profound inhibitory effect on keratinocyte proliferation in culture (Moses et al., 1985; Reiss and Sartorelli, 1987), but the data are controversial with respect to differentiation. It has been reported to be an enhancer (Mansbridge and Hanawalt, 1988; Matsumoto et al., 1990), inhibitor (Matsumoto et al., 1990) of keratinocyte differentiation, or to have no differentiative effect at all (Coffey et al., 1988), depending on the culture conditions. The expression patterns of TGFβ isoforms in the epidermis during skin development and in response to exogenous stimuli which perturb homeostasis (Akhurst et al., 1988; Flanders et al., 1989; Glick et al., 1989; Pelton et al., 1990, 1991; Millan et al., 1991; Levine et al., 1993), have suggested that these factors are important regulators of epidermal growth and homeostasis. To obtain further insight into the biological functions of TGFβ1 *in vivo*, two independent laboratories generated transgenic mice which utilized keratin promoters to drive TGFβ1 expression predominantly in the suprabasal differentiative compartment of the epidermis (Selleheyer et al., 1993; Cui et al., 1995).

Selleheyer et al. (1993) targeted bioactive TGFβ1 expression to the epidermis of transgenic mice, using a vector consisting of a human keratin 1(K1) gene promoter driving a bioactive porcine TGFβ1 cDNA, and showed that all of the founder mice died within 24 hours after birth due to inhibition of epidermal development. Histological and immunohistochemical studies on the mice showed complete inhibition of DNA synthesis in the epidermis and a reduced number of hair follicles in the skin.

Cui et al. (1995), using a similar gene promoter equivalent to human keratin 10 (K10), successfully generated viable lines of transgenic mice which constitutively expressed bioactive TGFβ1 in suprabasal keratinocytes. Surprisingly, and in contrast with the first study (Sellheyer et al., 1993), these mice showed a 2 to 3- fold increase in epidermal DNA labeling index over control mice, but did not have any obvious morphological or histological abnormalities. These results suggested that TGFβ1 not only increased keratinocyte proliferation *in vivo* (either directly or indirectly), but increased keratinocyte turnover in a compensatory fashion. Interestingly, when skin hyperplasia was induced in the transgenic epidermis by topical application of 12-tetradecanoyl-phorbol-13-acetate (TPA), the transgene product acted as a negative regulator of cell growth, and epidermal proliferation was strikingly attenuated in transgenic mice compared to control mice. Recombinant TGFβ1 was, therefore, acting as a stimulator of keratinocyte proliferation in quiescent adult mouse epidermis, but as a potent inhibitor of proliferation of hyperplastic keratinocytes. Studies on expression of the type II receptor (TβRII) were entirely consistent with this unexpected observation. In quiescent epidermis, TβRII expression was barely detectable, but increased rapidly in response to induction of hyperplasia by TPA (Cui et al., 1995). Taken together with earlier studies showing increased TGFβ1 expression in suprabasal keratinocytes in response to induction of epidermal hyperplasia (Akhurst et al., 1988; Krieg et al., 1991; Fowlis et al., 1992; Esherick et al., 1993), this data suggests that TGFβ1 and TβRII act in concert to regulate epidermal homeostasis *in vivo*.

The disparate phenotypes observed between the two studies of Sellheyer et al., (1993) and Cui et al. (1995), might have been due to several factors, such as the use of different keratin gene promoters, mouse strain-specific difference in expression levels of TβRII or, more likely, the different expression levels of recombinant TGFβ1 achieved in the transgenic epidermis. The K10 promoter drove gene expression exclusively in the suprabasal compartment (Cui et al., 1995), whereas, the K1 promoter used by Selleheyer et al. (1993) was additionally expressed in a subset of basal keratinocytes (Greenhalgh et al., 1993). Furthermore, TGFβ1 protein expression was probably much higher when driven from the K1-TGFβ1 than the K10-TGFβ1 transgene, because specific sequence modifications had been made in the K1-TGFβ1 transgene to optimize high level gene expression (Greenhalgh et al., 1993). In the K10-TGFβ1 transgenics, very low levels of recombinant TGFβ1, constitutively synthesized in suprabasal keratinocytes, might have acted in a predominantly autocrine fashion to promote suprabasal differentiation terminal differentiation and indirectly increase basal keratinocyte proliferation. However, in the study of Selleheyer et al., (1993), recombinant TGFβ1 constitutively expressed in both basal and suprabasal compartments, and at much higher levels, might have resulted in severe inhibition of epidermal proliferation, leading to neonatal lethality.

Ectopic TGFβ1 Expression in the Mammary Gland

Mammary gland development occurs predominantly postnatally, making it accessible to detailed study. There is considerable evidence to suggest that TGFβs are involved in mammary gland development. All three TGFβ isoforms are expressed, with overlapping patterns, in both the epithelium of the actively growing mammary end buds during branching morphogenesis and the epithelium of quiescent ducts (Robinson et al., 1991). Silberstein and Daniel (1987) first demonstrated the reversible inhibitory effects of TGFβ on murine mammary gland development *in vivo* by implanting slow-release pellets containing TGFβ adjacent to developing mammary epithelial end buds. Recently, two reports, using independent transgenic models, further confirmed the general inhibitory effects of TGFβ on mammary gland development. However, in the two transgenic models, the detailed phenotypes differed, again providing insight into the complexity of TGFβ responses *in vivo*.

Jhappan et al. (1993) produced transgenic mice expressing TGFβ1 from the mouse whey-acidic protein (WAP) gene promoter to target TGFβ1 transgene expression to the alveoli during pregnancy-induced alveolar development. Females from such transgenic lines exhibited significant inhibition of lobuloalveolar formation and were unable to suckle their young due to defective lactation. In a second study, Pierce et al. (1993) targeted TGFβ1 expression to the mammary ductal epithelial cells of the transgenic mice using the mouse mammary tumor virus (MMTV) promoter. These transgenic mice exhibited a different phenotype in which there was noticeable hypoplasia of the mammary ductal tree during peripubescent mammary development, but no distinguishable effect on alveolar outgrowths occurring during pregnancy. The MMTV transgenic females were able to feed their pups despite a reduced mammary ductal network. The two phenotypes observed in the two studies might be accounted for by the utilization of different promoters as well as different levels of transgene expression. Whereas the WAP promoter is expressed only in alveolar epithelial cells predominantly during pregnancy, the MMTV promoter is expressed throughout the whole process of mammary development in both ductal epithelium and alveoli. MMTV-TGFβ1 transgene expression suppressed mammary ductal development during puberty, but had no effects on lobuloalveolar outgrowth during pregnancy, when mammary epithelial cells become resistant to TGFβ1 (Pierce et al., 1993). However, it is likely that the WAP gene promoter drove higher expression levels of TGFβ1 during lobuloalveolar development and that these higher levels might have overcome the threshold of alveolar resistance to TGFβ1, resulting in inhibition of lobuloalveolar outgrowth and failure of lactation.

TGFβ receptor expression levels within the developing and pregnant mammary gland have not yet been reported. One might expect, in parallel to the studies on the keratin-promoter transgenic mice (Cui et al., 1995), that receptor levels in the weakly responding alveolar epithelium would be lower than in the responsive mammary end buds. The former cell type would, therefore, require much higher levels of TGFβ1 expression to exert a negative growth effect. It has been reported that TGFβ has different threshold levels for different biological activity effects *in vitro* and *in vivo* (Yang and Moses, 1990; Joyce et al., 1990). These threshold effects could be mediated, in part, by levels of expression of TβRII or other cell surface TGFβ binding proteins (see Massague's chapter).

B. TGFβ Gene "Knock-Out" Mice

An exceedingly powerful tool for addressing the function of a specific gene *in vivo* is gene "knockout" technology to generate mice with individual genetic elements ablated. Homologous recombination can be utilized to specifically mutate a gene of choice in mouse embryonal stem (ES) cells. These cells are derived from mouse blastocysts and have the unique property of being totipotential. If injected into host blastocysts, they can contribute to all tissues of the embryo, including the germ line. Thus "designer" lines of mice can be generated with heterozygous or homozygous mutations made to order. All three TGFβ genes have now been "knocked out" in ES cells, and lines of mice heterozygous for these knockouts have been established (T. Doetschman, personal communication). The first TGFβ1 gene to be successfully targeted was TGFβ1 (Shull et al., 1992; Kulkarni et al., 1993), and there is now extensive literature on the phenotype(s) of homozygous TGFβ1 knockout mice (Kulkarni and Karlsson, 1993; Letterio, et al., 1994; Geiser et al.,

1993; Shull and Doetschman, 1994; Hines et al., 1994; Kulkarni et al., 1994, 1995; Dickson et al., 1995).

Extensive descriptive studies from a number of laboratories had shown that the TGFβ1 gene is widely expressed during mammalian development (reviewed in Akhurst, 1994). It was initially surprising, therefore, that 50–60% of mice, which completely lacked a functional TGFβ1 gene, reached parturition with no overt developmental abnormalities. However, a homozygous TGFβ1 knockout line was not established, because these animals developed a fatal wasting syndrome, commencing at about 2 weeks of age, caused by multifocal inflammation. The TGFβ1 null mice died by about 3–4 weeks of age, and the phenotype was originally described as an autoimmune-like disorder (Shull et al., 1992; Kulkarni et al., 1993), possibly related to the immunosuppressive functions of TGFβ1 *in vitro* and *in vivo* (see Section IV).

The pathological nature of the postnatal lethal phenotype has now been studied in detail. Inflammation was most severe in the lungs and heart, and death was most likely due to cardiopulmonary failure, although several other organ systems were also severely affected (Kulkarni et al., 1995). The first histopathological lesions appeared around 8 days *postpartum* (p.p), when increased adhesion of leukocytes to vascular endothelium was observed. This increased cell adhesion to endothelial cells and their extracellular matrix appeared to be due to elevated expression of certain integrin molecules on the leukocyte cell surface, namely LFA-1 (integrin $\alpha_L\beta_2$) and VLA-4 (integrin $\alpha_4\beta_2$) (Hines et al., 1994). This was supported by the fact that inflammation in the homozygous TGFβ1 null animals could be ameliorated by injection of soluble synthetic fibronectin peptides, which presumably acted by competing out these excess fibronectin receptors (Hines et al., 1994). Geiser et al. (1993) also showed that there was elevated expression of MHC class I and II molecules in multiple tissues of the TGFβ1 null animals. The combination of elevated cell adhesion and MHC expression could account for the inflammation in TGFβ1 null animals, and there may be no necessity to invoke any autoimmune mechanism to explain the pathology (Kulkarni et al., 1995). However, recent studies have suggested that mitogen-challenged lymphoid cells from TGFβ1 null mice are defective in interleukin-2 (Il-2) synthesis and secretion, suggesting T-cell anergy. Additionally, immunosuppressive agents have been shown to reduce the TGFβ1 null pathology (Letterio et al., 1994; Kulkarni et al., 1995), and TGFβ1 itself is therapeutic in treating experimentally induced autoimmune disease in mice (Kuruvilla et al., 1991; Johns et al., 1991; Ross and Bona, 1993), suggesting that TGFβ1 might play an important endogenous immunosuppressive role *in vivo*.

The fact that developmentally normal TGFβ1 null pups can be born to heterozygous TGFβ1 null females raised the question of whether TGFβ1 is at all necessary for normal mammalian development. Letterio et al. (1994) demonstrated that TGFβ1 can cross the placenta intact and suggested that some null fetuses might receive adequate levels of maternal TGFβ1 (from their heterozygous mothers) to rescue them from intrauterine death. They managed to achieve one pregnancy in a homozygous null mother impregnated by a heterozygous father, which went to term. All four homozygous TGFβ1 null offspring had severe cardiac abnormalities and died shortly after birth (Letterio et al., 1994). Whether the cardiac defects were the primary or secondary phenotype caused by lack of TGFβ1 is still debatable.

Apart from the appearance of developmental abnormalities in TGFβ1 null pups which underwent gestation in a homozygous null mother, the number of TGFβ1 null homozygous animals born to heterozygous intercrosses was 50% less than expected from Mendelian predictions. This suggests that, either there are reproductive abnormalities in heterozygous animals, leading to under-representation of TGFβ1 null gametes (Shull and Doetschman, 1994), or that a significant fraction of TGFβ1 null conceptuses do die *in utero* (Dickson et al., 1995).

Dickson et al. (1995) specifically examined the cause of TGFβ1 null embryo lethality in heterozygous intercrosses from a mixed genetic background. They found that, although normal Mendelian (1:2:1) genotype ratios of wild-type, heterozygous, and null conceptuses were present at 8.5 and 9.5 days of gestation, by 11.5 days of gestation the number of viable homozygous TGFβ1 null embryos was reduced to 50% of expected levels, and there was also a 20% loss of heterozygotes, as had been previously reported at birth (Shull and Doetschman, 1994). The cause of death, which occurred specifically at 9.5 to 10.5 days, appeared to have been restricted to the extraembryonic yolk sac (Dickson et al., 1995). The vascular network, which develops from yolk sac blood islands, had failed to form properly in some homozygous and heterozygous TGFβ1 null conceptuses. Additionally and/or alternatively, many of the yolk sacs were severely anemic. The embryos had, thus, been starved of oxygen and nutrients prior to the establishment of the chorioallantoic placental circulation. The nonspecific phenotypes of the TGFβ1 null embryos *per se*, namely, edema and necrosis, were consistent with embryonic death being secondary to yolk sac defects (Dickson et al., 1995).

The specific cellular endothelial defects in the null mice appeared to be lack of endothelial tube formation and the establishment of weak endothelial connections, probably due to inadequate endothelial differentiation. Thus the few vessels which *did* form were weak. This would be consistent with TGFβ1 acting as an inducer of endothelial differentiation (Madri et al., 1988) and with the fact that TGFβ can increase expression of cell-adhesion molecules, extracellular matrix synthesis, and tight junction formation by endothelial cells (Madri et al., 1992). It was interesting that, although TGFβ1 is a potent endothelial cell-growth inhibitor, there appeared to be no endothelial cell hyperplasia in the null animals. This was also apparent in the yolk-sac hematopoietic defect, namely, anemia, which was an erythroid *hypo*plasia, despite the fact that TGFβ1 has been reported to be a potent inhibitor of hematopoiesis (Ohta et al., 1987). Thus, in these early embryos, it seems likely that the primary function of TGFβ1 *in vivo* is as a modulator of cellular differentiation and phenotype and not, as might have been predicted from *in vitro* studies, as a potent negative growth regulator.

The study of Dickson et al. (1995) suggested that there was no selective loss of TGFβ1 null gametes or embryos prior to 9.5 days of gestation (see Shull and Doetschman, 1994), but that all embryos, which were to die before birth, died between 9.5 and 10.5 days of gestation. Furthermore, the causative hematopoietic and endothelial defects were completely consistent with the fact that the first high-level TGFβ1 gene expression observed during embryogenesis is within hemangioblasts of the blood islands and their hematopoietic and endothelial descendants (Akhurst et al., 1990). The work of Dickson et al. (1995) was performed on animals of a mixed genetic background. It is now quite clear that the percentage of TGFβ1 null prenatal lethality depends on the genetic background of the mice (H. Su and C. Biron, personal communication). If TGFβ thresholds are important to different developmental reproductive processes, it is possible that the cause of homozygous TGFβ1 null embryo loss might depend on genetic background.

The finding of yolk-sac endothelial defects in TGFβ1 null animals is intriguing, because the first human hereditary disorder reported to be associated with regulating of TGFβ bioactivity is also characterized by vascular dysplasia. Hereditary hemorrhagic telangiectasia type I is caused by lesions in the endoglin gene (McAllister et al., 1994), the endothelial equivalent of the TGFβ type III receptor (see Massague's chapter).

The genes for TGFβ2 and TGFβ3 have also been knocked out in transgenic animals (T. Doetschman, personal communication). At present, homozygous TGFβ2 null animals have not yet been bred. Homozygous TGFβ3 null animals do reach parturition with no intrauterine death, but they *all* die within hours of birth because they cannot simultaneously breathe and suckle due to posterior clefting of the secondary palate (Proetzel et al., 1995; Kaartinen et al., 1995). Similar to the TGFβ1 prenatal lethality, this phenotype might have been predicted from descriptive studies on gene expression. During mammalian embryogenesis, the TGFβ3 gene is expressed at highest levels within the midline epithelial seam of the prefusion and fusing secondary palate (Fitzpatrick et al., 1990; Pelton et al., 1990). Detailed studies on the TGFβ3 knockout mice suggest that the cellular defect is one of epithelial cell migration out of the midline seam (Proetzel et al., 1995). Similar to the TGFβ1 knock out mice, this growth factor appears to affect primarily cellular phenotype, possibly via effects on the ECM, rather than cellular proliferation.

It is clear that there are many modulators of TGFβ biosynthesis, activation and activity (see Sections III and IV). This is reflected by the fact that the percentage of TGFβ1 null animals, which die *in utero*, depends on the genetic background onto which the genetic lesion is bred (C. Biron and H. Su, personal communication). Similarly, the severity of the TGFβ3 null phenotype depends on the genetic background of the mice (Proetzel et al., 1995). Presumably, genetic modifiers of the TGFβ knockout phenotypes could be genes which regulate TGFβ biosynthesis or the ability of the cell type to respond to TGFβ.

C. TGFβ in Wound Healing and Tissue Repair

The fact that platelets are the most concentrated source of TGFβ (Assoian et al., 1983) and that TGFβ can stimulate ECM synthesis, activate macrophages/monocytes, and suppress immune cell responses (see Section IV), indicates that TGFβ may play an important role in wound healing. Immunohistochemical studies following skin injury have shown that, within hours of injury, TGFβ1 expression was induced extracellularly at the site of the wound (Kane et al., 1991; Levin et al., 1993), suggesting that it could be a primary initiator of the wound healing response in the skin. TGFβ1 expression was also increased within keratinocytes of the leading edge of the migrating epithelial sheet (Kane et al., 1991). Here, TGFβ might induce fibronectin and integrin synthesis, and, thus, indirectly assist epithelial integrity and cell migration.

Early studies showed that a local injection of TGFβ resulted in rapid induction of granulation tissue (Sporn et al., 1983; Roberts et al., 1986). Therefore, TGFβ has been extensively studied as a potential enhancer of wounding healing. A single treatment with TGFβ1, purified from human platelets, significantly accelerated the healing of incisional skin lesions in rats, with a concomitant increase of collagen deposition at the wound site (Mustoe et al., 1987). Furthermore, in TGFβ-treated pig excisional wounds, *in situ* hybridization revealed enhancement in the levels of mRNA for matrix molecules including collagen I and III (Quaglino et al., 1991). Finally, application of recombinant human TGFβ1 to dorsal excisional punch lesions in pigs and to ulcer wounds in rabbit ear accelerated wound healing and increased the blood flow through the wound site (Beck et al., 1990a, 1990b). Similar effects of exogenous TGFβ on gastrointestinal incisional wounds (Mustoe et al., 1990) and partial thickness wounds (Jones et al., 1991) have been observed.

Systemic treatment with the anticancer drug, adriamycin, with glucocorticoids, or radiation treatment, often causes impaired wound healing. Data from several studies has demonstrated that administration of TGFβ to such wounds is effective in accelerating and improving the impaired healing. It was noted that TGFβ expression was conspicuously reduced in adriamycin-impaired wounds compared to normal wounds. Topical application of TGFβ could replace this healing deficit (Salomon et al., 1990). A single dose of TGFβ during wound healing in adriamycin-treated rats could return the breaking strength of the wound to near normal levels (Curtsinger et al., 1989). Likewise, in an irradiated guinea pig model (Bernstein et al., 1991) and in glucocorticoid-treated animal models (Pierce et al., 1989; Beck et al., 1991a; Slavin et al., 1992), TGFβ is beneficial to the wound-healing process. In various studies, application of TGFβ has ranged from a single dose at the time of wounding to repeated application throughout the healing period. Multiple applications proved to be most effective (Beck et al., 1991a).

In addition to TGFβ1, other TGFβ isoforms are also reported to affect cutaneous wound healing. Ksander et al. (1990) demonstrated that exogenous application of TGFβ2 increased the tensile strength of guinea pig dermal wounds as effectively as TGFβ1. Moreover, TGFβ2 and β3 appear to be dramatically induced in the migrating epidermis by 24 hours after excisional wounding in pigs (Levine et al., 1993).

Although TGFβ almost certainly is pivotal as an endogenous regulator of the initiation of cutaneous wound healing and repair, it is also believed to have detrimental effects at later stages, leading to scar formation. In excessive cases this may result in impairment of tissue function and growth. Using a rat incisional wound model, Shah and colleagues (Shah et al., 1992) demonstrated that injection of neutralizing antibodies to TGFβ during the healing process tended to reduce scarring. Furthermore, application of TGFβ neutralizing antibodies or decorin, a natural inhibitor of TGFβ activity, can protect against scarring in experimentally induced glomerulonephritis (Border et al., 1990, 1992).

In contrast to adult wounds, fetal wounds heal perfectly without scarring. However, injection of TGFβ into the fetal wound site can induce fibrosis (Krummel et al., 1988). Thus, it has been speculated that fetal healing might be distinct from that of the adult, in that TGFβ might not be induced at the fetal wound site (Roberts and Sporn, 1990). Whitby and Ferguson (1991) compared TGFβ expression immunohistochemically in lip wounds of 16-day gestational age fetuses, neonates, and adults. They found much reduced levels of TGFβ at the fetal wound site in comparison to neonatal or adult. They postulated that the difference of TGFβ expression between adult and fetal wound healing might account for lack of scar formation in the latter. Recently, using a combination of *in situ* hybridization and immunohistochemistry, a detailed time study of TGFβ RNA and protein expression was undertaken during the healing process following excision the the limb bud of an 11.5-day gestational age mouse fetus (Martin et al., 1993). In contrast to the observations of Whitby and Ferguson (1991), both TGFβ1 transcripts and protein were rapidly induced (1–3 hours postwounding) within the epithelial cells of the wound margin and mesenchyme of the wound. The major difference between adult wound healing and fetal wound healing in this limb bud system, was that, in the latter, TGFβ1 protein was rapidly (within 18 hours) cleared from the wound site (Martin et al., 1993), whereas, in an adult wound, TGFβ1 protein can still be present 5 days to one week postinjury (Kane et al., 1991; Levine et al., 1993). This observation emphasizes the possible importance of TGFβ1 in initiating wound closure in the fetus as well as the adult. Rapid clearance of TGFβ1 from the fetal wound, before complete closure of the wound, might be due to absence of an inflammatory response in the fetus and/or because of the distinct molecular phenotype of fetal tissue related to its increased plasticity. Nevertheless, the rapid removal of a potential scarring agent might well contribute to the perfection of fetal wound healing.

TGFβ is involved in the cutaneous healing process and in many other repair processes, such as promoting bone fracture healing (Roberts and Sporn, 1990), closure of defective skull bones (Beck et al., 1991b), and retinal defects (Glaser et al., 1992). Furthermore TGFβ1 or TGFβ2, administered intravenously, inhibit the early phase of liver regeneration following partial hepatectomy (Russell et al., 1988), suggesting that TGFβ may function as an effector of an inhibitory paracrine loop to prevent uncontrolled hepatocyte proliferation. Taken together, it is clear that TGFβ has a homeostatic action in the process of injury repair, either by diminishing the extent of injury or increasing the efficiency of repair.

D. Role of TGFβ in Multistage Epithelial Carcinogenesis

The outgrowth of a malignant tumor from a normal cell is a multistage process, requiring the participation of different classes of gene products, such as oncogenes and tumor suppressor genes (Fearon and Vogelstein, 1990; Bishop, 1991), and involving aberrations in regulating positive and negative growth regulators (Sporn and Roberts, 1985). TGFβ is a potent inhibitor of epithelial proliferation, and can affect epithelial differentiation, and, in some cases, induce apoptosis (see Sections IV and VI). One might expect, therefore, that deregulation of TGFβ metabolism could play an important role in multistage epithelial carcinogenesis.

Data collected from many laboratories has been confusing. In some cases TGFβ inhibited tumor cell proliferation (Valverius et al., 1989; Arteaga et al., 1990; Wu et al., 1992), whereas in others it stimulated tumour growth (Hubbs et al., 1989; Welch et al., 1990; Krieg et al., 1991; Arteaga et al., 1993). These contrary findings have promoted speculation that TGFβ may have biphasic effects during epithelial carcinogenesis, working as a tumor inhibitor at early stages of tumor promotion, when growth inhibitory responses are present (Haddow et al., 1991; Manning et al., 1991), but as a stimulator of carcinogenesis at later stages in tumors which have lost the negative growth responses to TGFβ, by contributing to stromal modification, immunosuppression, angiogenesis, invasion, and metastasis (see below).

TGFβ as a Tumor Suppressing Factor

TGFβ is a potent growth inhibitor of almost all normal epithelial cells (Moses et al., 1985; Masui et al., 1986; Barnard et al., 1989). Its expression is induced in mouse epidermis in response to various tumor promoters, including TPA (Akhurst et al., 1988; Esherick et al., 1993) and benzo[*a*]pyrene (Sherman et al., 1993), suggesting a function as an endogenous regulator of normal epithelial homeostasis and/or a modulator of tumor promotion (Akhurst et al., 1988; Parkinson and Balmain, 1990). *In vitro* studies have shown that TGFβ can inhibit proliferation of several human and rodent transformed and carcinoma cell lines (Manning et al., 1991, Zugmaier et al., 1989; Arteaga et al., 1990; Haddow et al., 1991, Baldwin and Korc, 1993), and these growth inhibitory effects can be reversed using anti-TGFβ antibodies (Arteaga et al., 1990). Wu et al. (1992) demonstrated that transfection of a TGFβ-responsive human colon carcinoma cell line, FET, with an antisense-expression vector for TGFβ1 enhanced its tumorigenicity *in vivo*, providing further evidence that TGFβ can act as an autocrine negative regulator of tumor growth *in vivo*.

A number of detailed studies have demonstrated that loss of the growth response to TGFβ, which occurs during multistage carcinogenesis, is a relatively late event in the tumorigenic pathway during skin, mammary, colon and lung carcinogenesis (Haddow et al., 1991; Zugmaier and Lippman, 1990; Manning et al., 1991). Thus, most benign tumours (Haddow et al., 1991; Manning et al., 1991) and even some carcinomas (Suardet et al., 1992; Baldwin and Korc, 1993) still retain growth sensitivity to TGFβ. Paradoxically, in many situations, TGFβ1 expression is elevated within these TGFβ-responsive tumors, both *in vitro* and *in vivo* (Derynck et al., 1987; Fowlis et al., 1992; Cui et al., 1994). It is, thus, possible that expression of this growth factor in tumour cells might suppress tumor outgrowth at early stages of carcinogenesis.

Direct evidence that TGFβ expression can suppress benign tumor outgrowth *in vivo* has come from recent studies on transgenic mice which express recombinant TGFβ in the epidermis (Cui, Fowlis, Akhurst, and Balmain, unpublished). Several transgenic lines of mice, which express recombinant TGFβ1 in suprabasal keratinocytes either constitutively (Cui et al., submitted) or in response to hyperplasia (Cui, Fowlis, Akhurst, and Balmain, unpublished) were subjected to a chemical carcinogenesis procedure, involving initiation with dimethylbenz[*a*]anthracene (DMBA), followed by repeated application of the tumor promoter, TPA, to the skin. The results revealed that there was a prominent suppression of benign tumor (papilloma) outgrowth in the transgenic mice compared with controls due to a prolonged tumor latency and significantly reduced tumor number per mouse. These studies provide the first direct evidence showing that TGFβ1 is a squamous tumor inhibitor *in vivo*.

Suppression of TGFβ Expression in Benign Skin Tumors Correlates with High Risk to Malignant Conversion

Most carcinomas arise from benign tumors (Fearon and Vogelstein, 1990; Yuspa, 1994), though there is probably a small subgroup of benign tumours which has a very high risk for malignant conversion, and which represents the true pool from which most carcinomas are derived (Glick et al., 1993; Yuspa, 1994). Many factors contribute to the high- or low-risk property of a benign tumor, including genetic changes in proto-oncogenes and tumor suppressor genes, (Kemp et al., 1993), and the manner in which tumours are induced (Hennings et al., 1985).

Two independent studies have shown an inverse correlation between TGFβ1 expression in benign tumors and the risk for malignant conversion. In the first study (Glick et al., 1993), benign tumors were induced using two different tumor initiation and promotion protocols, which gave rise to two classes of benign tumor with distinct risks (high and low) for malignant conversion. In the second study (Cui et al., 1994), mice with a null mutation in the *p53* gene were utilized alongside control mice. Following standard tumor initiation and promotion protocols, *p53* null mice generate benign skin tumors which have an unprecedentedly high-risk for malignant conversion. In both studies (Glick et al., 1993; Cui et al., 1994), the high-risk tumor group showed very much reduced TGFβ1 protein expression compared to the low-risk group. This downregulation of TGFβ1 protein expression apparently occurred posttranscriptionally or translationally (Fowlis et al., 1992, Glick et al., 1993), because TGFβ1 RNA levels were the same in both groups. TGFβ was clearly acting as a negative growth regulator in the low-risk tumors, because tumour cell proliferation, as assessed by bromodeoxyuridine labeling (Glick et al., 1993) or antiproliferating cell nuclear antigen (PCNA) staining (Cui et al., 1994), was markedly decreased in areas of high TGFβ1 immunostaining. The attenuation of TGFβ1 protein expression in the high-risk tumors probably contributed to malignant tumor outgrowth. Absence of TGFβ1 expression in benign tumors of the mouse is an important prognostic marker for skin cancer development.

Recently, studies using keratinocytes cultured from TGFβ1 null mice (Kulkarni et al., 1993) have provided further evidence linking loss of TGFβ1 expression and enhanced rate of malignant progression (Glick et al., 1994). After tumour initiation by transfection with a v-*ras*^{Ha} oncogene, cultured keratinocytes from homozygous or heterozygous TGFβ1 null or wild-type mice were grafted onto athymic mice. Grafts from the homozygous TGFβ1 null keratinocytes rapidly developed to dysplastic papillomas containing multifocal carcinomas, whereas those of the other genotypes formed well-differentiated papillomas.

Taken together, all of these results clearly demonstrate that TGFβ1 can inhibit tumor cell proliferation and can also suppress malignant conversion. The induction of TGFβ1 gene expression by tumor promoters (Akhurst et al., 1988; Krieg et al., 1991; Fowlis et al., 1992; Sherman et al., 1993) must, therefore, be an endogenous function of the cell in an attempt to prevent tumor outgrowth.

A Stimulator of Malignant Progression

When tumor cells progress to become TGFβ-resistant, TGFβ production may be increased because there is no longer a selection for downregulating production of this negative growth regulator. For example, production of TGFβ1 in various human colorectal carcinoma cell lines is inversely correlated with their responsiveness to TGFβ1 (Suardet et al., 1992). Malignant skin carcinomas, which are no longer responsive to TGFβ, exhibited increased TGFβ1 mRNA expression (Krieg et al., 1991). Once negative growth control is ablated, TGFβ production could indeed have a selective advantage to tumor growth via a number of mechanisms (Welch et al., 1990; Samuel et al., 1992; Gorsch et al., 1992; Arteaga et al., 1993; Hsu et al., 1994). TGFβ is known to be immunosuppressive and could serve such a role in a local manner to enhance tumor growth (Torre–Amione et al., 1990). Some of the mechanisms for stimulating tumor growth, invasion, and metastasis, may be identical to those by which TGFβ promotes developmental processes and stimulates wound healing, that is, modulation of ECM production, including stimulation of a desmoplastic response and promotion of angiogenesis. Welch et

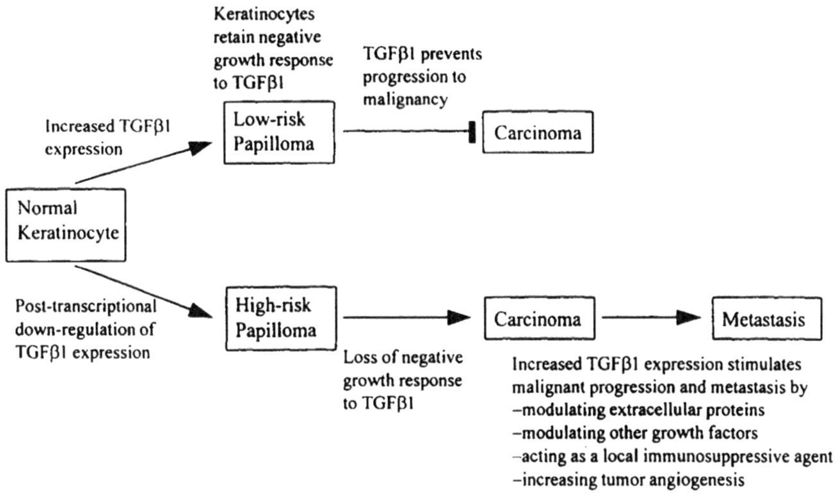

Figure 3. The role of TGFβ1 in multistage skin carcinogenesis.

al. (1990) demonstrated that TGFβ increased the number of surface lung metastases 2- to 3-fold after inoculating a TGFβ-resistant mammary adenocarcinoma clone into syngeneic rats. This increase in invasive potential was accompanied by a 2- to 6-fold increase in type IV collagenolytic activity and a 2.4-fold increase in heparanase activity.

Thus, TGFβ may act in a bifunctional manner during multistage carcinogenesis, acting as a growth inhibitor of normal keratinocytes and reducing the outgrowth of low-risk papillomas, but stimulating (high-risk) tumor progression to the malignant stage (Figure 3), when negative growth constraints are released.

VII. CONCLUSIONS

TGFβs clearly are important in many normal physiological processes *in vivo*, including the regulation of several developmental systems, tissue homeostasis, and wound healing. Aberrations in the control of TGFβ production and responses to this growth factor are probably centrally important in establishing malignant growth.

ABBREVIATIONS

CDK: cyclin-dependent kinase
CRE: cAMP-responsive element
DMBA: dimethylbenz[*a*]anthrancene
dpc: day post coitum
ECM: extracellular matrix

ES cell:	embryonal stem cell
FGF:	fibroblast growth factor
GM-CSF:	granulocyte/macrophage-colony stimulating factor
IGF:	insulin-like growth factor
IL:	interleukin
K1/K10:	cytokeratin 1 / cytokeratin 10
LAP:	latency-associated peptide
LTBP:	latent TGFβ-binding protein
man-6-p:	mannose-6-phosphate
MMTV:	mouse mammary tumor virus
NGF:	nerve growth factor
PAI:	plasminogen activator inhibitor
PCNA:	proliferating cell nuclear antigen
PDGF:	platelet-derived growth factor
p.p.:	*postpartum*
TβRII:	transforming growth factor type II receptor
TGFβ:	transforming growth factor β
TIMP:	tissue inhibitor of metalloproteinases
TNFα:	tumor necrosis factor α
TPA:	12-tetradecanoyl-phorbol-13-acetate
uPA:	urokinase-type plasminogen activator
UTR:	untranslated region
WAP:	whey-acidic protein

REFERENCES

Abbott, B. D. & Birnbaum, L. S. (1990). Retinoic acid-induced alterations in the expression of growth factors in the embryonic mouse palatal shelves. Teratology 42, 597–610.

Akhurst, R. J., Fee, F., & Balmain, A. (1988). Localized production of TGF-beta mRNA in tumour promoter-stimulated mouse epidermis. Nature 331, 363–365.

Akhurst, R. J., Lehnert, S. A., Faissner, A. J., & Duffie, E. (1990). TGF-β in murine morphogenetic processes: the early embryo and cardiogenesis. Development 108, 645–656.

Akhurst, R. J. (1994). The transforming growth factor β family in vertebrate embryogenesis. In: Growth Factors and Signal Transduction in Development (Nilsen–Hamilton, M., ed.), Wiley–Liss, New York, pp. 97–122.

Antonelli–Orlidge, A., Saunders, K. B., Smith, S. R., & D'Amore, P. A. (1989). An activated form of transforming growth factor β is produced by cocultures of endothelial cells and pericytes. Proc. Natl. Acad. Sci. USA 86, 4544–4548.

Anzano, M. A., Roberts, A. B., Meyers, C. A., Komoriya A., Lamb, L. C., Smith, J. M., & Sporn, M. (1982). Synergistic interaction between two classes of transforming growth factors from murine sarcoma cells. Cancer Res. 42, 4776–4778.

Archer, S. J., Bax, A., Roberts, A. B., Sporn, M. B., Ogawa, Y., Plez, K. A., Weatherbee, J., Tsang, M., Lucas, R., Zheng, B., Wanker, J., & Torchia, D. A. (1993). Transforming growth factor β1: Secondary structure as determined by heteronuclear magnetic resonance spectroscopy. Biochem. 32, 1164–1171.

Arrick, B. A., Lee, A. L., Grendell, R. L., & Derynck, R. (1991). Inhibition of translation of transforming growth factor β3 mRNA by its 5'-untranslated region. Mol. Cell. Biol. 11, 4306–4313.

Arrick, B. A., Grendell, R. L., & Griffin, L. A. (1994). Enhanced translational efficiency of a novel transforming growth factor β3 mRNA in human breast cancer cells. Mol. Cell. Biol. 14, 619–628.

Arteaga, C. L., Coffey Jr., Dugger, T. C., McCutchen, C. M., Moses, H. L., & Lyons, R. M. (1990). Growth stimulation of human breast cancer cells with antitransforming growth factor β antibodies: Evidence for negative autocrine regulation by transforming growth factor β. Cell Growth Differentiation 1, 367–374.

Arteaga, C. L., Carty–Dugger, T., Moses, H. L., Hurd, S. D., & Pietenpol, J. A. (1993). Transforming growth factor β1 can induce estrogen-independent tumorigenicity of human breast cancer cells in athymic mice. Cell Growth Differentiation 4, 193–201.

Assoian, R. K., Komoriya, A., Meyers, C. A., Miller, D. M., & Sporn, M. B. (1983). Transforming growth factor β in human platelets. Identification of a major storage site, purification, and characterization. J. Biol. Chem. 258, 7155–7160.

Assoian, R. K., Fleurdelys, B. E., Stevenson, H. C., Miller, P. J., Madtes, D. K., Raines, E. W., Ross, R., & Sporn, M. B. (1987). Expression and secretion of type β transforming growth factor by activated human macrophages. Proc. Natl. Acad. Sci. USA 84, 6020–6024.

Baldwin, R. L. & Korc, M. (1993). Growth inhibition of human pancreatic carcinoma cells by transforming growth factor β1. Growth Factors 8, 23–34.

Barnard, J. A., Beauchamp, R. D., Coffey, R. J., & Moses, H. L. (1989). Regulation of intestinal epithelial cell growth by transforming growth factor type β. Proc. Natl. Acad. Sci. USA 86, 1578–1582.

Barton, D. E., Foellmer, B. E., Du, J., Tamm, J., Derynck, R., & Francke, U. (1988). Chromosomal mapping of genes for transforming growth factors β 2 and β 3 in man and mouse: Dispersion of TGF-β gene family. Oncogene Res. 3, 323–331.

Basson, C. T., Kocher, O., Basson, M. D., Asis, A., & Madri, J. A. (1992). Differential modulation of vascular cell integrin and extracellular matrix expression *in vitro* by TGF-β1 correlates with reciprocal effects on cell migration. J. Cell Physiol. 153, 118–128.

Battegay, E. J., Raines, E. W., Seifert, F. A., Bowen–Pope, D. F., & Ross, R. (1990). TGF-β induces bimodal proliferation of connective tissue cells via complex control of an autocrine PDGF loop. Cell 63, 515–524.

Beck, L. S., Chen, T. L., Hirabayashi, S. E., DeGuzman, L., Lee, W. P., McFatridge, L. L., Xu, Y., Bates, R. L., & Ammann, A. J. (1990a). Accelerated healing of ulcer wounds in the rabbit ear by recombinant human transforming growth factor β1. Growth Factors 2, 273–282.

Beck, L. S., Chen, T. L., Mikalauski, P., & Ammann, A. J. (1990b). Recombinant human transforming growth factor β1 (rhTGF-beta 1) enhances healing and strength of granulation skin wounds. Growth Factors 3, 267–275.

Beck, L. S., DeGuzman, L., Lee, W. P., Xu, Y., McFatridge, L. A., & Amento, E. P. (1991a). TGF-β1 accelerates wound healing: Reversal of steroid-impaired healing in rats and rabbits. Growth Factors 5, 295–304.

Beck, L. S., DeGuzman, L., Lee, W. P., Xu, Y., McFatridge, L. A., Gillett, N. A., & Amento, E. P. (1991b). TGF-β induces bone closure of skull defects. J. Bone Mineral Res. 6, 1257–1265.

Bernstein, E. F., Harisiadis, L., Salomon, G., Norton, J., Solberg, S., Uitto, J., Glatstein, E., Glass, J., Talbot, T., & Russo, A. (1991). Transforming growth factor β improves healing of radiation-impaired wounds. J. Invest. Dermatol 97, 430–434.

Bishop, J. M. (1991). Molecular themes in oncogenesis. Cell 64, 235–248.

Border, W. A., Okuda, S., Languino, L. R., Sporn, M. B., & Ruoslahti, E. (1990). Suppression of experimental glomerulonephritis by antiserum against transforming growth factor β1. Nature 346, 371–374.

Border, W. A., Noble, N. A., Yamamoto, T., Harper, J. R., Yamaguchi, Y., Pierschbacher, M. D., & Ruoslahti, E. (1992). Natural inhibitor of transforming growth factor β protects against scarring in experimental kidney disease. Nature 360, 361–364.

Brown, P. D., Wakefield, L. M., Levinson, A. D., & Sporn, M. B. (1990). Physiochemical activation of recombinant latent transforming growth factor β1, 2, and 3. Growth factor 3, 35–43.

Brunner, A. M., Marquardt, H., Malacko, A. R., Lioubin, M. N., & Purchio, A. F. (1989). Site-directed mutagenesis of cysteine residues in the pro region of the transforming growth factor β1 precursor. Expression and characterization of mutant proteins. J. Biol. Chem. 264, 13660–13664.

Buchman, V. L., Sporn, M. B., & Davies, A. M. (1994). Role of transforming growth factor β isoforms in regulating the expression of nerve growth factor and neurotrophin-3 mRNA levels in embryonic cutaneous cells at different stages of development. Development 120, 1621–1629.

Bursch, W., Oberhammer, F., Jirtle, R. L., Askari, M., Sedivy, R., Grasl–Kraupp, B., Purchio, A. F., & Schulte–Hermann, R. (1993). Transforming growth factor β1 as a signal for induction of cell death by apoptosis. Br. J. Cancer 67, 531–536.

Burt, D. W. & Paton, I. R. (1992). Evolutionary origins of the TGFβ gene family. DNA Cell Biol. 11, 497–510.

Chakrabarty, S., Jan, Y., Brattain, M. G., Tobon, A., & Varani, J. (1989). Diverse cellular responses elicited from human colon carcinoma cells by transforming growth factor β1. Cancer Res. 49, 2112–2117.

Cheifetz, S., Hernandez, H., Laiho, M., ten Dijke, P., Iwata, K. K., & Massague, J. (1990). Distinct transforming growth factor β (TGF-β) receptor subsets as determinants of cellular responsiveness to three TGF-β isoforms. J. Biol. Chem. 265, 20533–20538.

Cheifetz, S., Bellon, T., Cales, C., Vera, S., Bernabeu, C., Massague, J., & Letarte, M. (1992). Endoglin is a component of the transforming growth factor β receptor system in human endothelial cells. J. Biol. Chem. 267, 19027–19030.

Chen, L. L., Dean, A., Jenkinson, T., & Mendelsohn, J. (1989). Effect of transforming growth factor β1 on proliferation and induction of hemoglobin accumulation in K-562 cells. Blood 74, 2368–2375.

Coffey, R. J., Sipes, N. J., Bascom, C. C., Graves–Deal, R., Pennington, C. Y., Weissman, B. E., & Moses, H. L. (1988). Growth modulation of mouse keratinocytes by transforming growth factors. Cancer Res. 48, 1596–1602.

Coletta, A. A., Wakefield, L. M., Howell, F. V., van Roozendaal, K. E., Danielpour, D., Ebbs, S. R., Sporn, M. B., & Baum, M. (1990). Anti-oestrogens induce secretion of active transformimg growth factor β in human fetal fibroblasts. Br. J. Cancer 62, 405–409.

Cui, W., Kemp, C. J., Duffie, E., Balmain, A., & Akhurst, R. J. (1994). Lack of TGFβ1 expression in benign skin tumours of *p53* ^{null} mice is prognostic for a high risk of malignant conversion. Cancer Res. 54, 5831–5836.

Cui, W., Fowlis, D. J., Cousins, F. M., Duffie, E., Bryson, S., Balmain, A., & Akhurst, R J. (1995). Concerted action of TGFβ1 and its type II receptor in control of epidermal homeostasis in transgenic mice. Genes and Dev. 9, 945–955.

Curtsinger, L. J., Petsch, J. D., Brown, G. L., von Fraunhofer, A., Ackerman, D., Polk, H. C. J., & Schultz, G. S. (1989). Reversal of adriamycin-impaired wound healing by transforming growth factor β. Surg. Bynecol. Obstet. 168, 517–522.

Czarniecki, C. W., Chiu, H. H., Wong, G. H. W., McCabe, S. M., & Palladino, M. A. (1988). Transforming growth factor β1 modulates the expression of class II histocompatability antigens on human cells. J. Immunol. 140, 4217–4223.

Daopin, S., Piez, K. A., Ogawa, Y., & Davie, D. R. (1992). Crystal structure of transforming growth factor β2. an unusual fold for the superfamily. Science 257, 369–373.

Denhez, F., Lafayatis, R., Kondaiah, P., Roberts, A. B., & Sporn, M. B. (1990). Cloning by polymerase chain reaction of a new mouse TGF-β, mTGF-β3. Growth Factors 3, 139–146.

Dennis, P. A. & Rifkin, D. B. (1991). Cellular activation of latent transforming growth factor β requires binding to the cation-independent mannose-6-phosphate/insulin-like growth factor receptor. Proc. Natl. Acad. Sci. USA 88, 580–584.

Derynck, R., Jarrett, J. A., Chen, E. Y., Eaton, D. H., Bell, J. R., Assoian, R. K., Roberts, A. B., Sporn, M. B., & Goeddel, D. V. (1985). Human transforming growth factor β cDNA sequence and expression in tumor cell lines. Nature 316, 701–705.

Derynck, R., Goeddel, D. V., Ullrich, A., Gutterman, J. U., Williams, R. D., Bringman, T. S., & Berger, W. H. (1987). Synthesis of messenger RNAs for transforming growth factors α and β and the epidermal growth factor receptor by human tumors. Cancer Res. 47, 707–712.

Dickson, M. C., Martin, J. S., Cousins, F. M., Kulkarni, A. B., Karlsson, S., & Akhurst, R. J. (1995). Transforming growth factor-beta 1 is essential for hematopoiesis and endothelial differentiation *in vivo*. Development 121, 1845–1854.

Esherick, J. S., DiCunto, F., Flanders, K. C., Missero, C., & Dotto, G. P. (1993). Transforming growth factor β1 induction is associated with TGF-β2 and β3 downmodulation in TPA-induced skin hyperplasia. Cancer Res. 53, 5517–5522.

Espevik, T., Figari, I. S., Ranges, G. E., & Palladino, M. A. J. (1988). Transforming growth factor β1 (TGF-β1) and recombinant human tumor necrosis factor-α reciprocally regulate the generation of lymphokine-activated killer cell activity. Comparison between natural porcine platelet-derived TGF-β1 and TGF-β2, and recombinant human TGF-β1. J. Immunol. 140, 2312–2316.

Fearon, E. R. & Vogelstein, B. (1990). A genetic model for colorectal tumorigenesis. Cell 61, 759–767.

Fiorelli, G., Ballock, R. T., Wakefield, L. M., Sporn, M. B., Gori, F., Masi, L., Frediani, U., Tanini, A., Bernabei, P. A., & Brandi, M. I. (1994). Role for autocrine TGF-β1 in regulating differentiation of a human leukemic cell line toward osteoclast-like cells. J. Cell. Physiol. 160, 482–490.

Fitzpatrick, D. R., Denhez, F., Kondaiah, P., & Akhurst, R. J. (1990). Differential expression of TGF-β isoforms in murine palatogenesis. Development 109, 585–595.

Flanders, K. C., Thompson, N. L., Cissel, D. S., Van Obberghen, S. E., Baker, C. C., Kass, M. E., Ellingsworth, L. R., Roberts, A. B., & Sporn, M. B. (1989). Transforming growth factor β1: Histochemical localization with antibodies to different epitopes. J. Cell Biol. 108, 653–660.

Flaumenhaft, R., Abe, M., & Rifkin, D. B. (1992). Basic fibroblast growth factor-induced activation of latent transforming growth factor β in endothelial cells: Regulation of plasminogen activator activity. J. Cell Biol. 118, 901–909.

Flaumenhaft, R., Abe, M., Sato, Y., Miyazono, K., Harpel, J., Heldin, C., & Rifkin, D. B. (1993). Role of latent TGF-β binding protein in the activation of latent TGF-β by cocultures of endothelial and smooth muscle cells. J. Cell Biol. 120, 995–1002.

Florini, J. R., Roberts, A. B., Ewton, D. Z., Falen, S. L., Flanders, K. C., & Sporn, M. B. (1986). Transforming growth factor β. A very potent inhibitor of myoblast differentiation, identical to the differentiation inhibitor secreted by Buffalo rat liver cells. J. Biol. Chem. 261, 16509–16513.

Fowlis, D. J., Flanders, K., Duffie, E., Balmain, A., & Akhurst, R.J. (1992). Discordant TGFβ1 RNA and protein localizations during chemical carcinogenesis of the skin. Cell Growth Differentiation 3, 81–91.

Frolik, C. A., Dart, L. L., Meyers, C. A., Smith, D. M., & Sporn, M. B. (1983). Purification and initial characterization of a type β transforming growth factor from human placenta. Proc. Natl. Acad. Sci. USA 80, 3676–3680.

Fujii, D., Brissenden, J. E., Derynck, R., & Francke, U. (1986). Transforming growth factor β gene maps to human chromosome 19 long arm and to mouse chromosome 7. Somat. Cell Mol. Genet. 12, 281–288.

Gamble, J. R. & Vadas, M. A. (1988). Endothelial adhesiveness for blood neutrophils is inhibited by transforming growth factor β. Science 242, 97–99.

Gamble, J. R. & Vadas, M. A. (1991). Endothelial cell adhesiveness for human T-lymphocytes is inhibited by transforming growth factor β1. J. Immunol. 146, 1149–1154.

Gatherer, D., ten Dijke, P., Baird, D. T., & Akhurst, R. J. (1990). Expression of TGFβ isoforms during first trimester human embryogenesis. Development 110, 445–460.

Geiser, A. G., Kim, S., Roberts, A. B., & Sporn, M. B. (1991). Characterization of the mouse transforming growth factor β1 promoter and activation by the Ha-ras oncogene. Mol. Cell. Biol. 11, 84–92.

Geiser, A., Letterio, J. J., Kulkarni, A. B., Karlsson, S., Roberts, A. B., & Sporn, M. B. (1993). TGF-β1 controls expression of major histocompatibility genes in the postnatal mouse: Aberrant histocompatibility antigen expression in the pathogenesis of the TGFβ1 null phenotype. Proc. Natl. Acad. Sci. USA 90, 9944–9948.

Gentry, L. E., Webb, N. R., Lim, G. J., Brunner, A. M., Ranchalis, J. E., Twardzik, D. R., Lioubin, M. N., Marquardt, H., & Purchio, A. F. (1987). Type 1 transforming growth factor β: Amplified expression and secretion of mature and precursor polypeptides in Chinese hamster ovary cells. Mol. Cell Biol. 7, 3418–3427.

Gentry, L. E. & Nash, B. W. (1990). The pro domain of pre-pro-transforming growth factor β1 when independently expressed is a functional binding protein for the mature growth factor. Biochem. 29, 6851–6857.

Glaser, B. M., Michels, R. G., Kuppermann, B. D., Sjaarda, R. N., & Pena, R. A. (1992). Transforming growth factor β2 for the treatment of full-thickness macular holes. Ophthalmology 99, 1162–1173.

Glick, A. B., Flanders, K. C., Danielpour, D., Yuspa, S. H., & Sporn, M. B. (1989). Retinoic acid induces transforming growth factor β2 in cultured keratinocytes and mouse epidermis. Cell Regul. 1, 87–97.

Glick, A. B., Kulkarni, A. B., Tennenbaum, T., Hennings, H., Flanders, K. C., O'Reilly, M., Sporn, M. B., Karlsson, S., & Yuspa, S. H. (1993). Loss of expression of transforming growth factor β in skin and skin tumors is associated with hyperproliferation and a high risk for malignant conversion. Proc. Natl. Acad. Sci. USA 90, 6076–6080.

Glick, A. B., Lee, M. M., Darwiche, N., Kulkarni, A. B., Karlsson, S., & Yuspa, S. H. (1994). Targeted deletion of the TGF-β1 gene causes rapid progression to squamous cell carcinoma. Genes Dev. 8, 2429–2440.

Gorsch, S. M., Memoli, V. A., Stukel, T. A., Gold, L. I., & Arrick, B. A. (1992). Immunohistochemical staining for transforming growth factor β1 associates with disease progression in human breast cancer. Cancer Res. 52, 6949–6952.

Gray, S. M. & Mason, A. J. (1990). Requirement for activin A and transforming growth factor β pro-region in homodimer assembly. Science 247, 1328–1330.

Graycar, J. L., Miller, D. A., Arrick, B. A., Lyons, R. M., Moses, H. L., & Derynck, R. (1989). Human transforming growth factor β3: Recombinant expression, purification, and biological activities in comparison with transforming growth factors β1 and β2. Mol. Endocrinol. 3, 1977–1986.

Greenhalgh, D. A., Rothnagel, J. A., Quintanilla, M. I., Orengo, C. C., Gagne, T. A., Bundman, D. S., Longley, M. A., & Roop, D. R. (1993). Induction of epidermal hyperplasia, hyperkeratosis, and papillomas in transgenic mice by a targeted v-Ha-ras oncogene. Mol. Carcinogen. 7, 99–110.

Haddow, S., Fowlis, D. J., Parkinson, K., Akhurst, R. J., & Balmain, A. (1991). Loss of growth control by TGFβ occurs at a late stage of mouse carcinogenesis and is independent of ras gene activation. Oncogene 6, 1465–1470.

Heimark, R. L., Twardzik, D. R., & Schwartz, S. M. (1986). Inhibition of endothelial regeneration by type β transforming growth factor from platelets. Science 233, 1078–1080.

Heino J. & Massague, J. (1989). Transforming growth factor-β switches the pattern of integrins expressed in MG-63 human osteosarcoma cells and causes a selective loss of cell adhesion to laminin. J. Biol. Chem. 36, 21806–21811.

Hennings, H., Spangler, E. F., Shores, R., Mitchell, P., Devor, D., Shamsuddin, A. K. M., Elgjo, K. M., & Yuspa, S. H. (1985). Induction of papillomas with a high probability of conversion to malignancy. Carcinogenesis 6, 1607–1610.

Hines, K. L., Kulkarni, A. B., McCarthy, J. B., Tian, H., Ward, J. M., Christ, M., McCartney–Francis, L., Furcht, L. T., Karlsson, S., & Wahl, S. M. (1994). Synthetic fibronectin peptides interrupt inflammatory cell infiltration in transforming growth factor β1 knockout mice. Proc. Natl. Acad. Sci. USA 91, 5187–5191.

Hsu, S., Huang, F., Hafez, M., Winawer, S., & Friedman, E. (1994). Colon carcinoma cells switch their response to transforming growth factor β1 with tumor progression. Cell Growth Differentiation 5, 267–275.

Hubbs, A. F., Hahn, F. F., & Thomassen, D. G. (1989). Increased resistance to transforming growth factor β accompanies neoplastic progression of rat tracheal epithelial cells. Carcinogenesis. 10, 1599–1605.

Ignotz, R. A. & Massague, J. (1986). Transforming growth factor β stimulates the expression of fibronectin and collagen and their incorporation into the extracellular matrix. J. Biol. Chem. 261, 4337–4345.

Ignotz, R. A., Heino, J., & Massague, J. (1989). Regulation of cell adhesion receptors by transforming growth factor β. Regulation of vitronectin receptor and LFA-1. J. Biol. Chem. 264, 389–392.

Jennings, J. C., Mohan, S., Linkhart, T. A., Widstrom, R., & Baylink, D. J. (1988). Comparison of the biological actions of TGF-β1 and TGF-β2: Differential activity in endothelial cells. J. Cell Physiol. 137, 167–172.

Jhappan, C., Geiser, A. G., Kordon, E. C., Bagheri, D., Hennighausen, L., Roberts, A. B., Smith, G. H., & Merlino, G. (1993). Targeting expression of a transforming growth factor β1 transgene to the pregnant mammary gland inhibits alveolar development and lactation. EMBO J. 12, 1835–1845.

Johns, L. D., Flanders, K. C., Ranges, G. E., & Sriram, S. (1991). Successful treatment of experimental allergic encephalomyelitis with transforming growth factor β1. J. Immunol. 147, 1792–1796.

Jones, S. C., Curtsinger, L. J., Whalen, J. D., Pietsch, J. D., Ackerman, D., Brown, G. L., & Schultz, G. S. (1991). Effect of topical recombinant TGF-β on healing of partial thickness injuries. J. Surg. Res. 51, 344–352.

Joyce, M. E., Roberts, A. B., Sporn, M. B., & Nolander, M. E. (1990). Transforming growth factor β and the initiation of chondrogenesis and osteogenesis in the rat femur. J. Cell Biol. 110, 2195–2207.

Jürgensmeier, J. M., Schmitt, C. P., Viesel, E., Höfler, P., & Bauer, G. (1994). Transforming growth factor β-treated normal fibroblasts eliminate transformed fibroblasts by induction of apoptosis. Cancer Res. 54, 393–398.

Kaartinen, V., Voncken, J. W., Shuler, C., Warburton, D., Bu, D., Heisterkamp, N., & Groffen, J. (1995). Abnormal lung development and cleft palate in mice lacking TGF-β3 indicates defects of epithelial-mesenchymal interaction. Nature Gen. 11, 415–421.

Kane, C. J., Hebda, P. A., Mansbridge, J. N., & Hanawalt, P. C. (1991). Direct evidence for spatial and temporal regulation of transforming growth factor β1 expression during cutaneous wound healing. J. Cell Physiol. 148, 157–173.

Kanzaki, T., Olofsson, A., Moren, A., Wernstedt, C., Hellman, U., Miyazono, K., Claesson–Welsh, L., & Heldin, C. (1990). TGF-β1 binding protein: A component of the large latent complex of TGF-β1 with multiple repeat sequences. Cell 61, 1051–1061.

Kato, Y., Iwamoto, M., Koike, T., Suzuki, F., & Takano, Y. (1988). Terminal differentiation and calcification in rabbit chondrocyte cultures grown in centrifuge tubes: Regulation by transforming growth factor β and serum factors. Proc. Natl. Acad. Sci. USA 85, 9552–9556.

Kehrl, J. H., Roberts, A. B., Wakefield, L. M., Jakowlew, S., & Sporn, M. B. (1986a). Transforming growth factor β is an important immunomodulatory protein for human B lymphocytes. J. Immunol. 137, 3855–3860.

Kehrl, J. H., Wakefield, L. M., Roberts, A. B., Jakowlew, S., Alvarez Mon, M., Derynck, R., Sporn, M. B., & Fauci, A. S. (1986b). Production of transforming growth factor β by human T-lymphocytes and its potential role in the regulation of T-cell growth. J. Exp. Med. 163, 1037–1050.

Kehrl, J. H., Thevenin, C., Rieckmann, P., & Fauci, A. S. (1991). Transforming growth factor β suppresses human B-lymphocyte Ig production by inhibiting synthesis and the switch from the membrane form to the secreted form of Ig mRNA. J. Immun. 146, 4016–4023.

Keller, J. R., Jaconsen, S. E. W., Sill, K., Ellingsworth, L., & Ruscetti, F. W. (1991). Stimulation of granulopoiesis by transforming growth factor β: Synergy with granulocyte-macrophage colony-stimulating factor. Proc. Natl. Acad, Sci. USA 88, 7190–7193.

Keller, J. R., Bartelmez, S. H., Sitnicka, E., Ruscetti, F. W., Mariaestella, O., Gooya, J. M., & Jacobsen, S. E. W. (1994). Distinct and overlapping direct effects of macrophage inhibitory protein-1α and transforming growth factor β on hematopoietic progenitor/stem cell growth. Blood 84, 2175–2181.

Kemp, C. J., Donehower, L. A., Bradley, A., & Balmain, A. (1993). Reduction of *p53* gene dosage does not increase initiation or promotion but greatly enhances malignant progression of chemically induced skin tumours. Cell 74, 813–822.

Kim, S., Glick, A., Sporn, M. B., & Roberts, A. B. (1989). Characterization of the promoter region of the human transforming growth factor β1 gene. J. Biol. Chem. 264, 402–408.

Kim, S., Park, K., Koeller, D., Kim, K. Y., Wakefield, L. M., Sporn, M. B., & Roberts, A. B. (1992). Posttranscriptional regulation of the transforming growth factor β1 gene. J. Biol. Chem. 267, 13702–13707.

Kingsley, D. M. (1994). The TGF-β superfamily: New members, new receptors, and new genetic tests of function in different organisms. Genes Dev. 8, 133–146.

Kojima, S., Nara, K., & Rifkin, D. B. (1993). Requirement for transglutaminase in the activation of latent transforming growth factor β in bovine endothelial cells. J. Cell Biol. 121, 439–448.

Kondaiah, P., Sands, M. J., Smith, J. M., Fields, A., Roberts, A. B., Sporn, M. B., & Melton, D. A. (1990). Identification of a novel transforming growth factor β (TGFβ5) mRNA in *Xenopus laevis*. J. Biol. Chem. 265, 1089–1093.

Kovacina, K. S., Steele Perkins, G., Purchio, A. F., Lioubin, M., Miyazono, K., Heldin, C. H., & Roth, R. A. (1989). Interactions of recombinant and platelet-transforming growth factor β1 precursor with the insulin-like growth factor II/mannose-6-phosphate receptor. Biochem. Biophys. Res. Commun. 160, 393–403.

Kozak, M. (1987). An analysis of 5′-noncoding sequences from 699 vertebrate messenger RNAs. Nucleic Acids Res. 15, 8125–8147.

Krieg, P., Schnapke, R., Furstenberger, G., Vogt, I., & Marks, F. (1991). TGF-β1 and skin carcinogenesis: Antiproliferative effect *in vitro* and TGF-β1 mRNA during epidermal hyperproliferation and multistage tumorigenesis. Molecular Carcinogen. 4, 129–137.

Krummel, T. M., Michna, B. A., Thomas, B. L., Sporn, M. B., Nelson, J. M., Salzberg, A. M., Cohen, I. K., & Diegelmann, R. F. (1988). Transforming growth factor β (TGF-β) induces fibrosis in a fetal wound model. J. Pediatr. Surg. 23, 647–652.

Ksander, G. A., Ogawa, Y., Chu, G. H., McMullin, H., Rosenblatt, J. S., & McPherson, J. M. (1990). Exogenous transforming growth factor β2 enhances connective tissue formation and wound strength in guinea pig dermal wounds healing by secondary intent. Ann. Surg. 211, 288–294.

Kulkarni, A. B., Huh, C., Becker, D., Geiser, A., Lyght, M., Flanders, K. C., Roberts, A. B., Sporn, M. B., Ward, J. M., & Karlsson, S. (1993). Transforming growth factor β1 null mutation in mice causes excessive inflammatory response and early death. Proc. Natl. Acad. Sci. USA 90, 770–774.

Kulkarni, A. B. & Karlsson, S. (1993). Transforming growth factor β1 (TGF-β1) knockout mice: A mutation in one cytokine causes dramatic inflammatory disease. Am. J. Pathol. 143, 3–9.

Kulkarni, A. B., Ward, J. M., Geiser, A. G., Letterio, J. J., Hines, K. L., Christ, M., D'Souza, R. N., Huh, C., Roberts, A. B., Sporn, M. B., McCartney-Francis, N., Wahl, S. M., Glick, A. B., Yuspa, S. H., Mackall, C., Gress, R. E., & Karlsson, S. (1994). TGFβ1 knockout mice: Immune dysregulation and pathology, 8th Symposium on molecular biology of hematopoiesis. In: Molecular Biology of Haematopoiesis (Abrahm, N. G., Shadduck, R. K., Levine, A. S., & Takaku, F., eds.), Vol. 3, pp. 749–757, Intercept Ltd., Andover, NH.

Kulkarni, A. B., Ward, J. M., Yaswen, L., Mackall, C. L., Bauer, S. R., Huh, C. G., Gress, R. E., & Karisson, S. (1995). Transforming growth factor β1 null mice: An animal model for inflammatory disorders. Am. J. Pathol. 146, 1–12.

Kuruvilla, A. P., Shah, R., Hochwald, G. M., Uggitt, H. D., Palladino, M. A., & Thorbecke, G. J. (1991). Protective effect of transforming growth factor β1 on experimental autoimmune diseases in mice. Proc. Natl. Acad. Sci. USA 88, 2918–2921.

Kyprianou, N. & Isaacs, J. T. (1989). Expression of transforming growth factor in the rat ventral prostate during castration-induced programmed cell death. Mol. Endocrinol. 3, 1515–1522.

Lafyatis, R., Lechleider, R., Kim, S., Jakowlew, S., Roberts, A. B., & Sporn, M. B. (1990). Structural and functional characterisation of the transforming growth factor β3 promoter. J. Biol. Chem. 265, 19128–19136.

Letterio, J. J., Gieser, A. G., Kulkarni, A. B., Roche, N. S., Sporn, M. B., & Roberts, A. B. (1994). Maternal rescue of the transforming growth factor β knockout. Science 264, 1936–1938.

Levine, J. H., Moses, H. L., Gold, L. I., & Nanney, L. B. (1993). Spatial and temporal patterns of immunoreactive transforming growth factor β1, β2, and β3 during excisional wound repair. Am. J. Pathol. 143, 368–380.

Li, M. L., Aggeler, J., Farson, D. A., Hatier, C., Hassell, J., & Bissell, M. J. (1987). Influence of a reconstituted basement membrane and its components on casein gene expression and secretion in mouse mammary epithelial cells. Proc. Natl. Acad. Sci. USA 84, 136–140.

Lin, H. Y., Wang, X., Ng, E. E., Weinberg, R. A., & Lodish, H. F. (1992). Expression cloning of the TGF-β type II receptor, a functional transmembrane serine/threonine kinase. Cell 68, 775–785.

Lin, J. & Chou, C. (1992). *In Vitro* apoptosis in the human hepatoma cell line induced by transforming growth factor β1. Cancer Res. 52, 385–388.

Lindholm, D., Castrén, E., Kiefer, R., Zafran, F., & Thoenen, H. (1992). Transforming growth factor β1 in the rat brain: Increase after injury and inhibition of astrocyte proliferation. J. Cell Biol. 117, 395–400.

Lopez, A. R., Cook, J., Deininger, P. L., & Derynck, R. (1992). Dominant negative mutants of transforming growth factor β1 inhibit the secretion of different transforming growth factor β isoforms. Mol. Cell. Biol. 12, 1674–1679.

Lyons, R. M., Gentry, L. E., Purchio, A. F., & Moses, H. L. (1990). Mechanism of activation of latent recombinant transforming growth factor β1 by plasmin. J. Cell Biol. 110, 1361–1367.

Madisen, L., Lioubin, M. N., Marquardt, H., & Purchio, A. F. (1990). High-level expression of TGF-β2 and the TGF-β2(414) precursor in Chinese hamster ovary cells. Growth factor 3, 129–138.

Madri, J. A., Williams, S. K., Wyatt, T., & Mezzio, C. (1983). Capillary endothelial cell cultures: Phenotypic modulation by matrix components. J. Cell Biol. 97, 153–165.

Madri, J. A., Pratt, B. M., & Tucker, A. M. (1988). Phenotypic modulation of endothelial cells by transforming growth factor β depends upon the composition and organization of the extracellular matrix. J. Cell Biol. 106, 1375–1384.

Madri, J. A., Bell, L., & Merwin, J. R. (1992). Modulation of vascular cell behavior by transforming growth factors β. Mol. Reprod. Dev. 32, 121–126.

Manning, A. M., Williams, A. C., Games, S. M., & Paraskeva, C. (1991). Differential sensitivity of human colonic adenoma and carcinoma cells to transforming growth factor β (TGFβ): Conversion of an adenoma cell line to a tumorigenic phenotype is accompanied by a reduced response to the inhibitory effects of TGFβ. Oncogene 6, 1471–1476.

Mansbridge, J. N. & Hanawalt, P. C. (1988). Role of transforming growth factor β in the maturation of human epidermal keratinocytes. J. Invest. Dermatol. 90, 336–341.

Martin, P., Dickson, M. C., Millan, F. A., & Akhurst, R. J. (1993). Rapid induction and clearance of TGFβ1 is an early response to wounding in the mouse embryo. Dev. Genet. 14, 225–236.

Maslen, C. L., Corson, G. M., Maddox, B. K., Glanville, R. W., & Sakai, L. Y. (1991). Partial sequence of a candidate gene for the Marfan syndrome. Nature 352, 334–337.

Massague, J., Cheifetz, S., Endo, T., & Nadal–Ginard, B. (1986). Type β transforming growth factor is an inhibitor of myogenic differentiation. Proc. Natl. Acad. Sci. USA 83, 8206–8210.

Massague, J. (1990). The transforming growth factor β family. Ann. Rev. Cell Biol. 6, 597–641.

Massague, J., Cheifetz, S., Boyd, F. T., & Andres, J. L. (1990). TGF β receptors and TGFβ binding proteoglycans: Recent progress in identifying their functional properties. Ann. NY Acad. Sci. USA 593, 59–72.

Massague, J. (1992). Receptors for the TGFβ family. Cell 69, 1067–1070.

Massague, J., Cheifetz, S., Laiho, M., Ralph, D. A., Weis, F. M. B., & Zentella, A. (1992). Transforming growth factor β. Cancer Surv. 12, 81–103.

Masui, T., Wakefield, L. M., Lechner, J. F., LaVeck, M. A., Sporn, M. B., & Harris, C. C. (1986). Type β transforming growth factor is the primary differentiation-inducing serum factor for normal human bronchial epithelial cells. Proc. Natl. Acad. Sci. USA 83, 2438–2442.

Matsumoto, K., Hashimoto, K., Hashiro, M., Yoshimasa, H., & Yoshikawa, K. (1990). Modulation of growth and differentiation in normal human keratinocytes by transforming growth factor β. J. Cell. Physiol. 145, 95–101.

McAllister, K. A., Grogg, D. W., Johnson, D. W., Gallione, C. J., Baldwin, M. A., Jackson, C. E., Helmbold, E. A., Markel, D. S., McKinnon, W. C., Murrell, J., McCormick, M. K., Pericak-Vance, M. A., Heutink, P., Oostra, B. A., Haitjema, T., Westerman, C. J. J., Porteous, M. E., Guttmacher, A. E., Letarte, M., & Marchuk, D. A. (1994). Endoglin, a TGFβ binding protein of endothelial cells, is the gene for hereditary hemorrhagic telangiectasia type 1. Nature Genet. 8, 345–351.

McDonald, N. Q. & Hendrickson, W. A. (1993). A structural superfamily of growth factors containing a cystine knot motif. Cell 73, 421–424.

Merwin, J. R., Anderson, J., Kocher, O., van Itallie, C., & Madri, J. A. (1990). Transforming growth factor β modulates extracellular matrix organization and cell–cell junctional complex formation during *in vitro* angiogenesis. J. Cell Physiol. 142, 117–128.

Merwin, J. R., Newman, W., Beall, D., Tucker, A., & Madri, J. A. (1991). Vascular cells respond differentially to transforming growth factors β1 and β2. Am. J. Pathol. 138, 37–51.

Millan, F. A., Kondaiah, P., Denhez, F., & Akhurst, R. J. (1991). Embryonic gene expression patterns of TGF-β1, 2, and 3 suggest different developmental functions *in vivo*. Development 111, 131–144.

Miller, D. M., Ogawa, Y., Iwata, K. K., ten Dijke, P., Purchio, A. F., Soloff, M. S., & Gentry, L. E. (1992). Characterization of the binding of transforming growth factor β1, β2, and β3 to recombinant β1-latency-associated peptide. Mol. Endocrinol. 6, 694–702.

Miyazono, K., Hellman, U., Wernstedt, C., & Heldin, C. H. (1988). Latent high molecular weight complex of transforming growth factor β1. Purification from human platelets and structural characterization. J. Biol. Chem. 263, 6407–6415.

Miyazono, K. & Heldin, C. H. (1989). Role for carbohydrate structures in TGFβ1 latency. Nature 338, 158–160.

Miyazono, K., Olofsson, A., Colosetti, P., & Heldin, C. (1991). A role of the latent TGFβ1-binding protein in the assembly and secretion of TGFβ1. EMBO J. 10, 1091–1101.

Miyazono, K., Ichijo, H., & Heldin, C. (1993). Transforming growth factor β: Latent forms, binding proteins, and receptors. Growth Factors 8, 11–22.

Moses, H. L., Branum, E. L., Proper, J. A., & Robinson, R. A. (1981). Transforming growth factor production by chemically transformed cells. Cancer Res. 41, 2842–2848.

Moses, H. L., Tucker, R. F., Leof, E. B., Coffey, R. J., Halper, J., & Shipley, G. D. (1985). Type β transforming growth factor is a growth stimulator and a growth inhibitor. In: Cancer Cells (Feramisco, J., Ozanne, B., & Stiles, C., eds.), pp. 65–71, Cold Spring Harbor Laboratory, New York.

Moulton, B. C. (1994). Transforming growth factor β stimulates endometrial stromal apoptosis *in vitro*. Endocrinol. 134, 1055–1060.

Muller, G., Behrens, J., Nussbaumer, U., Bohlen, P., & Birchmeier, W. (1987). Inhibitory action of transforming growth factor β on endothelial cells. Proc. Natl. Acad. Sci. USA 84, 5600–5604.

Mustoe, T. A., Pierce, G. F., Thomason, A., Gramates, P., Sporn, M. B., & Deuel, T. F. (1987). Accelerated healing of incisional wounds in rats induced by transforming growth factor β. Science 237, 1333–1336.

Mustoe, T. A., Landes, A., Cromack, D. T., Mistry, D., Griffin, A., Deudl, T. F., & Pierce, G. F. (1990). Differential acceleratiion of healing of surgical incisions in the rabbit gastrointestinal tract by platelet-derived growth factor and transforming growth factor, type β. Surgery 108, 324–329.

Noda, M. & Rodan, G. A. (1987). Type β transforming growth factor (TGFβ) regulation of alkaline phosphatase expression, and other phenotype-related mRNAs in osteoblastic rat osteosarcoma cells. J. Cell Physiol. 133, 426–437.

Noma, T., Glick, A. B., O'Reilly, M. A., Miller, J., Roberts, A. B., & Sporn, M. B. (1991). Molecular cloning and structure of the human transforming growth factor β2 gene promoter. Growth Factors 4, 247–255.

Nugent, M. A. & Edelman, E. R. (1992). Transforming growth factor β1 stimulates the production of basic fibroblast growth factor-binding proteoglycans in Balb/c3t3 cells. J. Biol. Chem. 267, 21256–21264.

O'Reilly, M. A., Geiser, A. G., Kim, S., Bruggeman, L. A., Luu, A. X., Roberts, A. B., & Sporn, M. B. (1992). Identification of an activating transcription factor (ATF) binding site in the human transforming growth factor β2 promoter. J. Biol. Chem. 267, 19938–19943.

Oberhammer, F., Bursch, W., Parzefall, W., Breit, P., Erber, E., Stadler, M., & Schulte–Hermann, R. (1991). Effect of transforming growth factor β on cell death of cultured rat hepatocytes. Cancer Res. 51, 2478–2485.

Oberhammer, F. A., Pavelka, M., Sharma, S., Tiefenbacher, R., Purchio, A. F., Bursch, W., & Schulte–Hermann, R. (1992). Induction of apoptosis in cultured hepatocytes and in regressing liver by transforming growth factor β1. Proc. Natl. Acad. Sci. USA 89, 5408–5412.

Ohta, M., Greenberger, J. S., Anklesaria, P., Bassols, A., & Massague, J. (1987). Two forms of transforming growth factor β distinguished by multipotential haematopoietic progenitor cells. Nature 329, 539–541.

Olson, E. N., Sternberg, E., Hu, J. S., Spizz, G., & Wilcox, C. (1986). Regulation of myogenic differentiation by type β transforming growth factor. J. Cell Biol. 103, 1799–1805.

Ortaldo, J. R., Mason, A. T., Oshea, J. J., Smyth, M. J., Falk, L. A., Kennedy, I. C. S., Longo, D. L., & Ruscetti, F. W. (1991). Mechanistic studies of transforming growth factor β inhibition of IL-2-dependent activation of CD-3 large granular lymphocyte functions: Regulation of IL-2Rβ (p75) signal transduction. J. Immun. 146, 3791–3798.

Parkinson, K. & Balmain, A. (1990). Chalones revisited—a possible role for TGFβ in tumor promotion. Carcinogenesis 20, 195–198.

Pelton, R. W., Hogan, B. L., Miller, D. A., & Moses, H. L. (1990). Differential expression of genes encoding TGFs β1, β2, and β3 during murine palate formation. Developmental Biology 141, 456–460.

Pelton, R. W., Saxena, B., Jones, M., Moses, H. L., & Gold, L. I. (1991). Immunohistochemical localization of TGFβ1, TGFβ2, and TGFβ3 in the mouse embryo: Expression patterns suggest multiple roles during embryonic development. J. Cell Biol. 115, 1091–1105.

Perrella, M. A., Yoshizumi, M., Fen, Z., Tsai, J., Hsieh, C., Kourembanas, S., & Lee, M. (1994). Transforming growth factor β1, but not dexamethasone, downregulates nitric oxide synthase mRNA after its induction by interleukin-1β in rat smooth muscle cells. J. Biol. Chem. 269, 14595–14600.

Pierce, G. F., Mustoe, T. A., Lingelbach, J., Masakowski, V. R., Gramates, P., & Deuel, T. F. (1989). Transforming growth factor β reverses the glucocorticoid-induced wound-healing deficit in rats: Possible regulation in macrophages by platelet-derived growth factor. Proc. Natl. Acad. Sci. USA 86, 2229–2233.

Pierce, D. F. J., Johnson, M. D., Matsui, Y., Robinson, S. D., Gold, L. I., Purchio, A. F., Daniel, C. W., Hogan, B. L. M., & Moses, H. L. (1993). Inhibition of mammary duct development but not alveolar outgrowth during pregnancy in transgenic mice expressing active TGF-β1. Genes Dev. 7, 2308–2317.

Pircher, R., Jullien, P., & Lawrence, D. A. (1986). β-Transforming growth factor is stored in human blood platelets as a latent high molecular weight complex. Biochem. Biophys. Res. Commun. 136, 30–37.

Prehn, J. H. M., Peruche, B., Unsicker, K., & Krieglstein, J. (1993). Transforming growth factor β1 prevents degeneration of primary neuronal cultures induced by cytotoxic hypoxia or glutamate. J. Neurochem. 60, 1665–1672.

Proetzel, G., Pawlowski, S. A., Wiles, M. V., Yin, M., Boivin, G. P., Howles, P. N., Ding, J., Ferguson, M. J., & Doetschman, T. (1995). Transforming growth factor-β3 is required for secondary palate fusion. Nature Gen. 11, 409–414.

Qian, S. W., Burmester, J. K., Merwin, J. R., Madri, J. A., Sporn, M. B., & Roberts, A. B. (1992). Identification of a structural domain that distinguishes the actions of the type 1 and 2 isoforms of transforming growth factor β on endothelial cells. Proc. Natl. Acad. Sci. USA 89, 6290–6294.

Quaglino, D., Nanney, L. B., Ditesheim, J. A., & Davidson, J. M. (1991). Transforming growth factor β stimulates wound healing and modulates extracellular matrix gene expression in pig skin incisional wound model. J. Invest. Dermatol. 97, 34–42.

Reiss, M. & Sartorelli, A. C. (1987). Regulation of growth and differentiation of human keratinocytes by type β transforming growth factor and epidermal growth factor. Cancer Res. 47, 6705–6709.

Ristow, H. J. (1986). BSC-1 growth inhibitor/type β transforming growth factor is a strong inhibitor of thymocyte proliferation. Proc. Natl. Acad. Sci. USA 83, 5531–5533.

Roberts, A. B., Anzano, M. A., Lamb, L. C., Smith, J. M., & Sporn, M. B. (1981). New class of transforming growth factors potentiated by epidermal growth factor. Proc. Natl. Acad. Sci. USA 78, 5339–5343.

Roberts, A. B., Anzano, M. A., Meyer, C. A., Wideman, J., Blacher, R., Pan, Y., Stein, S., Lehrman, S. R., Smith, J. M., Lamb, L. C., & Sporn, M. B. (1983). Purification and properties of a type β transforming growth factor from bovine kidney. Biochem. 22, 5692–5698.

Roberts, A. B., Sporn, M. B., Assoian, R. K., Smith, J. M., Roche, N. S., Wakefield, L. M., Heine, U. I., Liotta, L. A., Falanga, V., Kehrl, J. H., et al. (1986). Transforming growth factor type β: Rapid induction of fibrosis and angiogenesis *in vivo* and stimulation of collagen formation *in vitro*. Proc. Natl. Acad. Sci. USA 83, 4167–4171.

Roberts, A. B. & Sporn, M. B. (1990). The transforming growth factor βs. In: Peptide Growth Factors and Their Receptors—Handbook of Experimental Pharmacology (Sporn, M. B. and Roberts, A. B., eds.), pp. 419–472, Springer-Verlag, Heidelberg.

Roberts, A. B., Vodovotz, Y., Roche, N. S., Sporn, M. B., & Nathan, C. F. (1992). Role of nitric oxide in antagonistic effects of transforming growth factor β and interleukin-1β on the beating rate of cultured cardiac myocytes. Mol. Endocrinol. 6, 1921–1930.

Robinson, S. D., Silberstein, G. B., Roberts, A. B., Flanders, K. C., & Daniel, C. W. (1991). Regulated expression and growth inhibitory effects of transforming growth factor β isoforms in mouse mammary gland development. Development 113, 867–878.

Romeo, D. S., Park, K., Roberts, A. B., Sporn, M. B., & Kim, S. (1993). An element of the transforming growth factor β1 5′-untranslated region represses translation and specifically binds a cytosolic factor. Mol. Endocrinol. 7, 759–766.

Rook, A. H., Kehrl, J. H., Wakefield, L. M., Roberts, A. B., Sporn, M. B., Burlington, D. B., Lane, H. C., & Fauci, A. S. (1986). Effects of transforming growth factor β on the functions of natural killer cells: Depressed cytolytic activity and blunting of interferon responsiveness. J. Immunol. 136, 3916–3920.

Rosa, F., Roberts, A. B., Danielpour, D., Dart, L. L., Sporn, M. B., & Dawid, I. B. (1988). Mesoderm induction in amphibians: The role of TGF-β2-like factors. Science 239, 783–785.

Ross, N. R. & Bona, C. (1993). Defining criteria for autoimmune diseases (Witebsky's postulates revisited). Immunol. Today 14, 426–430.

Rotello, R. J., Leiberman, R. C., Purchio, A. F., & Gerschenson, L. E. (1991). Coordinated regulation of apoptosis and cell proliferation by transforming growth factor β1 in cultured uterine epithelial cells. Proc. Natl. Acad. Sci. USA 88, 3412–3415.

Russell, W. E., Coffey, R. J., Jr., Ouellette, A. J., & Moses, H. L. (1988). Type β transforming growth factor reversibly inhibits the early proliferative response to partial hepatectomy in the rat. Proc. Natl. Acad. Sci. USA 85, 5126–5130.

Salomon, G. D., Kasid, A., Bernstein, E., Buresh, C., Director, E., & Norton, J. A. (1990). Gene expression in normal and doxorubicin-impaired wounds: Importance of transforming growth factor β. Surgery 108, 318–322.

Samuel, S. K., Hurta, R. A. R., Kondaiah, P., Khalil, N., Turley, E. A., Wright, J. A., & Greenberg, A. H. (1992). Autocrine induction of tumor protease production and invasion by a metallothionein-regulated TGF-β1. EMBO J. 11, 1599–1605.

Sato, Y. & Rifkin, D. B. (1989). Inhibition of endothelial cell movement by pericytes and smooth muscle cells: Activation of a latent transforming growth factor β1-like molecule by plasmin during coculture. J. Cell Biol. 109, 309–315.

Schlunegger, M. P. & Grutter, M. G. (1992). An unusual feature revealed by the crystal structure at 2.2Å resolution of human transforming growth factor β2. Nature 258, 430–434.

Scotto, L. & Assoian, R. K. (1993). A GC-rich domain with bifunctional effects on mRNA and protein levels: Implications for control of transforming growth factor β1 expression. Mol. Cell. Biol. 13, 3588–3597.

Sellheyer, K., Bickenbach, J. R., Rothnagel, J. A., Bundman, D., Longley, M. A., Krieg, T., Roche, N. S., Roberts, A. B., & Roop, D. R. (1993). Inhibition of skin development by overexpression of transforming growth factor β1 in the epidermis of transgenic mice. Proc. Natl. Acad. Sci. USA 90, 5237–5241.

Sha, X., Yang, L., & Gentry, L. E. (1991). Identification and analysis of discrete functional domains in the pro region of pre-pro-transforming growth factor β1. J. Cell Biol. 114, 827–839.

Shah, M., Foreman, D. M., & Ferguso, M. W. J. (1992). Control of scarring in adult wounds by neutralizing antibody to TGFβ Lancet 339, 213–214.

Sherman, J. H., Miller, M., Albert, R. E., & Baxter, S. (1993). Dose- and time-dependent expression of transforming growth factor β1 mRNA and protein in mouse epidermis and papillomas after repeated topical application of benzo[*a*]pyrene. Mol. Carcinog. 8, 264–271.

Shull, M. M., Ormsby, I., Kier, A. B., Pawlowski, S., Diebold, R. J., Yin, M., Allen, R., Sidman, C., Proetzel, G., Calvin, D., Annunziata, N., & Doetschman, T. (1992). Targeted disruption of the mouse transforming growth factor β1 gene results in multifocal inflammatory disease. Nature 359, 693–699.

Shull, M. M. & Doetschman, T. (1994). Transforming growth factor β1 in reproduction and development. Mol. Reprod. Devel. 39, 239–246.

Silberstein, G. B. & Daniel, C. W. (1987). Reversible inhibition of mammary gland growth by transforming growth factor β. Science 237, 291–293.

Slavin, J., Nash, J. R., & Kingsnorth, A. N. (1992). Effect of transforming growth factor β and basic fibroblast growth factor on steroid-impaired healing intestinal wounds. Br. J. Surg. 79, 69–72.

Sporn, M. B., Roberts, A. B., Shull, J. H., Smith, J. M., Ward, J. M., & Sodek, J. (1983). Polypeptide transforming growth factors: Isolation from bovine sources and use for wound healing *in vivo*. Science 219, 1329–1331.

Sporn, M. B. & Roberts, A. B. (1985). Autocrine growth factors and cancer. Nature (Lond.) 313, 745–747.

Sporn, M. B. & Roberts, A. B. (1992). Transforming growth factor β: Recent progress and new challenges. J. Cell Biol. 119, 1017–1021.

Suardet, L., Gaide, A., Calmes, J., Sordat, B., Givel, J., Eliason, J. F., & Odartchenko, N. (1992). Responsiveness of three newly established human colorectal cancer cell lines to transforming growth factors β1 and β2. Cancer Res. 52, 3705–3712.

ten Dijke, P., Geurts van Kessel, A. H. M., Foulkes, J. G., & Le Beau, M. M. (1988). Transforming growth factor type β3 maps to human chromosome 14, region q23–q24. Oncogene 3, 721–724.

Torre–Amione, G., Beauchamp, R. D., Koeppen, H., Park, B. H., Schreiber, H., Moses, H. L., & Rowley, D. A. (1990). A highly immunogenic tumor transfected with a murine transforming growth factor type β1 cDNA escapes immune surveillance. Proc. Natl. Acad. Sci. USA 87, 1486–1490.

Torti, F. M., Torti, S. V., Larrick, J. W., & Ringold, G. M. (1989). Modulation of adipocyte differentiation by tumor necrosis factor and transforming growth factor β. J. Cell Biol. 108, 1105–1113.

Tsuji, T., Okada, F., Yamaguchi, K., & Nakamura, T. (1990). Molecular cloning of the large subunit of transforming growth factor type β masking protein and expression of the mRNA in various tissues. Proc. Natl. Acad. Sci. USA 87, 8835–8839.

Valverius, E. M., Walker–Jones, D., Bates, S. E., Stampfer, M. R., Clark, R., McCormick, F., Dickson, R. B., & Lippman, M. E. (1989). Production of and responsiveness to transforming growth factor β in normal and oncogene-transformed human mammary epithelial cells. Cancer Res. 49, 6269–6274.

Wahl, S. M., Hunt, D. A., Wakefield, L. M., McCartney–Francis, N., Wahl, L. M., Roberts, A. B., & Sporn, M. B. (1987). Transforming growth factor type β induces monocyte chemotaxis and growth factor production. Proc. Natl. Acad. Sci. USA 84, 5788–5792.

Wahl, S. M., Hunt, D. A., Wong, H. L., Dougherty, S., McCartney–Francis, N., Ellingsworth, L., Schmidt, J. A., Hall, G., Roberts, A. B., et al. (1988). Transforming growth factor β is a potent immunosuppressive agent that inhibits IL-1-dependent lymphocyte proliferation. J. Immunol. 140, 3026–3032.

Wahl, S. M. (1992). Transforming growth factor β (TGFβ) in inflammation: A cause and a cure. J. Clin. Immunol. 12, 61–74.

Welch, D. R., Fabra, A., & Nakajima, M. (1990). Transforming growth factor β stimulates mammary adenocarcinoma cell invasion and metastatic potential. Proc. Natl. Acad. Sci. USA 87, 7678–7682.

Whitby, D. J. & Ferguson, M. W. J. (1991). The extracellular matrix of lip wounds in feral, neomatal, and adult mice. Development 112, 651–668.

Wu, S. P., Theodoresou, D., Kerbel, R. S., Willson, J. K., Mulder, K. M., Humphrey, L. E., & Brattain, M. G. (1992). TGFβ1 is an autocrine-negative growth regulator of human colon carcinoma FET cells *in vivo* as revealed by transfection of an antisense expression vector. J. Cell Biol. 116, 187–196.

Wyllie, A. H. (1985). The biology of cell death in tumors. Anticancer Res. 5, 131–136.

Yang, E. Y. & Moses, H. L. (1990). Transforming growth factor β1-induced changes in cell migration, proliferation, and angiogenesis in the chicken chorioallantoic membrane. J. Cell Biol. 111, 731–741.

Yuspa, S. H. (1994). The pathogenesis of squamous cell cancer: Lessons learned from studies of skin carcinogenesis—33rd G. H. A. Clowes Memorial Award Lecture. Cancer Res. 54, 1178–1189.

Zentrella, A. & Massague, J. (1992). Transforming growth factor β induces myoblast differentiation in the presence of mitogens. Proc. Natl. Acad. Sci. USA 89, 5176–5180.

Zugmaier, G., Ennis, B. W., Deschauer, B., Katz, D., Knabbe, C., Wilding, G., Daly, P., Lippman, M. E., & Dickson, R. B. (1989). Transforming growth factors type β1 and β2 are equipotent growth inhibitors of human breast cancer cell lines. J. Cell. Physiol. 141, 353–361.

Zugmaier, G. & Lippman, M. E. (1990). Effects of TGF-β on normal and malignant mammary epithelium. Ann. NY. Acad. Sci. 593, 272–275.

THE TRANSFORMING GROWTH FACTOR β SUPERFAMILY

Alexandra C. McPherron and Se-Jin Lee

	Abstract	357
I.	Introduction	358
II.	Sequences and Structures	358
III.	Müllerian Inhibiting Substance (MIS)	369
IV.	Inhibins/Activins	371
V.	Bone Morphogenetic Proteins	376
VI.	Nodal, GDNF, and Other Growth/Differentiation Factors Related to TGFβ	380
VIII.	Conclusions	383
	References	384

ABSTRACT

The TGFβ superfamily encompasses a large group of growth and differentiation factors affecting a wide range of biological processes in both vertebrates and invertebrates. Although the various members of this family are all structurally related, they are quite divergent from TGFβ and from each other with respect to their amino acid

Growth Factors and Cytokines in Health and Disease
Volume 1B, pages 357–393.

ISBN: 0-7623-0091-4

sequences and the specific activities that they possess *in vitro* and *in vivo*. Like TGFβ, many of these factors are important in regulating cell differentiation processes during embryonic development and in maintaining tissue homeostasis in adult animals. Because these molecules are potent extracellular regulators of cell growth and differentiation, understanding their biology will be important for deciphering the complex regulatory network of cell–cell interactions that take place during development and for understanding how such regulatory mechanisms can go awry in the pathogenesis of human diseases and how the biological properties of these factors can be exploited for clinical applications.

I. INTRODUCTION

Since the elucidation of the primary structure of transforming growth factor β (TGFβ) in 1985, a large number of growth and differentiation factors structurally related to TGFβ have been identified in invertebrate and vertebrate species. In mammals, twenty-three distinct members of the TGFβ superfamily have been reported in the literature to date. In addition to the three isoforms of TGFβ, the mammalian family members include Müllerian inhibiting substance (MIS), the inhibins and activins, the bone morphogenetic proteins (BMPs), nodal, glial cell line-derived neurotrophic factor (GDNF), and a number of additional factors of unknown function identified on the basis of their sequence homology to other family members. The various members of this gene family encode secreted factors that are similar to TGFβ in their overall structure and in their ability to regulate cell differentiation processes during embryonic development and/or in adult tissues. Although many of these factors share some structural and functional similarity to TGFβ, the members of this superfamily are only distantly related to TGFβ at the primary amino acid sequence level and are quite disparate from each other and from TGFβ with respect to the specific activities that they possess *in vitro* and the specific biological roles that they play *in vivo*.

Here we will review the current state of knowledge of the mammalian members of the TGFβ superfamily focusing primarily on their *in vivo* roles. Additional detail regarding their biological activities can be found in a number of recent reviews describing individual family members or subgroups thereof (for example, see Vale et al., 1991; Dye et al., 1992; Wozney, 1992; Lee and Donahoe, 1993; Hillier and Miró, 1993; Josso et al., 1993; Behringer, 1994; Kingsley, 1994; Moore et al., 1994; Urist, 1994). In addition, chapters describing TGFβ itself and receptors for members of this superfamily can be found elsewhere in this volume.

II. SEQUENCES AND STRUCTURES

Each member of the TGFβ superfamily has a primary structure depicted schematically in Figure 1. The overall structure of each family member contains several features shared by all members of this superfamily. First, each sequence contains a

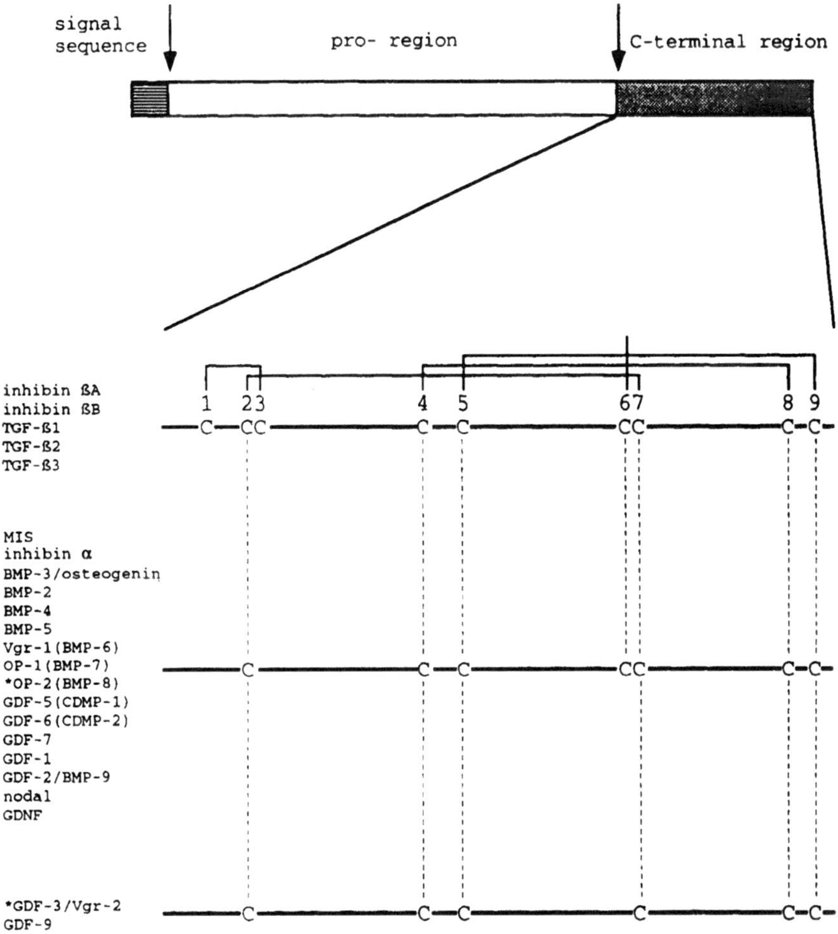

Figure 1. Schematic diagram of the primary structure of TGFβ superfamily members. Arrows point to sites of proteolytic processing. The cysteines in the C-terminal region are denoted by "C"s and numbered relative to their positions in mature TGFβ. Lines connecting the numbered cysteines represent the intra- and intermolecular disulfide bridges found in the mature TGFβ2 dimer. The individual factors that conform to the diagrammed structures are listed to the left of the individual structures. The asterisks designate the factors that contain an extra cysteine residue in the C-terminal region.

core of hydrophobic amino acids near the N-terminus that functions as a signal sequence for secretion. Second, each family member appears to be made in a precursor form that is proteolytically processed. Third, the C-terminal regions of all family members show some degree of homology to each other at the primary amino acid sequence level and contain six to nine cysteine residues separated by a characteristic spacing. And fourth, the active species in each case appears to be a disulfide-linked dimer of C-terminal fragments; in most cases, the active form has been shown to be a homodimer of C-terminal fragments, but heterodimers between different family members have also been detected, and in certain cases, these heterodimers have been shown to possess biological activities distinct from the corresponding homodimers.

Each member of this family is believed to be synthesized in a prepro- form that must undergo at least two proteolytic processing events to generate the biologically active molecule. Following cleavage of the signal peptide at the N-terminus, a second cleavage event removes the pro- region, which generally spans approximately 70 to 75% of the length of the precursor protein. It is this second processing event that liberates the mature C-terminal dimer, which is sufficient in most cases for full biological activity. In nearly all cases where the N-terminal amino acid of the mature protein has been determined, the cleavage site has been shown to conform to the sequence RxxR, and all family members contain at least one RxxR or dibasic sequence approximately 110 to 140 amino acids from the C-terminus. Based on the sequence specificity of this proteolytic cleavage, it has been presumed that subtilisin-like proteases (such as furin) are involved in carrying out this processing event *in vivo* (for review, see Barr, 1991). However, two lines of evidence have suggested that a second class of proteases may also play a role in regulating the maturation of at least certain members of this family. First, highly purified preparations of bone morphogenetic proteins have been shown to contain not only molecules that are members of the TGFβ superfamily, but also an unrelated molecule, BMP-1, which is also capable of inducing bone formation *in vivo* and which shows sequence homology to a known metalloprotease (Wozney et al., 1988). And second, genetic studies in *Drosophila* have suggested that the *decapentaplegic* gene product, which is a member of the TGFβ superfamily most homologous to two bone morphogenetic proteins (BMP-2 and BMP-4) (Padgett et al., 1987; Wozney et al., 1988), may interact with the *tolloid* gene product (Ferguson and Anderson, 1992), which also contains a metalloprotease domain (Shimell et al., 1991). Based on these observations, it seems likely that these metalloprotease-like proteins may play a role in regulating the activities of particular TGFβ family members, perhaps by directly carrying out the proteolytic processing event that generates the active C-terminal dimers.

In general, it is only in the C-terminal region that all of the family members are related at the primary sequence level. An alignment of the C-terminal regions of all known mammalian family members is shown in Figure 2, and a tabulation of amino acid sequence homologies between the C-terminal regions of family members compared pairwise is shown in Figure 3. Based on these sequence comparisons, certain family members can be divided into subgroups consisting of highly related factors. At least four subgroups are readily apparent from these sequence comparisons: one subgroup consisting of BMP-2 and BMP-4; a second subgroup consisting of BMP-5, Vgr-1 (Vg-related-1) (also known as BMP-6), OP-1 (osteogenic pro-

```
                                1        23                                    4   5
MIS                 PGRAQRSAGATAADGPCALRELSVDL----RAERSVLIPETYQANNCQGVCGWPQSDRNPRY--
inhibin α           LRLLQRPPEEPAAHANCHRVALNISF-QELGWERWIVYPPSFIFHYCHGGCGLHIPPNLSLPV-
inhibin ßA          RRRRGLECDGKV--NICCKKQFFVSF-KDIGWNDWIIAPSGYHANYCEGECPSHIAGTSGSSL-
inhibin ßB          IRKRGLECDGRT--NLCCRQQFFIDF-RLIGWNDWIIAPTGYYGNYCEGSCPAYLAGVPGSAS-
BMP-3/osteogenin    QTLKKARRKQWIEPRNCARRYLKVDF-ADIGWSEWIISPKSFDAYYCSGACQFPMPKSLKP---
BMP-2               EKRQAKHKQRKRLKSSCKRHPLYVDF-SDVGWNDWIVAPPGYHAFYCHGECPFPLADHLNS---
BMP-4               RSPKHHSQRARKKNKNCRRHSLYVDF-SDVGWNDWIVAPPGYQAFYCHGDCPFPLADHLNS---
BMP-5               RMSSVGDYNTSEQKQACKKHELYVSF-RDLGWQDWIIAPEGYAAFYCDGECSFPLNAHMNA---
Vgr-1(BMP-6)        RGSGSSDYNGSELKTACKKHELYVSF-QDLGWQDWIIAPKGYAANYCDGECSFPLNAHMNA---
OP-1(BMP-7)         RMANVAENSSSDQRQACKKHELYVSF-RDLGWQDWIIAPEGYAAYYCEGECAFPLNSYMNA---
OP-2(BMP-8)         LPGIFDDVHGSHGRQVCRRHELYVSF-QDLGWLDWVIAPQGYSAYYCEGECSFPLDSCMNA---
GDF-5(CDMP-1)       SATRQGKRPSKNLKARCSRKALHVNF-KDMGWDDWIIAPLEYEAFHCEGLCEFPLRSHLEP---
GDF-6(CDMP-2)       FASRHGKRHGKKSRLRCSKKPLHVNF-KELGWDDWIIAPLEYEAYHCEGVCDFPLRSHLEP---
GDF-7               GGGGAGRGHGRRGRSRCSRKSLHVDF-KELGWDDWIIAPLDYEAYHCEGVCDFPLRSHLEP---
GDF-1               PRRDAEPVLGGGPGGACRARRLYVSF-REVGWHRWVIAPRGFLANYCQGQCALPVALSGSGGPP
GDF-2/BMP-9         GPLLARRKRSTGASSHCQKTSLRVNF-EDIGWDSWIIAPKEYDAYECKGGCFFPLADDVTP---
GDF-3/Vgr-2         KRRAAISVPKGFCRNFCHRHQLFINF-QDLGWHKWVIAPKGFMANYCHGECPFSMTTYLNS---
GDF-9               FNLSEYFRQFLLPQNECELHDFRLSF-SQLKWDNWIVAPHRYNPRYCKGDCPRAGRHRYGS---
nodal               WGRRQRRHHLPDRSQLCRRVKFQVDF-NLIGWGSWIIYPKQYNAYRCEGECPNPVGEEFHP---
GDNF                NSRGKGRRGQRGKNRGCVLTAIHLNV---TDLGLGYETKEELIFRYCSGSCDAAETTYDKI---
TGF-ß1              RALDTNYCFSSTE-KNCCVRQLYIDFRKDLGWK-WIHEPKGYHANFCLGPCPYIWSLD------
TGF-ß2              RALDAAYCFRNVQ-DNCCLRPLYIDFKRDLGWK-WIHEPKGYNANFCAGACPYLWSSD------
TGF-ß3              RALDTNYCFRNLE-ENCCVRPLYIDFRQDLGWK-WVHEPKGYYANFCSGPCPYLRSAD------
```

```
                                       67                                          8 9
MIS                 -GNHVVLLLKMQA--RGAALARPPCCV--PTAYAGKLLISLSEER--ISAHHVPNMVATECGCR
inhibin α           -PGAPPTPAQPYS----LLPGAQPCCAALPGTMRPLHVRTTSDGGYSFKYETVPNLLTQHCACI
inhibin ßA          -SFHSTVINHYRMRGHSPFANLKSCCV--PTKLRPMSMLYY-DDGQNIIKKDIQNMIVEECGCS
inhibin ßB          -SFHTAVVNQYRMRGLNPGT-VNSCCI--PTKLSTMSMLYF-DDEYNIVKRDVPNMIVEECGCA
BMP-3/osteogenin    -SNHATIQSIVRA-VGVVPGIPEPCCV--PEKMSSLSILFF-DENKNVVLKVYPNMTVESCACR
BMP-2               -TNHAIVQTLVNS---VNSKIPKACCV--PTELSAISMLYL-DENEKVVLKNYQDMVVEGCGCR
BMP-4               -TNHAIVQTLVNS---VNSSIPKACCV--PTELSAISMLYL-DEYDKVVLKNYQEMVVEGCGCR
BMP-5               -TNHAIVQTLVHL--MFPDHVPKPCCA--PTKLNAISVLYF-DDSSNVILKKYRNMVVRSCGCH
Vgr-1(BMP-6)        -TNHAIVQTLVHL--MNPEYVPKPCCA--PTKLNAISVLYF-DDNSNVILKKYRNMVVRACGCH
OP-1(BMP-7)         -TNHAIVQTLVHF--INPETVPKPCCA--PTQLNAISVLYF-DDSSNVILKKYRNMVVRACGCH
OP-2(BMP-8)         -TNHAILQSLVHL--MKPNAVPKACCA--PTKLSATSVLYY-DSSNNVILRKARNMVVKACGCH
GDF-5(CDMP-1)       -TNHAVIQTLMNS--MDPESTPPTCCV--PTRLSPISILFI-DSANNVVYKQYEDMVVESCGCR
GDF-6(CDMP-2)       -TNHAIIQTLMNS--MDPGSTPPSCCV--PTKLTPISILYI-DAGNNVVYNEYEEMVVESCGCR
GDF-7               -TNHAIIQTLLNS--MAPDAAPASCCV--PARLSPISILYI-DAANNVVYKQYEDMVVEACGCR
GDF-1               ALNHAVLRALMHA--AAPGAADLPCCV--PARLSPISVLFF-DNSDNVVLRQYEDMVVDECGCR
GDF-2/BMP-9         -TKHAIVQTLVHL--EFPTKVGKACCV--PTKLSPISILYKDDMGVPTLKYHYEGMSVAECGCR
GDF-3/Vgr-2         -SNYAFMQALMHM---ADPKVPKAVCV--PTKLSPISMLYQ-DSDKNVILRHYEDMVVDECGCG
GDF-9               -PVHTMVQNIIYE--KLDSSVPRPSCV--PAKYSPLSVLTI-EPDGSIAYKEYEDMIATRCTCR
nodal               -TNHAYIQSLLKR--YQPHRVPSTCCA--PVKTKPLSMLYV--DNGRVLLEHHKDMIVEECGCL
GDNF                -LKNLSRNRRLVS-----DKVGQACCR--PIAFDDDLSFL--DDNLVYHIL--RKHSAKRCGCI
TGF-ß1              -TQYSKVLALYNQ--HNPGASAAPCCV--PQALEPLPIVYY-VGRKPKV-EQLSNMIVRSCKCS
TGF-ß2              -TQHSRVLSLYNT--INPEASASPCCV--SQDLEPLTILYY-IGKTPKI-EQLSNMIVKSCKCS
TGF-ß3              -TTHSTVLGLYNT--LNPEASASPCCV--PQDLEPLTILYY-VGRTPKV-EQLSNMVVKSCKCS
```

Figure 2. Alignment of the C-terminal sequences of human MIS (Cate et al., 1986), human inhibin α, human inhibin βA, human inhibin βB (Mason et al., 1986), human BMP-3/osteogenin, human BMP-2, human BMP-4 (Wozney et al., 1988), human BMP-5, human Vgr-1 (BMP-6), human OP-1 (BMP-7) (Celeste et al., 1990; Özkaynak, 1990), human OP-2 (BMP-8) (Özkaynak et al., 1992), human GDF-5 (CDMP-1), human GDF-6 (CDMP-2) (Chang et al., 1994), murine GDF-7 (Storm et al., 1994), human GDF-1 (Lee, 1991), human GDF-2/BMP-9 (Wozney and Celeste, 1993), murine GDF-3/Vgr-2 (McPherron and Lee, 1993), human GDF-9 (McGrath et al., 1995), murine nodal (Zhou et al., 1993b), human GDNF (Lin et al., 1993), human TGFβ1 (Derynck et al., 1985), human TGFβ2 (de Martin et al., 1987), and human TGFβ3 (Derynck et al., 1988; ten Dijke et al., 1988). The positions corresponding to the nine cysteine residues in TGFβ are shaded and numbered.

	MIS	inhibin α	inhibin ßA	inhibin ßB	BMP-3/osteogenin	BMP-2	BMP-4	BMP-5	Vgr-1(BMP-6)	OP-1(BMP-7)	OP-2(BMP-8)	GDF-5(CDMP-1)	GDF-6(CDMP-2)	GDF-7	GDF-1	GDF-2/BMP-9	GDF-3/Vgr-2	GDF-9	nodal	GDNF	TGF-ß1	TGF-ß2	TGF-ß3
MIS	100	18	24	25	30	27	27	24	24	27	27	27	26	25	34	24	22	21	22	18	28	23	25
inhibin α	-	100	26	25	29	22	22	24	25	24	26	24	25	26	23	22	25	26	23	20	23	22	24
inhibin ßA	-	-	100	63	36	42	41	43	44	43	42	40	43	41	37	40	42	30	39	18	41	37	36
inhibin ßB	-	-	-	100	37	42	42	37	41	42	37	37	38	36	35	33	41	30	37	22	35	34	37
BMP-3/osteogenin	-	-	-	-	100	48	47	43	44	42	41	47	44	46	42	42	42	29	41	17	32	32	32
BMP-2	-	-	-	-	-	100	92	61	61	60	55	57	54	57	42	52	53	34	42	23	35	34	36
BMP-4	-	-	-	-	-	-	100	59	60	58	55	57	55	57	43	52	50	35	40	22	34	33	35
BMP-5	-	-	-	-	-	-	-	100	91	88	74	52	54	52	46	54	50	31	40	23	34	35	36
Vgr-1(BMP-6)	-	-	-	-	-	-	-	-	100	87	75	51	53	52	46	54	53	31	43	21	35	37	39
OP-1(BMP-7)	-	-	-	-	-	-	-	-	-	100	74	51	53	53	47	52	50	30	41	22	34	38	38
OP-2(BMP-8)	-	-	-	-	-	-	-	-	-	-	100	50	51	53	47	49	55	27	44	24	30	35	38
GDF-5(CDMP-1)	-	-	-	-	-	-	-	-	-	-	-	100	82	80	46	51	48	32	42	22	33	34	37
GDF-6(CDMP-2)	-	-	-	-	-	-	-	-	-	-	-	-	100	82	43	54	46	32	43	22	34	35	37
GDF-7	-	-	-	-	-	-	-	-	-	-	-	-	-	100	48	51	45	33	45	24	36	35	38
GDF-1	-	-	-	-	-	-	-	-	-	-	-	-	-	-	100	39	50	28	35	18	33	32	33
GDF-2/BMP-9	-	-	-	-	-	-	-	-	-	-	-	-	-	-	-	100	48	33	44	24	32	33	33
GDF-3/Vgr-2	-	-	-	-	-	-	-	-	-	-	-	-	-	-	-	-	100	32	40	24	36	31	32
GDF-9	-	-	-	-	-	-	-	-	-	-	-	-	-	-	-	-	-	100	31	20	22	25	24
nodal	-	-	-	-	-	-	-	-	-	-	-	-	-	-	-	-	-	-	100	19	28	32	31
GDNF	-	-	-	-	-	-	-	-	-	-	-	-	-	-	-	-	-	-	-	100	19	17	21
TGF-ß1	-	-	-	-	-	-	-	-	-	-	-	-	-	-	-	-	-	-	-	-	100	74	78
TGF-ß2	-	-	-	-	-	-	-	-	-	-	-	-	-	-	-	-	-	-	-	-	-	100	82
TGF-ß3	-	-	-	-	-	-	-	-	-	-	-	-	-	-	-	-	-	-	-	-	-	-	100

Figure 3. Amino acid homologies among the mammalian TGFβ superfamily members. Numbers represent percent amino acid identities between each pair calculated from the first cysteine conserved in all family members (cysteine 2 of the C-terminal region shown in Figure 2). Shaded boxes represent homologies among highly related members within particular subgroups.

tein-1) (also known as BMP-7), and OP-2 (also known as BMP-8); a third subgroup consisting of GDF-5 (growth/differentiation factor-5) (also known as CDMP-1 [cartilage-derived morphogenetic protein-1]), GDF-6 (also known as CDMP-2), and GDF-7; and a fourth subgroup consisting of the three TGFβ isoforms (TGFβ1, TGFβ2, and TGFβ3). Within each of these four subgroups, the family members are approximately 70 to 90% identical when compared pairwise.

These four subgroups, however, encompass only about half of the known family members, and, because all of the sequences appear to form a continuum, any classification schemes used to further subdivide the various factors based on homology alone have been somewhat arbitrary. In general, when family members are compared to those outside of a given subgroup, the degree of sequence homology is only in the range of 20 to 50%, and certain family members, such as MIS, inhibin α, GDF-9, and GDNF, are only distantly related to all other family members. The high degree of sequence divergence among members of this superfamily is also evident in examining the distribution of conserved amino acids within the C-terminal region (Figure 2). In particular, there are only seven invariant amino acids scattered throughout the entire region. Six of these invariant amino acids correspond to cysteine residues, and the seventh is an invariant glycine residue located approximately halfway into the C-terminal region. Hence, although a number of laboratories have been attempting to identify new family members either by a degenerate PCR strategy or by low stringency screening with known sequences, the high degree of sequence divergence and the distribution of conserved amino acids have made a thorough search essentially impossible. Based on these considerations, it is virtually certain that this superfamily will turn out to encompass considerably more factors than are presently known.

Certain family members, like the TGFβs, contain nine cysteine residues each in the C-terminal region. Based on the three-dimensional structure of TGFβ2 (Daopin et al.,1992; Schlunegger and Grütter, 1992), it is known that eight of the cysteine residues form four intramolecular disulfide bridges. In the crystal structure, each TGFβ2 monomer forms a structure that resembles an open left hand with the four "fingers" formed by antiparallel β-sheets. A disulfide bond joining the first and third cysteine residues in TGFβ2 anchors an α-alpha helix of the amino-terminal "thumb" to the β-sheet of the first "finger". Three additional disulfide bonds are formed at the "palm" of the hand by cysteines two and seven, cysteines four and eight, and cysteines five and nine. These three disulfide bonds form a cystine knot structure in which one disulfide bridge passes through the ring formed by the other two; this cystine knot motif is also seen in platelet-derived growth factor and nerve growth factor (for review, see McDonald and Hendrickson, 1993), which also contain four extended antiparallel β-strands and also function as dimers. In addition to the four intramolecular disulfide bonds formed by eight of the nine cysteine residues in TGFβ2, one intermolecular disulfide is formed by the remaining cysteine residue (cysteine six). This sole covalent intermolecular linkage joins the two monomers so that the "palm" regions face each other but are oppositely oriented.

Although the three-dimensional structures of other family members have not yet been reported, the conservation of the cysteine residues suggests that they will have an overall fold similar to TGFβ2. In addition to the TGFβ isoforms, inhibin βA and βB also contain nine similarly spaced cysteine residues in the C-terminal region (Mason et al., 1985; Forage et al., 1986). Although most other family members contain only seven cysteine residues in this region, because the two missing cysteine residues correspond to cysteines one and three in TGFβ2, it seems likely that the most significant deviation from the overall fold would be a possible effect on the interaction between the "thumb" and the first "finger". OP-2 (Özkaynak et al., 1992) and GDF-3/Vgr-2 (Jones et al., 1992; McPherron and Lee, 1993), each contain an additional cysteine residue in the C-terminal region whose position does not correspond to any of the nine cysteine residues in TGFβ2; whether these additional cysteines are involved in disulfide bond formation is not known. Two family members, GDF-9 and GDF-3/Vgr-2 (Jones et al., 1992; McPherron and Lee, 1993), lack the cysteine residue (cysteine six) that forms the sole intermolecular disulfide bond in TGFβ2. Although GDF-3/Vgr-2 contains an additional cysteine residue in the "thumb" region that could be involved in forming an intermolecular linkage, GDF-9 contains no other cysteine residues and, therefore, is unlikely to form a disulfide-linked dimer. Whether either of these molecules forms dimers at all has not yet been established, but noncovalent interactions, such as hydrophobic interactions, between monomers almost certainly play a major role in stabilizing the TGFβ2 dimer.

In this respect, several studies have produced conflicting results regarding the importance of intermolecular disulfide bond formation for biological activity in other family members. In particular, for the inhibins and activins, "monomeric" forms of α and βA subunits are present at high levels in follicular fluid (Knight et al., 1989; Robertson et al., 1989), and naturally-occurring "monomeric" βA subunits are only 2 to 5-fold less active than disulfide-linked βA dimers (Robertson et al., 1992). In contrast, mutants in which cysteine six of the βA subunit was changed either to alanine or serine by site-directed mutagenesis were only 1% as active as native dimeric βA (Hüsken-Hindi et al., 1994; Mason, 1994), and an analogous mutation in TGFβ1 in which cysteine six was changed to serine virtually abolished the ability of TGFβ1 to inhibit growth of mink lung epithelial cells (Brunner et al., 1992; Amatayakul-Chantler et al., 1994). However, the effect of these mutations on other activities of these proteins was less severe; that is, the cysteine to serine mutation in the inhibin βA subunit reduced receptor binding by only 5-fold (Hüsken–Hindi et al., 1994), and the same mutation in TGFβ1 had only a 5-fold effect on its ability to induce expression from the plasminogen activator inhibitor-1 promoter and a partial or no effect on receptor binding (Amatayakul–Chantler et al., 1994). Hence, only certain activities may require forming a disulfide-linked dimer. Whether the activities of either inhibin βA or TGFβ1 subunits in these assays were mediated by truly monomeric forms of these proteins or by noncovalently-linked dimers was not determined conclusively in any of these studies.

Although much is known about the structure and function of the C-terminal regions of these family members, very little is known about the role of the pro-regions. In general, there is little discernible relationship among the pro-regions of different family members at the primary sequence level. Even between family members belonging to the same subgroup, the degree of homology in the pro- region is significantly lower than in the C-terminal region. For example, BMP-2 and BMP-4, the two most highly related factors in the C-terminal region (92% amino acid sequence identity), are only approximately 50% homologous in the pro-region. In addition, for any given molecule, the pro- region tends to be less conserved through evolution than the C-terminal region. Although the function of the pro- region is not entirely understood, several types of experiments have suggested that at least one function may be to direct proper folding, dimerization, and processing of the C-terminal region. Experiments using hybrid constructs in which the pro- region of one family member was substituted with that of another have shown that the pro- regions are capable of acting in a heterologous manner.

For example, the pro- region of BMP-2 is capable of directing the efficient processing and assembly of C-terminal BMP-4 dimers in a hybrid BMP-2/BMP-4 construct (Hammonds et al., 1991). Although this result might be expected, given the high degree of sequence homology between BMP-2 and BMP-4, the BMP-2 pro- region is also capable of functioning with more distantly related C-terminal regions, such as GDF-1 (M. Baxter and S.-J. Lee, unpublished results) and *Xenopus* Vg-1 (Thomsen and Melton, 1993), in hybrid constructs.

The role of the pro- region in directing folding and assembly has been investigated most thoroughly for inhibin βA and TGFβ1 (Gray and Mason, 1990). In particular, when inhibin βA or TGFβ1 was expressed in mammalian cells, secreted dimers of inhibin βA or TGFβ1 subunits were detected only when the pro- region was coexpressed. Remarkably, expression of secreted dimers could be at least partially reconstituted by coexpressing the pro- region in a separate construct, and in this respect, the pro- region of TGFβ1 could partially substitute for the pro- region of inhibin βA when coexpressed with the C-terminal region of inhibin βA. Hence, to some extent, the pro- region is capable of functioning with a heterologous factor and also acting in *trans* to direct the assembly of C-terminal dimers.

In addition to affecting the folding and assembly of the C-terminal region, the pro-region may also directly influence biological activity. For example, purified MIS consists mainly of a disulfide-linked dimer of full-length uncleaved molecules that can be cleaved *in vitro* to yield a tightly bound complex of C-terminal dimer and pro- region (Pepinsky et al., 1988). When the pro- region and C-terminal region were purified separately from such a mixture, the C-terminal region by itself was biologically active, but full potency could be generated only upon reconstituting the C-terminal region with the pro- region (Wilson et al., 1993). Hence, the pro-region of MIS appears to be capable of potentiating the activity of the C-terminal dimer even though the pro- region alone is inactive. In the case of TGFβ, the pro-region appears to have an inhibitory effect on the biological activity of the C-terminal region (Gentry and Nash, 1990). Indeed, TGFβ appears to exist *in vivo* in a latent form in which the C-terminal dimer is held in a complex with other proteins, including the pro- region (Miyazono et al., 1988). Biologically active TGFβ can be generated *in vitro* only by dissociating the C-terminal dimer from this latent complex, such as by treatment with acid or base, chaotropic agents, or plasmin (Lawrence et al., 1985; Lyons et al., 1988). In fact, cells transfected with TGFβ1 mutants, in which either of two cysteine residues in the pro- region was changed to serine, secreted mature disulfide-linked C-terminal dimers that did not require acid activation for full biological activity (Brunner et al., 1989). Whether other members of this superfamily also exist in such latent complexes is not known.

Although all the members of this superfamily share a common overall structure, they appear to be quite divergent with respect to their primary amino acid sequences and their biological functions. Many of these factors appear to influence multiple biological processes, and conversely, many biological processes appear to be influenced by multiple members of this family. Tables 1 and 2 summarize the data

Table 1. Embryonic and Adult Expression Patterns of the Mammalian TGFβ Superfamily Members and Their Known *In Vitro* and *In Vivo* activities (see text for references)

Family Member	*Expression Patterns*	*Activities*
MIS	developing testes: adult ovaries, testes	induction of Müllerian duct regression; inhibition of oocyte maturation; inhibition of lung production of DSPC
inhibin α	blastocyst; developing ovaries, testes, adrenal glands; placenta; adult ovaries, testes, brain, spinal cord, pituitary gland, adrenal glands, kidneys, spleen, pancreas.	inhibin: inhibition of FSH production; regulation of steroidogenesis by Leydig and thecal cells; inhibition of spermatogonial proliferation and oocyte maturation; regulation of monocyte function; stimulation of DNA synthesis by thymocytes
inhibin βA	early cleavage stage embryo; developing ovaries, testes, heart, skin, bones, intervertebral disks, tendons, whisker follicles, hair bulbs, teeth, brain, adrenal glands; placenta, deciduum; adult ovaries, testes, brain, spinal cord, adrenal glands, spleen, bone marrow, endothelium, regenerating liver, pancreas, eyes.	activin: stimulation of FSH production; regulation of steroidogenesis by Leydig, thecal, and adrenal cells; regulation of neuronal survival and differentiation; stimulation of spermatogonial proliferation, oocyte maturation, erythroid differentiation, cytokine production by monocytes, insulin secretion, and vascular smooth muscle cell proliferation; potentiation of BMP-induced bone formation; inhibition of vascular endothelial and adrenal cell proliferation; inhibition of DNA synthesis by thymocytes and stimulated hepatocytes; induction of hepatocyte cell death
inhibin βB	early cleavage stage embryo, blastocyst; developing ovaries, testes, brain, salivary glands, adrenal glands; placenta; adult ovaries, testes, pituitary gland, brain, adrenal glands, pancreas.	
BMP-3/ osteo-genin	perichondrium and periosteum of developing bones; developing lungs, kidneys, brain, intestine; adult lungs, kidneys, brain, intestine, prostate gland.	induction of cartilage formation.
BMP-2	apical ectodermal ridge, interdigital mesenchyme, perichondrium and osteogenic zone of developing limbs; prevertebrae; developing heart, blood, vessels, hair and whisker follicles, teeth, craniofacial mesenchyme, otic vesicles, branchial pouches, kidneys; placenta, amnion, deciduum; adult hair follicles, prostate gland.	induction of cartilage and bone formation; inhibition of early limb mesenchymal cell proliferation; regulation of dental mesenchymal cell differentiation; regulation of neural differentiation.

(*continued*)

Table 1. (Continued)

Family Member	*Expression Patterns*	*Activities*
BMP-4	apical ectodermal ridge and mesenchyme of early developing limbs; developing facial processes, whisker follicles, teeth, branchial pouches and arches, gut mesenchyme, heart, diencephalon, otic vesicles; allantois, amnion, deciduum; adult bones during fracture healing; adult lungs, kidneys, liver, spleen, prostate gland.	induction of cartilage and bone formation; regulation of dental mesenchymal cell differentiation; inhibition of hair cell proliferation; induction of apoptosis of neural crest cells; regulation of neural differentiation and monocyte chemotaxis.
BMP-5	condensing mesenchyme, perichondrium and periosteum of developing skeletal structures; developing lungs, liver, heart, ureter, renal pelvis, bladder, intestine, brain, meninges; adult lungs, heart, liver, external ears.	induction of cartilage and bone formation; regulation of neural differentiation.
Vgr-1 (BMP-6)	blastocyst; hypertrophic cartilage of developing limbs, vertebrae, ribs, and head; developing skin, palate, tongue, heart, brain, spinal cord, meninges, liver, lungs, kidneys; placenta, amnion, yolk sac, deciduum; adult ovaries, testes, brain, meninges, muscle, kidneys, adrenal glands, uterus, cervix, vagina, lungs, esophagus, stomach, prostate.	inhibition of keratinocyte proliferation; induction of neural differentiation.
OP-1 (BMP-7)	postnatal kidneys, bladder, brain, adrenal glands, heart, lungs, spleen, intestine.	induction of cartilage and bone formation; regulation of neural differentiation; protection against reperfusion injury following myocardial ischemia.
OP-2 (BMP-8)	?	?
GDF-5 (CDMP-1)	precartilaginous mesenchyme and perichondrium in developing limbs; placenta; adult uterus, oviducts, brain, thymus, heart, lungs, kidneys, adrenal glands.	?
GDF-6 (CDMP-2)	hypertrophic cartilage in developing limbs; adult cricoid and articular cartilage.	?
GDF-7	?	?
GDF-1	developing brain, spinal cord; adult brain, spinal cord.	?
GDF-2 /BMP-9	?	?
GDF-3/ Vgr-2	developing long bones, calvaria, ribs, and bones of the head; adult bone marrow, spleen, thymus, adipose tissue.	?
GDF-9	oocytes	?
nodal	primitive node	?
GDNF	developing brain, spinal cord, muscle.	promotion of dopaminergic and motor neuron survival.

Table 2. Mouse and Human Map Positions and Descriptions of the Phenotypes Resulting from Mutations of the Mammalian TGFβ Superfamily Members

Family Member	*Map Position*		*Null Phenotype*
	Mouse	*Human*	
MIS	10	19p13.3-13.2	persistence of Müllerian duct structures in males; Leydig cell hyperplasia
inhibin α	1	2q33-qter	elevated FSH and activin levels; development of gonadal stromal and adrenal tumors; cachexia.
inhibin βA	13	7p15-14	?
inhibin βB	1	2cen-q13	slightly elevated FSH levels; increased gestational time in mutant mothers; inability to nurse; failure of eyelid fusion in some genetic backgrounds.
BMP-3/osteogenin	5	4p14-q21	?
BMP-2	2	20p12	?
BMP-4	14	14	?
BMP-5	9	6	(*short ear*) reduction in sizes of external ears, the manubrium at the top of the sternum, and the ventral process of the sixth cervical vertebra; absence of one pair of ribs; altered shapes of the xyphoid process and thyroid and larynx cartilages; impaired ability to repair rib fractures; tendency to develop liver granulomas, hydroureter, hydronephrosis, and lung cysts.
Vgr-1 (BMP-6)	13	6	?
OP-1 (BMP-7)	?	20	?
OP-2 (BMP-8)	?	?	?
GDF-5 (CDMP-1)	2	*20p11-q13	(*brachypodism*) shortened long bones, metacarpals and metatarsals; fused proximal and medial phalanges.
GDF-6 (CDMP-2)	4	?	?
GDF-7	12	*2pter-p23	?
GDF-1	?	?	?
GDF-2/BMP-9	?	?	?
GDF-3/Vgr-2	6	?	?
GDF-9	?	?	?
nodal	?	?	(413-d) embryonic lethality; absence of mesoderm formation; overproliferation of ectoderm.
GDNF	?	?	?

Notes: Mouse map positions for MIS (King et al., 1991), inhibins α, βA, and βB (Barton et al., 1989), BMP-3/osteogenin, BMP-2, BMP-4, and Vgr-1 (Dickinson et al., 1990), BMP-5 (Kingsley et al., 1992), GDF-5 (CDMP-1), GDF-6 (CDMP-2), and GDF-7 (Storm et al., 1994), and GDF-3/Vgr-2 (Jones et al., 1992), and human map positions for MIS (Cohen-Haguenauer et al., 1987), inhibins α, βA, and βB (Barton et al., 1989), BMP-3/osteogenin (Tabas et al., 1991), BMP-2 (Tabas et al., 1991; Rao et al., 1992), BMP-4 (Tabas et al., 1993), BMP-5, Vgr-1 (BMP-6), and OP-1 (BMP-7) (Hahn et al., 1992) are shown. Human map positions, based solely on the corresponding mouse locations, are designated by an asterisk (see text for references describing mutant phenotypes).

currently available regarding the expression patterns, biological activities, map positions, and mutant phenotypes for each member of the superfamily. In the sections below, we will discuss each factor individually with particular focus on the functions that these factors are likely to perform *in vivo*.

III. MÜLLERIAN INHIBITING SUBSTANCE (MIS)

The existence of a male hormone capable of suppressing the development of female reproductive organs was first proposed by Lillie (1916) to explain the abnormal sexual differentiation of the freemartin calf. The freemartin calf is sterile because it lacks a uterus, oviducts, and the upper part of the vagina, structures that normally develop from the Müllerian ducts. In addition, the ovaries of the freemartin calf lack germ cells and have a morphology resembling that of seminiferous tubules (for review, see Jost et al., 1973). Lillie proposed that the freemartin effect resulted from the exposure of the female embryo to a hormone produced by the male twin through shared chorionic circulation. Castration experiments carried out by Jost (1947) in rabbits demonstrated that this circulating hormone was produced by the testis and was distinct from testosterone. Jost named this factor anti-Müllerian hormone based on its ability to cause regression of the Müllerian ducts. This factor is more commonly known today as Müllerian inhibiting substance (MIS). The development of an *in vitro* assay in which regression of the Müllerian duct could be monitored in cultured embryonic rat genital ridge tissue (Picon, 1969) eventually led to the purification and cloning of MIS, revealing its structural similarity to TGFβ (Cate et al., 1986).

Several lines of evidence suggest that MIS is, in fact, important in regulating the development of the Müllerian duct. In particular, MIS has been shown to be expressed at high levels by Sertoli cells of the developing testis at the time during which the Müllerian duct is normally regressing (for review, see Josso et al., 1993; Lee and Donahoe, 1993; Behringer, 1994), and MIS protein has been detected in the sera of male fetuses but not in the sera of female fetuses, consistent with an endocrine role for MIS in male sex differentiation (Josso et al., 1993). MIS continues to be expressed at high levels by the testis after birth until the onset of puberty, suggesting that MIS may also perform additional functions in regulating male sex development. MIS is also expressed in the ovary, suggesting that MIS may be involved in regulating reproductive function in females as well. Specifically, MIS has been shown to be expressed by granulosa cells, and because the levels of MIS expression appear to be inversely correlated with follicular maturation, it has been proposed that MIS plays a role in regulating oocyte maturation (for review, see Josso et al., 1993; Lee and Donahoe, 1993).

The activities of MIS *in vitro* also suggest that MIS is important in sex differentiation and development. Specifically, MIS is the only factor known to be capable of inducing Müllerian duct regression in embryonic genital ridge explant cultures (for review, see Josso et al., 1993). Furthermore, MIS can induce germ cell depletion and formation of seminiferous tubelike structures in organ cultures of fetal rat ovaries (Vigier et al., 1987). These activities are consistent with MIS being critical for regulating regression of the Müllerian duct *in vivo* and for inducing the phenotypes in the freemartin calf. In addition to these effects on fetal tissues, MIS can also influence maturation of germ cells in gonads of postnatal animals; that is, MIS has been shown capable of blocking meiotic progression of rat oocytes *in vitro* (Takahashi et al., 1986) as well as stimulating germ cell development in cultured neonatal testes (Zhou et al., 1993a).

Effects of MIS on reproductive development have also been observed in transgenic mouse experiments in which a metallothionein I promoter was used to express MIS ectopically in females (Behringer et al., 1990). In particular, expression of MIS in these animals completely blocked the development of Müllerian duct structures, resulting in females with a blind vagina and no uterus or oviducts. In the highest producing transgenic lines, the adult females also lacked ovaries. At birth, ovaries were present in these animals and contained few oocytes, but by the 16th day, the ovaries were completely devoid of germ cells and developed cordlike structures resembling seminiferous tubules similar to those seen in the ovaries of freemartin calves. Overexpression of MIS also led to abnormalities in male sexual development. For example, some, but not all, male transgenic mice in the highest producing lines had vaginal openings and mammary gland development along with underdeveloped male sex structures, suggesting that high levels of MIS may have interfered with androgen production. In addition, some of the male transgenic mice also had undescended testes.

Mutant studies have confirmed the essential role of MIS in regulating male sex differentiation. Homozygous mutant male mice carrying a targeted disruption of the MIS gene were pseudohermaphrodites; that is, these animals had female sex structures, including a uterus, oviducts, and a vagina, in addition to male sex structures (Behringer et al., 1994). Although the female sex structures appeared to be somewhat abnormal (the uterus tended to be smaller and had fewer endometrial glands and stroma, the oviducts failed to coil, and the vagina ended in the urethra rather than outside the body), the persistence of Müllerian duct-derived structures in animals that were genetically male certainly suggests that MIS is essential in regulating the regression of the Müllerian duct *in vivo*. MIS appears to play a similar role in humans as well, as several male patients with persistent Müllerian duct syndrome (PMDS) carry mutations in the MIS gene (Knebelmann et al., 1991; Carré–Eusèbe et al., 1992; Imbeaud et al., 1994; Loeff et al., 1994). The phenotype of homozygous MIS-deficient male mice also suggests that MIS may play a direct role in regulating testicular function. Specifically, although these animals appeared to be normal in producing viable sperm, the testes of mutant males appeared to be partially abnormal in that Leydig cell hyperplasia was observed in some of the mutant mice, and the presence of a Leydig cell tumor was noted in one mutant male. Taken together with the feminization of the external genitalia observed in transgenic males overexpressing MIS (Behringer et al., 1990), these results suggest that MIS may be a negative regulator of Leydig cell proliferation *in vivo*.

Although several lines of evidence have suggested that MIS may perform additional functions *in vivo*, the abnormalities in the mutant mice were restricted to male sex differentiation. For example, as discussed earlier, the expression pattern of MIS in the ovary and the activity of MIS on oocytes both *in vitro* and *in vivo* have suggested that MIS may be important in regulating female reproductive development; however, homozygous MIS-deficient females appeared to be normal in reproductive development and function. In addition, it has been proposed that MIS is involved in regulating lung development. In particular, binding sites for MIS have been detected in the developing lung (Catlin et al., 1992), and MIS has been shown capable of inhibiting production of disaturated phosphatidylcholine (DSPC), a component of surfactant, *in vitro*, when incubated with minced lung cultures (Catlin et al., 1988) and *in vivo*, when injected into rat fetuses (Catlin et al., 1990). The lungs of the MIS-deficient mice, however, also appeared to be normal (Behringer et al., 1994). Hence, if MIS does perform additional functions *in vivo*, either the activity of MIS is redundant with that of other regulatory molecules or the effects of MIS deficiency in the mutant animals were too subtle to be appreciated.

IV. INHIBINS/ACTIVINS

Like MIS, the inhibins were also originally identified as hormones synthesized by the testis. McCullagh (1932) proposed the name inhibin to describe an aqueous testicular hormone capable of preventing pituitary hypertrophy following castration. It is now known that the gonads of both sexes produce one or more substances that can inhibit secretion of follicle-stimulating hormone (FSH) by the anterior pituitary gland (for review, see Vale et al., 1991; Dye et al., 1992). The purification of inhibin from follicular fluid followed by the molecular cloning of inhibin revealed that its structure is a disulfide-linked heterodimer composed of α and β subunits, each related in primary structure to TGFβ (Mason et al., 1985; Forage et al., 1986; Mayo et al., 1986). Subsequently, a substance having the opposite effect of inhibin, that is, a substance capable of stimulating pituitary secretion of FSH, was also identified in follicular fluid, and this substance, termed activin, was shown to be a dimer consisting solely of β subunits (Ling et al., 1986; Vale et al., 1986). Because details regarding the identification and purification of the inhibins and activins, their patterns of expression in the gonads, their hormonal regulation, and their effects on pituitary and gonadal cells can be found in a number of excellent recent reviews (Vale et al., 1991; Dye et al., 1992; Hillier and Miró, 1993; Moore et al., 1994), we will focus primarily on *in vivo* studies and possible extragonadal activities of these proteins.

Both α and β subunits have structures typical of all members of the TGFβ superfamily. In mammals, two different β subunits, βA and βB, have been identified

(Mason et al., 1985; Forage et al., 1986). In the C-terminal region the two β subunits show a high degree of amino acid sequence homology to each other (approximately 60% identity) whereas inhibin α is only distantly related to the β subunits (approximately 25% identical) and to all other members of the superfamily. Each of the β subunits, like the TGFβs, contains nine cysteine residues in the C-terminal region, whereas the α subunit contains only seven cysteine residues. Multiple forms of the inhibins and activins utilizing different combinations of subunits have been detected. Both α-βA and α-βB heterodimers have been found *in vivo*, and these have been termed inhibin A and inhibin B, respectively. Similarly, all possible dimeric combinations of β subunits, βA-βA, βA-βB, and βB-βB, have also been detected, and these have been termed activin A, activin AB, and activin B, respectively. Because the two β subunits appear functionally equivalent, the presence or absence of the α subunit determines whether the dimer has inhibin- or activin-like activity. In nearly all cases where both inhibins and activins have been tested in the same assay, the two molecules have opposite biological effects (for review, see Vale et al., 1991). The diametrically opposed activities of the inhibins and activins constitute the most dramatic example in the TGFβ superfamily where heterodimers and homodimers of different family members can have profoundly different biological activities.

In both males and females, inhibins and activins are believed to affect the differentiation and function of cells in the gonad indirectly by regulating pituitary secretion of FSH. These proteins were first purified based on their ability to modulate FSH secretion by cultured anterior pituitary cells. However, the expression patterns of the inhibins and activins within the gonad and the activities of these molecules on gonadal cells in culture suggest that they may also directly affect reproductive function in both males and females by regulating the proliferation and differentiation of both support cells and germ cells. In males, both α and β subunits are expressed by Sertoli and Leydig cells of the testis, and the inhibins and/or activins are capable of regulating steroidogenesis by Leydig cells and proliferation of spermatogonia *in vitro* (for review, see Dye et al., 1992; Moore et al., 1994). In females, both α and β subunits are expressed by granulosa cells and thecal cells of the ovary, and the inhibins and/or activins are capable of regulating steroidogenesis in stromal cells (for review, see Dye et al., 1992; Hillier and Miró et al., 1993) and maturation of oocytes in culture (O et al., 1989; Itoh et al., 1990).

Inhibins and activins are also synthesized at many extragonadal sites and have activities unrelated to reproduction (for review, see Dye et al., 1992). For example, activin A may play a role in regulating hematopoiesis. Activin A was first implicated as a possible hematopoietic factor by the finding that erythroid differentiation factor (EDF), which was originally identified based on its ability to induce hemoglobin synthesis in Friend erythroleukemia cells, was identical to activin A (Eto et al., 1987; Murata et al., 1988). It has since been shown that activin A is expressed in the bone marrow and is capable of enhancing erythropoietin-induced colony formation by erythroid and multipotential progenitor cells (Yu et al., 1987;

Broxmeyer et al., 1988; Meunier et al., 1988; Shiozaki et al., 1989, 1992), perhaps indirectly by stimulating cytokine production by monocytes (Yu et al., 1989; Nakao et al., 1991). In this regard, activin can stimulate production of interleukin-1, interleukin-6, and tumor necrosis factor α by monocytes (Yamashita et al., 1993), and inhibin can inhibit production of interferon-γ by monocytes and stimulate monocyte chemotaxis (Petraglia et al., 1991). Activin and inhibin have also been shown to have effects on DNA synthesis by rat thymocytes (Hedger et al., 1989).

Inhibin subunits are also expressed in neural tissue (Meunier et al., 1988; Roberts et al., 1991) and possess activities related to neural function. For example, activin A can function *in vitro* as a survival molecule for embryonic neural retina cells and P19 teratoma cells (which can differentiate into neuronal-like cells), and both activin A and inhibin A can promote the survival of the B50 neural cell line (Schubert et al., 1990). In contrast, activin is expressed in retinal pigment epithelial cells and can inhibit the growth of these cells in culture (Jaffe et al., 1994). Activin can also affect the differentiation of neural cells. For example, activin can inhibit differentiation of several neuroblastoma cell lines and can counteract the effect of retinoic acid on the differentiation of P19 cells in culture (Hashimoto et al., 1990). It has also been shown that activin A is secreted by choroid cells of the eye and is capable of inducing somatostatin synthesis in ciliary ganglion neurons *in vitro* (Coulombe et al., 1993). Finally, activin A can alter the pattern of neurotransmitter synthesis by sympathetic neurons in culture (Fann and Patterson, 1994).

Inhibin and activin have been reported to have activities in a variety of other cell types as well. Inhibin subunits are expressed in pancreatic islet cells (Ogawa et al., 1993; Yasuda et al., 1993a), and activin can stimulate insulin secretion by these cells (Totsuka et al., 1988). Activin is also expressed in the adrenal gland (Crawford et al., 1987; Meunier et al., 1988) and can affect mitogenesis and steroidogenesis in fetal adrenal cells (Spencer et al., 1990, 1992; Nishi et al., 1992). Activin is produced by vascular endothelial cells and can inhibit endothelial cell growth (McCarthy and Bicknell, 1993) and stimulate vascular smooth muscle cell growth *in vitro* (Kojima et al., 1993). Activin is upregulated in its expression following partial hepatectomy (Yasuda et al., 1993b) and possesses activities suggesting that it is a negative regulator of hepatocyte growth; in particular, activin can inhibit epidermal growth factor (EGF) and hepatocyte growth factor (HGF) induced proliferation of hepatocytes (Yasuda et al., 1993b) and can induce death of hepatocytes both *in vitro* and *in vivo* (Schwall et al., 1993). Activin has also been reported to have a number of activities related to bone formation. Specifically, activin is expressed in bone and can potentiate bone morphogenetic protein-induced bone formation (Roberts et al., 1991; Ogawa et al., 1992), stimulate DNA and collagen synthesis in osteoblasts (Centrella et al., 1991), and enhance the formation of osteoclast-like cells in bone marrow cultures (Sakai et al., 1993).

Finally, inhibin subunits are expressed during mouse embryogenesis beginning at very early cleavage stages (Manova et al., 1992; van den Eijnden–van Raaij et al., 1992; Albano et al., 1993; Lu et al., 1993). Although the specific roles that activin plays in early embryos are not known, studies carried out with *Xenopus* embryos have suggested that activin may be involved in regulating early embryonic inductive events. In particular, activin is capable of inducing the formation of mesoderm and anterior structures in animal cap explant cultures (Asashima et al., 1990; Smith et al., 1990; Thomsen et al., 1990). Although an *in vivo* role for activin in regulating early inductive events is supported by the finding that a dominant negative activin receptor can inhibit mesoderm formation (Hemmati–Brivanlou and Melton, 1992), neither of the β subunits appears to be expressed at the appropriate time during embryonic development as would be expected for the endogenous inducer (Thomsen et al., 1990). Whether activin constitutes all or part of the endogenous mesoderm inducer in *Xenopus* and whether activin plays a role in patterning early mammalian embryos are not known.

At least certain *in vivo* roles for inhibin and activin have been substantiated by genetic experiments in the mouse. Mice lacking inhibin but not activin function were generated by targeted disruption of the inhibin α gene (Matzuk et al., 1992, 1994). As expected, both male and female homozygous mutant mice had abnormally high circulating levels of FSH; because circulating activin levels were also elevated in these mice, the increased pituitary secretion of FSH may have resulted either from decreased inhibition by inhibin or from increased stimulation by activin. Unexpectedly, however, these mice had other abnormalities. In particular, although both male and female animals appeared normal at birth, every animal examined eventually developed gonadal tumors. In females, tumors appeared to be derived from granulosa and thecal cells and developed at 4 to 13 weeks of age. In males, the tumors appeared to be derived from stromal cells and were usually seen by 5 weeks of age. Although the development of these tumors led to sterility in these mutant animals, germ cell maturation appeared to be normal at early stages of gonadal development in both sexes. The development of tumors in these mice was believed to result from altered inhibin/activin levels rather than from abnormally high levels of circulating FSH because FSH-secreting pituitary tumors in humans do not appear to predispose these patients to the development of gonadal tumors (Galway et al., 1990).

Inhibin-deficient animals also developed a cachexia-like wasting syndrome that included hepatocellular necrosis (Matzuk et al., 1994). The effect on liver tissue most likely resulted from elevated activin levels because the infusion of activin is known to induce death of hepatocytes in normal mice (Schwall et al., 1993). Because the cachexia-like syndrome eventually led to death, the possible effects of inhibin on other tissues could not be directly examined in these mutant animals. However, the mutant mice could be kept alive longer by removing their gonads before 6 weeks of age, and these gonadectomized animals subsequently developed adrenal cortical tumors. Hence, inhibin appears to perform a tumor suppressor function not only for gonadal cells but also for adrenal cells. In this respect, activin and/or inhibin are known to have direct effects on differentiation and/or proliferation of both gonadal stromal cells (for review, see Dye et al., 1992; Hillier and Miró,

1993; Moore et al., 1994) and adrenal cells (Spencer et al., 1990, 1992; Nishi et al., 1992).

Gene disruption experiments have also been carried out to investigate the *in vivo* role for the βB gene (Vassalli et al., 1994). Although mutant male mice appeared to be normal in reproductive function, female mice that were homozygous for a mutation in the βB gene showed a severe impairment of reproductive function. In particular, mutant females showed an increase in the duration of gestation, with most animals giving birth 1–2 days later than their heterozygous and wild-type litter mates and some animals failing to deliver altogether. In addition, homozygous mutant mothers were unable to adequately nurse their newborn pups, and histological examination of the mammary glands revealed an increase in glandular tissue relative to adipose tissue as well as an increase in milk accumulation. Although the underlying physiological basis for these reproductive abnormalities was not determined, the authors suggested that both the increased gestation time and the inability to nurse could potentially be explained by alterations in oxytocin expression by the endometrium and/or the pituitary gland. In this respect, localization studies in the central nervous system have detected β- but not α-chain immunoreactivity in neural pathways involved in regulating the secretion of oxytocin by the pituitary gland (Sawchenko et al., 1988), and administration of antibodies directed against the β subunit blocks milk release (Vale et al., 1991).

Because these mutant mice lacked inhibin B as well as two forms of activin (activin B and activin AB), it was not possible to attribute the observed phenotypes to the loss of a specific inhibin or activin form. However, the homozygous mutant animals were also found to have slightly elevated serum levels of FSH relative to their heterozygous litter mates, suggesting that the lack of inhibin B had slightly more of an effect than the loss of the activin forms, at least with respect to pituitary regulation. Except for the defects in reproductive function, the only other phenotype observed in the homozygous mutant animals was a frequent failure (only in certain genetic backgrounds) of eyelid fusion during embryogenesis which led to the development of eye lesions as the animals aged.

Although numerous *in vivo* roles for the βB gene have been suggested by the various *in vitro* activities of the inhibins and activins and the widespread expression pattern of the βB gene, several compensating factors may have contributed to the relatively restricted phenotypes in the βB mutant mice. First, the disruption of the βB gene would result in the loss of both inhibin and activin forms, which generally have counteracting biological effects. Second, the βB gene disruption would not eliminate the production of inhibin A and activin A, which may have redundant activities with the forms of these proteins that are eliminated in the mutant animals. And finally, βA subunit levels were found to be upregulated in these mutant mice, suggesting the existence of a compensatory mechanism for inhibin βB deficiency; this compensatory mechanism operated at the translational level because inhibin βA mRNA levels were not elevated in the mutant mice.

V. BONE MORPHOGENETIC PROTEINS

The existence of bone morphogenetic proteins (BMPs) was first recognized by Urist (1965), who showed that extracts prepared from demineralized bone could induce local de novo bone formation when implanted subcutaneously into rodents. The ultimate purification of bone morphogenetic proteins was made possible by the demonstration that the osteoinductive activity could be separated from the bone matrix and, then, reconstituted in an active form with the collagenous matrix prior to implantation (Sampath and Reddi, 1981). Using oligonucleotide probes based on peptide sequences obtained from highly purified protein preparations, Wozney et al. (1988) isolated genes encoding four distinct proteins capable of inducing bone formation in this assay. Three of these, BMP-2 (originally designated BMP-2a), BMP-3 (also called osteogenin; Luyten et al., 1989), and BMP-4 (originally designated BMP-2b), showed clear homology to both TGFβ and the other known members of the TGFβ superfamily. Since this original report, a number of additional factors have been identified based on their homology to the BMPs and to other members of the TGFβ superfamily, and at least some of these also appear to be involved in regulating bone formation *in vivo*.

Based on the predicted amino acid sequences of the mature molecules, the BMPs and BMP-like molecules can be classified into at least three distinct subgroups of highly related factors. One subgroup consists of BMP-2 and BMP-4 (Wozney et al., 1988), which show 92% amino acid sequence identity to each other in the C-terminal region. A second subgroup consists of BMP-5 (Celeste et al., 1990), Vgr-1 (BMP-6) (Lyons et al., 1989a; Celeste et al., 1990), OP-1 (BMP-7) (Celeste et al., 1990; Özkaynak et al., 1990), and OP-2 (BMP-8) (Özkaynak et al., 1992), which are approximately 70–90% identical to each other. A third subgroup consists of GDF-5 (CDMP-1), GDF-6 (CDMP-2), and GDF-7 (Chang et al., 1994; Storm et al., 1994), which are 80–82% identical to each other. In general, when the sequences of these factors are compared even between subgroups, they are more highly related to each other than they are to other members of the TGFβ superfamily.

Some or all of these proteins are believed to be important in regulating bone formation *in vivo*. Although BMP-2 (Wozney et al., 1988; Wang et al., 1990), BMP-3/osteogenin (Wozney et al., 1988; Luyten et al., 1989), BMP-4 (Hammonds et al., 1991), BMP-5 (Cox et al., 1991; D'Alessandro et al., 1991), and OP-1 (Sampath et al., 1992) are the only factors in this group that have been reported, to date, to be active in inducing cartilage and/or bone formation in the rat implant assay, it seems likely that at least some of the other highly homologous factors will also be shown to possess osteogenic activity. Based on the remarkable ability of these proteins to induce endochondral bone formation de novo, it has been hypothesized, since their initial identification, that these factors may be key regulators of bone development and repair *in vivo*. By a variety of morphological and biochemical criteria, the cascade of cellular events triggered by the administration of these proteins in the rat implant assay is remarkably similar to the normal sequence of events that occur during endochondral bone formation in embryonic development and during fracture healing (for review, see Reddi, 1981). The fact that these proteins can recapitulate the entire normal process of bone formation is certainly consistent with the hypothesis that this is their normal *in vivo* role.

The expression patterns of many of these factors during embryogenesis also suggest that they may play a role in regulating bone development. In particular, *in situ* hybridization studies carried out with mammalian embryos have demonstrated the presence of BMP-2, BMP-3/osteogenin, BMP-4, BMP-5, Vgr-1, GDF-5, and GDF-6 transcripts in developing skeletal structures. For example, several of these factors have been shown to be expressed at different stages of limb development during mouse embryogenesis. BMP-2 mRNA has been detected in the apical ectodermal ridge during early limb development, in the interdigital mesenchyme at later stages, and in the perichondrium and osteogenic zone at still later stages (Lyons et al., 1990). BMP-4 transcripts have also been detected early in limb development, both in the apical ectodermal ridge and in the mesenchyme at day 10.5 postcoitum (p.c.) (Jones et al., 1991). In contrast, at later stages of limb development, GDF-5 is expressed in the precartilaginous mesenchyme and perichondrium (Storm et al., 1994), BMP-5 is also expressed in the perichondrium (King et al., 1994), and Vgr-1 is expressed by cells of the hypertrophic cartilage (Lyons et al., 1989b). GDF-6 has also been shown to be expressed in hypertrophic cartilage during human embryogenesis (Chang et al., 1994). Hence, the expression patterns of these factors suggest that development of the long bones may be regulated, at least in part, by the sequential actions of multiple members of this family.

In addition to the roles that these factors may play in regulating long bone development, their expression patterns suggest that they may be involved in regulating other aspects of bone and cartilage development. For example, during mouse embryogenesis, BMP-2 transcripts have been detected in the condensing mesenchymal cells of prevertebrae in day 12.5 embryos (Lyons et al., 1989b), BMP-5 transcripts have been detected in many developing skeletal structures, including the genital tubercle, external ears, sternum, trachea, vertebrae, ribs, and thyroid, hyoid, and nasal cartilages (King et al., 1994), and Vgr-1 transcripts have been detected not only in the developing limbs but also in other skeletal structures, including vertebrae, ribs, and head, throughout embryonic development (Lyons et al., 1989b). Similarly, in human embryos, BMP-3/osteogenin transcripts have been detected in the perichondrium and periosteum at specific sites in the developing skeleton (Vukicevic et al., 1994).

Some of these factors are also expressed in mature bone. In particular, the sequences of BMP-2 (Wozney et al., 1988), BMP-3/osteogenin (Wozney et al., 1988; Luyten et al., 1989), BMP-5 (Celeste et al., 1990), Vgr-1 (Lyons et al., 1989a; Celeste et al., 1990), and OP-1 (Sampath et al., 1990) all contain segments identical to the sequences of tryptic peptides derived from purified protein preparations having bone inductive activity. Because the starting material for purification was an extract prepared from homogenized adult bone, it seems likely that some of these factors are present in adult bone. cDNA clones encoding BMP-2, BMP-4, BMP-5, and OP-1 were all originally isolated from libraries prepared from an osteosarcoma cell line. Although the precise roles that these factors play in adult bone tissue are not known, one possibility is that they may be involved in regulating bone repair or bone remodeling. In this respect, at least one of these molecules has been shown to be expressed transiently during fracture healing; specifically, BMP-4 mRNA was not detected in ribs prior to fracture but appeared to be localized in proliferating periosteum, medullary cavity, and nearby muscles early in the fracture healing process (Nakase et al., 1994).

The essential roles that these factors play *in vivo* in regulating bone repair and/or skeletal development have been established in two cases in which mouse mutants known to have skeletal defects were found to carry mutations in a BMP or BMP-like gene. One of these mutants is the *short ear (se)* mouse, which has a number of skeletal abnormalities, including a reduction in the sizes of the external ears, the manubrium at the top of the sternum, and the ventral process of the sixth cervical vertebra, an alteration in the shape of the xyphoid process and thyroid and larynx cartilages, and a reduction in the number of ribs (12 pairs instead of the normal 13) (Green and Green, 1942, 1946; Green, 1951). In addition, *se* mice have a defect in their ability to repair rib fractures (Green, 1958). A positional cloning strategy used to home in on the *se* locus on mouse chromosome 9 revealed that the region deleted or rearranged in *se* mutants corresponded to the BMP-5 gene (Kingsley et al., 1992), and, subsequently, one *se* mutant allele was shown to be a nonsense mutation (King et al., 1994). The second mutant is the *brachypodism* (*bp*) mouse, which has skeletal abnormalities restricted to the limbs. In particular, these mice have shortened long bones, metacarpals, and metatarsals and fused proximal and medial phalanges (Landauer, 1952; Grüneberg and Lee, 1973). The proximity of the map location of the GDF-5 gene to the *bp* locus on chromosome 2 (Storm et al., 1994) and the cell non-autonomous nature of the *bp* phenotype (Owens and Solursh, 1982; Malinina et al., 1984) suggested GDF-5 as a candidate gene for *bp*, and sequence analysis revealed frameshift mutations preceding the mature C-terminal portion of the GDF-5 molecule in three different *bp* alleles (Storm et al., 1994). Based on the phenotypes of *se* and *bp* mice, it seems clear that BMP-like molecules are essential for normal skeletal development and for fracture healing. Moreover, because the phenotypes of *se* and *bp* mice differ in the specific skeletal structures that are altered, it seems likely that the development of the vertebrate skeleton requires the composite actions of multiple BMP-like signaling molecules, each regulating the development of a unique spectrum of skeletal structures.

In addition to the roles that these factors play in regulating bone development, the expression patterns of many of these molecules suggest that they may be involved in influencing a variety of other developmental processes as well. For example, during mouse embryogenesis, BMP-2 mRNA has been detected not only in developing bone and cartilage but also in developing heart and blood vessels, hair and whisker follicles, and tooth buds (Lyons et al., 1989b, 1990; Vainio et al., 1993). Similarly, BMP-4 mRNA transcripts have been detected in the mesoderm of the posterior primitive streak in day 8.5 mouse embryos and in a variety of tissues at later stages, including the developing heart and blood vessels, whisker follicles, diencephalon, otic vesicle, and mesoderm surrounding the developing gut and lung buds (Jones et al., 1991) as well as in dental mesenchyme (Vainio et al., 1993). In contrast, Vgr-1 expression during mouse embryogenesis is localized predominantly to developing squamous epithelia in a number of tissues and to the developing nervous system, most notably to cells in the roof plate and those adjacent to the floor plate (Lyons et al., 1989b; Jones et al., 1991). *In situ* hybridization studies with human embryos have detected BMP-3/osteogenin transcripts in developing kidney, lung, and intestine (Vukicevic et al., 1994). Many of these factors also appear to be expressed in a variety of nonskeletal tissues in adult animals (Lyons et al., 1989a; Özkaynak et al., 1991, 1992; Harris et al., 1994; Storm et al., 1994).

In addition to their widespread expression patterns, the activities of these molecules *in vitro* and *in vivo* also suggest that the BMPs and BMP-like factors may regulate a wide range of differentiation processes. A number of biological activities, for example, have been reported for BMP-2 and BMP-4, and in many of these cases, the biological activities were predicted from their expression patterns during embryonic development. For example, BMP-2 and BMP-4 are known to be expressed in the developing tooth bud (Lyons et al., 1990; Vainio et al., 1993; Heikinheimo, 1994), and beads soaked with either protein can induce changes in gene expression and in the morphology of dental mesenchyme explants similar to those induced by dental epithelium (Vainio et al., 1993). BMP-2 and BMP-4 also appear to be involved in regulating epithelial-mesenchymal interactions during early limb development. Specifically, both BMP-2 and BMP-4 are expressed in the apical ectodermal ridge (Lyons et al., 1990; Jones et al., 1991), and BMP-2 can inhibit proliferation of the underlying mesenchymal cells (Niswander and Martin, 1993). Similarly, mRNA localization studies have suggested roles for BMP-2 and BMP-4 in the development of whisker follicles (Lyons et al., 1989b, 1990; Jones et al., 1991), and transgenic mouse experiments in which BMP-4 was expressed ectopically in hair follicles have shown that BMP-4 can inhibit the proliferation of hair cells (Blessing et al., 1993). BMP-4 is expressed in the primitive streak region during gastrulation in the mouse (Jones et al., 1991), and, although the precise roles that BMP-4 plays during these early stages are not known, mRNA injection experiments in *Xenopus* embryos have suggested that BMP-4 may be involved in patterning ventral mesoderm (for review, see Harland, 1994). BMP-4 has been shown to possess a number of additional activities influencing the differentiation of neural cells as well. For example, BMP-4 has been shown to be expressed in specific rhombomeres during chick hindbrain development, and BMP-4 has been shown capable of inducing apoptosis of neural crest cells derived from these rhombomere segments in explant cultures (Graham et al., 1994). BMP-4 can also induce extension of neurites and expression of VGF, a neuronal-specific mRNA, in rat pheochromocytoma PC12 cells (Paralkar et al., 1992).

Other BMPs and BMP-like molecules have also been demonstrated to have activities unrelated to bone induction. For example, Vgr-1 can inhibit the proliferation of undifferentiated keratinocytes in culture, consistent with its expression in squamous epithelium (Drozdoff et al., 1994). OP-1 has been shown to have a protective effect against myocardial ischaemia reperfusion injury and to be capable of inhibiting the binding of activated polymorphonuclear leukocytes to vascular endothelium (Lefer et al., 1992). BMP-3/osteogenin and BMP-4 have been shown to be chemoattractants for human blood monocytes *in vitro* (Cunningham et al., 1992). And a number of factors, including OP-1, BMP-2, BMP-4, BMP-5, and Vgr-1, can induce expression of the neural cell adhesion molecules N-CAM and L1 and formation of cell aggregates in a neuroblastoma-glioma hybrid cell line, NG108-15 (Perides et al., 1992, 1994).

The phenotypes of mice carrying mutations in these genes have also provided some information regarding the roles that these factors may play in addition to regulating skeletal development. In particular, *short ear* mice have an increased frequency of lung cysts, liver granulomas, hydronephrosis, and hydroureter (Green, 1968), suggesting that BMP-5 performs important regulatory functions in these tissues; BMP-5 is known to be expressed in many of these tissues (Özkaynak et al., 1992; King et al., 1994). In contrast, the abnormalities in *brachypodism* mice appear to be restricted to the skeleton (Landauer, 1952) despite the fact that GDF-5 is expressed in a number of different tissues in adult animals (Storm et al., 1994); specifically, *brachypodism* mice show no abnormalities in female reproductive function even though GDF-5 is expressed at highest levels in the uterus in adult mice. Clearly, the availability of mice carrying mutations in other BMPs and BMP-like genes will be essential for determining whether a particular factor plays an essential role in regulating a given biological process or whether different family members may be carrying out redundant functions.

VI. NODAL, GDNF, AND OTHER GROWTH/DIFFERENTIATION FACTORS RELATED TO TGFβ

In addition to MIS, the inhibins/activins, and the bone inducing factors, a number of additional molecules have been identified that are clearly members of this family of growth and differentiation factors. Based on their predicted sequences, these molecules appear to belong to subgroups different than those discussed above, and for that reason, they will be described separately here. However, because expression patterns and biological activities of many of these factors have only been partially characterized, it is possible that some of these additional factors may perform functions similar to those carried out by the factors already discussed.

Nodal was identified based on a retroviral insertional mutation in mice that led to embryonic lethality. Specifically, embryos homozygous for the 413-d proviral insertion showed hyperplasia of embryonic and extraembryonic ectoderm, failed to form mesoderm from ecotoderm, and failed to undergo gastrulation (Conlon et al., 1991; Iannaccone et al., 1992). Experiments in which mutant embryonic stem cells were injected into wild-type blastocysts showed that the mutant cells could contribute to all somatic tissues, suggesting that the gene mutated by the 413-d insertion normally functions cell nonautonomously (Conlon et al., 1991). Analysis of the 413-d proviral integration site revealed that the mutated gene was a new member of the TGFβ superfamily, and this gene was given the name nodal (Zhou et al., 1993b). By *in situ* hybridization to mouse embryos, nodal was shown to be expressed specifically in the primitive node, which is believed to serve as an organizing center for axial patterning. Based on the expression pattern of nodal and the phenotype of 413-d mutants, it has been hypothesized that nodal is essential in mesoderm formation and/or in axial development during mouse embryogenesis.

Glial cell line-derived neurotrophic factor (GDNF) is a recently described family member identified based on its ability to promote survival of embryonic midbrain dopaminergic neurons in culture (Lin et al., 1993). GDNF was purified from the conditioned medium of the rat B49 glial cell line, and the isolation of the GDNF gene using partial amino acid sequences derived from the purified protein revealed that GDNF is a distantly related member of this family. The sequence of GDNF is quite unusual compared to other family members in the high degree of sequence divergence of its C-terminal region (less than 20% homology to the other family members) are in the relatively small size of its pro- region (fewer than 60 amino acids in length). GDNF has been shown to have potent survival activity on two types of neurons. First, GDNF can promote the survival of midbrain dopaminergic neurons both in culture (Lin et al., 1993) and in grafts of ventral mesencephalic tissue containing the developing substantia nigra (Strömberg et al., 1993). Second, GDNF is the most potent factor known for promoting the survival of motor neurons both in dissociated spinal cord neuronal cultures and in the facial nerve following axotomy (Henderson et al., 1994). The expression pattern of GDNF is also consistent with GDNF's importance in supporting the survival of these neuronal populations *in vivo*. Specifically, GDNF mRNA has been detected in regions of the brain containing dopaminergic innervation, such as the striatum, ventral pallidum, olfactory tubercule, piriform cortex, and cingulate cortex (Schaar et al., 1993; Strömberg et al., 1993; Springer et al., 1994) and in regions innervated by or in the vicinity of motor neurons, such as in developing skeletal muscle, in Schwann cells, and in the ventral roots adjacent to the spinal cord (Henderson et al., 1994; Springer et al., 1994). Hence, GDNF is likely to function as a neurotrophic factor for both of these neuronal populations *in vivo*. Although GDNF has been shown to be inactive on certain other types of neurons, such as GABAergic neurons and serotonergic neurons (Lin et al., 1993), the expression of GDNF in other parts of the nervous system, such as in the hippocampus, dorsal root ganglia, thalamus, pineal gland, recessus, and cerebellar anlage, suggests that it may influence other neuronal populations as well (Strömberg et al., 1993; Springer et al., 1994).

GDF-2/BMP-9 was identified by two groups independently on the basis of its homology to known members of the family (S.–J. Lee, unpublished results; Wozney and Celeste, 1993). Although neither the expression pattern nor the biological activity of GDF-2/BMP-9 has yet been reported, GDF-2/BMP-9 is highly homologous (80% identical) to a chicken gene, dorsalin-1, that appears to play a role in the development of the nervous system (Basler et al., 1993). In particular, dorsalin-1 has been shown to be expressed in the dorsal third of the neural tube and to have two activities related to specification of dorsal cell fates. First, dorsalin-1 can induce migration of neural crest cells from dorsal neural explants, and second, dorsalin-1 can inhibit motor neuron differentiation in intermediate neural plate explants. These results suggest that dorsalin-1 may be involved in regulating dorsoventral patterning during nervous system development. Whether GDF-2/BMP-9 is the ortholog of chicken dorsalin and performs a similar function in mammals is not known.

Another family member that appears to play a role in the development of the nervous system is GDF-1, which was originally isolated from a day 8.5 mouse embryo cDNA library (Lee, 1990). Although the expression pattern of GDF-1 during early embryogenesis has not been characterized, at later stages of development and in adult animals, the expression of GDF-1 appears to be nervous system-specific (Lee, 1991). The transcript expressed in the nervous system is quite unusual in that it appears to be bicistronic. The downstream cistron encodes GDF-1, which contains all of the hallmarks characteristic of TGFβ family members, and the upstream cistron (designated UOG-1) encodes a protein that appears to contain multiple membrane-spanning domains. Although the function of UOG-1 is not known, two yeast genes have recently been identified that show sequence homology to UOG-1. One of these, *LAG1*, has been shown to play a role in regulating the life span of yeast cells (D'mello et al., 1994). The other, *YKL156*, was identified as an open reading frame on chromosome XI (Boyer et al., 1993). Whether additional molecules related to UOG-1 exist in mammals and whether UOG-1 or UOG-1-like molecules play a role in the biology of GDF-1 or other TGFβ family members are not known.

Finally, two additional factors, GDF-9 and GDF-3/Vgr-2, were also identified based on their homology to known family members. The expression of GDF-9 has been shown to be highly tissue and cell type specific. In particular, GDF-9 appears to be expressed exclusively in the ovaries in adult animals (McPherron and Lee, 1993), and within the ovary, GDF-9 expression is restricted to oocytes (McGrath et al., 1995). The expression of GDF-3/Vgr-2 is also tissue-restricted. During mouse embryogenesis, GDF-3/Vgr-2 expression peaks at midgestation and is localized to developing bone throughout the embryo (Jones et al., 1992). In adult animals GDF-3/Vgr-2 is expressed at highest levels in bone marrow, spleen, thymus, and adipose tissue (McPherron and Lee, 1993). As discussed earlier, the sequences of these two factors are unusual in that they are missing the cysteine residue that is known, in the case of TGFβ2, to form the sole intermolecular disulfide bond holding the dimer together covalently.

VIII. CONCLUSIONS

The TGFβ superfamily encompasses a large group of signaling molecules involved in regulating the growth and differentiation of a wide range of cell types. Nearly two dozen members of this superfamily have been identified in mammals alone, and homologous factors have been identified in other vertebrates, including amphibians, birds, and fish, and in invertebrates, such as insects and echinoderms. Based on the range of biological processes influenced by this group of molecules and the high degree of sequence divergence among the known members of this superfamily, it seems virtually certain that there are a large number of additional family members yet to be identified both in mammals and in other organisms.

The members of this superfamily are related in their overall structure and in their ability to regulate cell growth and differentiation. Most, if not all, of these factors are likely to be critical in regulating embryonic development. In fact, the development of virtually every organ system appears to be regulated by members of this family, and many developmental processes seem to be regulated by multiple members of this family acting in concert or in sequence. These factors also appear essential for maintaining tissue homeostasis in adult animals and for mediating tissue repair or regeneration. Because the functions of many of these factors are only beginning to be elucidated and because it is likely that only a fraction of the total superfamily has been discovered, it seems clear that the full range of biological processes influenced by members of this superfamily is only beginning to be appreciated.

Because the members of this superfamily are potent regulators of cell differentiation and proliferation, a number of these molecules are being explored for potential applications in the treatment of human disease. For example, certain members of this family may have applications in promoting tissue repair. In particular, the ability of the bone morphogenetic proteins to induce endochondral bone formation *in vivo* has suggested applications for these proteins in fracture healing (for description, see Alper, 1994); the BMPs have shown efficacy in repairing bony defects both in animal models and in humans, and two molecules, BMP-2 and OP-1, are currently being evaluated in clinical trials. Moreover, the BMPs have been shown to be effective in inducing dentine formation in animal models, and OP-1 is currently being evaluated in clinical trials as a potential alternative to root canal surgery (for description, see Alper, 1994). Other family members may have applications in oncology. As discussed earlier, genetic studies have indicated that inhibin α and MIS may function as tumor suppressor genes; indeed, MIS has been shown to inhibit growth of a variety of human tumor cells, including those of Müllerian duct origin, when grown in culture or when transplanted into immunodeficient mice (for review, see Lee and Donahoe, 1993). The ability of GDNF to promote the survival of dopaminergic neurons and motor neurons has suggested its use in the treatment of certain neurodegenerative diseases, such as Parkinson's disease and amyotrophic lateral sclerosis, or in promoting nerve regeneration, such as in the repair of spinal cord injuries.

It is already clear that this family of growth and differentiation factors holds enormous promise for many clinical applications. Some of these applications are beginning to be realized while others are only beginning to be explored. Only after all of the members of this superfamily have been identified and the biological activities of these factors have been elucidated will it be possible to sense the full impact of this group of proteins on clinical practice. Nevertheless, it seems likely that, in the next century, we will see numerous examples where an understanding of the basic biology of these factors has led to advances in treating human disease by directly using these molecules as therapeutic agents.

REFERENCES

Albano, R. M., Groome, N., & Smith, J. C. (1993). Activins are expressed in preimplantation mouse embryos and in ES and EC cells and are regulated on their differentiation. Development 117, 711–723.

Alper, J. (1994). Boning up: Newly isolated proteins heal bad breaks. Science 263, 324–324.

Amatayakul-Chantler, S., Qian, S. W., Gakenheimer, K., Böttinger, E. P., Roberts, A. B., & Sporn, M. B. (1994). [Ser^{77}]transforming growth factor-β1: Selective biological activity and receptor binding in mink lung epithelial cells. J. Biol. Chem. 269, 27687–27691.

Asashima, M., Nakano, H., Shimada, K., Kinoshita, K., Ishii, K., Shibai, H., & Ueno, N. (1990). Mesodermal induction in early amphibian embryos by activin A (erythroid differentiation factor). Roux's Arch. Dev. Biol. 198, 330–335.

Barr, P. J. (1991). Mammalian subtilisins: The long-sought dibasic processing endoproteases. Cell 66, 1–3.

Barton, D. E., Yang-Feng, T. L., Mason, A. J., Seeburg, P. H., & Francke, U. (1989). Mapping genes for inhibin subunits α, β_A, and β_B on human and mouse chromosomes and studies of *jsd* mice. Genomics 5, 91–99.

Basler, K., Edlund, T., Jessell, T. M., & Yamada, T. (1993). Control of cell pattern in the neural tube: Regulation of cell differentiation by *dorsalin-1*, a novel TGFβ family member. Cell 73, 687–702.

Behringer, R. R. (1994). In: Current Topics in Developmental Biology (R. A. Pedersen, ed.), Vol. 29, pp. 171–187, Academic Press, San Diego.

Behringer, R. R., Cate, R. L., Froelick, G. J., Palmiter, R. D., & Brinster, R. L. (1990). Abnormal sexual development in transgenic mice chronically expressing Müllerian inhibiting substance. Nature 345, 167–170.

Behringer, R. R., Finegold, M. J., & Cate, R. L. (1994). Müllerian-inhibiting substance function during mammalian sexual development. Cell 79, 415–425.

Blessing, M., Nanney, L. B., King, L. E., Jones, C. M., & Hogan, B. L. M. (1993). Transgenic mice as a model to study the role of TGF-β-related molecules in hair follicles. Genes Dev. 7, 204–215.

Boyer, J., Pascolo, S., Richard, G.-F., & Dujon, B. (1993). Sequence of a 7.8 kb segment on the left arm of yeast chromosome XI reveals four open reading frames, including the *CAP1* gene, an intron-containing gene and a gene encoding a homolog to the mammalian *UOG-1* gene. Yeast 9, 279–287.

Broxmeyer, H. E., Lu, L., Cooper, S., Schwall, R. H., Mason, A. J., & Nikolics, K. (1988). Selective and indirect modulation of human multipotential and erythroid hematopoietic progenitor cell proliferation by recombinant human activin and inhibin. Proc. Natl. Acad. Sci. USA 85, 9052–9056.

Brunner, A. M., Lioubin, M. N., Marquardt, H., Malacko, A. R., Wang, W.–C., Shapiro, R. A., Neubauer, M., Cook, J., Madisen, L., & Purchio, A. F. (1992). Site-directed mutagenesis of glycosylation sites in the transforming growth factor β1 (TGFβ1) and TGFβ2 (414) precursors and of cysteine residues within mature TGFβ1: Effects on secretion and bioactivity. Mol. Endocrinol. 6, 1691–1700.

Brunner, A. M., Marquardt, H., Malacko, A. R., Lioubin, M. N., & Purchio, A. F. (1989). Site-directed mutagenesis of cysteine residues in the pro region of the transforming growth factor-β1 precursor. J. Biol. Chem. 264, 13660–13664.

Carré–Eusèbe, D., Imbeaud, S., Harbison, M., New, M. I., Josso, N., & Picard, J.–Y. (1992). Variants of the anti-Müllerian hormone gene in a compound heterozygote with the persistent Müllerian duct syndrome and his family. Hum. Genet. 90, 389–394.

Cate, R. L., Mattaliano, R. J., Hession, C., Tizard, R., Farber, N. M., Cheung, A., Ninfa, E. G., Frey, A. Z., Gash, D. J., Chow, E. P., Fisher, R. A., Bertonis, J. M., Torres, G., Wallner, B. P., Ramachandran, K. L., Ragin, R. C., Manganaro, T. F., MacLaughlin, D. T., & Donahoe, P. K. (1986). Isolation of the bovine and human genes for Müllerian inhibiting substance and expression of the human gene in animal cells. Cell 45, 685–698.

Catlin, E. A., Ezzell, R. M., Donahoe, P. K., Manganaro, T. F., Ebb, R. G., & MacLaughlin, D. T. (1992). Müllerian inhibiting substance binding and uptake. Dev. Dyn. 193, 295–299.

Catlin, E. A., Manganaro, T. F., & Donahoe, P. K. (1988). Müllerian inhibiting substance depresses accumulation *in vitro* of disaturated phosphatidylcholine in fetal rat lung. Am. J. Obstet. Gynecol. 159, 1299–1303.

Catlin, E. A., Powell, S. M., Manganaro, T. F., Hudson, P. L., Ragin, R. C., Epstein, J., & Donahoe, P. K. (1990). Sex-specific fetal lung development and Müllerian inhibiting substance. Ann. Rev. Respir. Dis. 141, 466–470.

Celeste, A. J., Iannazzi, J. A., Taylor, R. C., Hewick, R. M., Rosen, V., Wang, E. A., & Wozney, J. M. (1990). Identification of transforming growth factor β family members present in bone-inductive protein purified from bovine bone. Proc. Natl. Acad. Sci., USA 87, 9843–9847.

Centrella, M., McCarthy, T. L., & Canalis, E. (1991). Activin-A binding and biochemical effects in osteoblast-enriched cultures from fetal-rat parietal bone. Mol. Cell. Biol. 11, 250–258.

Chang, S. C., Hoang, B., Thomas, J. T., Vukicevic, S., Luyten, F. P., Ryba, N. J. P., Kozak, C. A., Reddi, A. H., & Moos, J., M. (1994). Cartilage-derived morphogenetic proteins. J. Biol. Chem. 269, 28227–28234.

Cohen–Haguenauer, O., Picard, J. Y., Mattei, M.–G., Severo, S., van Long, N., de Tand, M.–F., Guerrier, D., Hors–Cayla, M.–C., Josso, N., & Frezal, J. (1987). Mapping of the gene for anti-Müllerian hormone to the short arm of human chromosome 19. Cytogenet. Cell Genet. 44, 2–6.

Cönlon, F. L., Barth, K. S., & Robertson, E. J. (1991). A novel retrovirally induced embryonic lethal mutation in the mouse: assessment of the developmental fate of embryonic stem cells homozygous for the 413.d proviral integration. Development 111, 969–981.

Coulombe, J. N., Schwall, R., Parent, A. S., Eckenstein, F. P., & Nishi, R. (1993). Induction of somatostatin immunoreactivity in cultured ciliary ganglion neurons by activin in choroid cell-conditioned medium. Neuron 10, 899–906.

Cox, K., Holtrop, M., D'Allesandro, J. S., Wang, E. A., Wozney, J. M., & Rosen, V. (1991). Histological and ultrastructural comparison of the *in vivo* activities of RHBMP 2 and RHBMP 5. J. Bone Min. Res. 6, S155.

Crawford, R. J., Hammond, V. E., Evans, B. A., Coghland, J. P., Haralambidis, J., Hudson, B., Penschow, J. D., Richards, R. I., & Tregear, G. W. (1987). Alpha-inhibin gene expression occurs in the ovine adrenal cortex, and is regulated by adrenocorticotropin. Mol. Endocrinol. 1, 699–706.

Cunningham, N. S., Paralkar, V., & Reddi, A. H. (1992). Osteogenin and recombinant bone morphogenetic protein 2B are chemotactic for human monocytes and stimulate transforming growth factor beta 1 mRNA expression. Proc. Natl. Acad. Sci., USA 89, 11740–11744.

D'Allessandro, J. S., Cox, K. A., Israel, D. I., LaPan, P., Moutsatsos, J. K., Nove, J., Rosen, V., Ryan, M. C., Wozney, J. M., & Wang, E. A. (1991). Purification, characterization and activities of recombinant human bone morphogenetic protein 5. J. Bone Min. Res. 6, S153.

D'mello, N. P., Childress, A. M., Franklin, D. S., Kale, S. P., Pinswasdi, C., & Jazwinski, S. M. (1994). Cloning and characterization of *LAG1*, a longevity-assurance gene in yeast. J. Biol. Chem. 269, 15451–15459.

Daopin, S., Piez, K. A., Ogawa, Y., & Davies, D. R. (1992). Crystal structure of transforming growth factor-β2: An unusual fold for the superfamily. Science 257, 369–373.

deMartin, R., Haendler, B., Hofer–Warbinek, R., Gaugitsch, H., Wrann, M., Schlusener, H., Seifert, J. M., Bodmer, S., Fontana, A., & Hofer, E. (1987). Complementary DNA for human glioblastoma-derived T-cell suppressor factor, a novel member of the transforming growth factor-β gene family. EMBO J. 6, 3673–3677.

Derynck, R., Jarrett, J. A., Chen, E. Y., Eaton, D. H., Bell, J. R., Assoian, R. K., Roberts, A. B., Sporn, M. B., & Goeddel, D. V. (1985). Human transforming growth factor β complementary DNA sequence and expression in normal and transformed cells. Nature 316, 701–705.

Derynck, R., Lindquist, P. B., Lee, A., Wen, D., Tamm, J., Graycar, J. L., Rhee, L., Mason, A. J., Miller, D. A., Coffey, R. J., Moses, H. L., & Chen, E. Y. (1988). A new type of transforming growth factor-β, TGF-β3. EMBO J. 7, 3737–3743.

Dickinson, M. E., Kobrin, M. S., Silan, C. M., Kingsley, D. M., Justice, M. J., Miller D. A., Ceci, J. D., Lock, L. F., Lee, A., Buchberg, A. M., Siracusa, L. D., Lyons, K. M., Derynck, R., Hogan, B. L. M., Copeland, N. G., & Jenkins, N. A. (1990). Chromosomal localization of seven members of the murine TGF-β superfamily suggests close linkage to several morphogenetic mutant loci. Genomics 6, 505–520.

Drozdoff, V., Wall, N. A., & Pledger, W. J. (1994). Expression and growth inhibitory effect of decapentaplegic Vg-related protein 6: Evidence for a regulatory role in keratinocyte differentiation. Proc. Natl. Acad. Sci. USA 91, 5528–5532.

Dye, R. B., Rabinovici, J., & Jaffe, R. B. (1992). Inhibin and activin in reproductive biology. Obstet. and Gynecol. Surv. 47, 173–185.

Eto, Y., Tsuji, T., Takezawa, M., Takano, S., Yokogawa, Y., & Shibai, H. (1987). Purification and characterization of erythroid differentiation factor (EDF) isolated from human leukemia cell line THP-1. Biochem. Biophys. Res. Commun. 142, 1095–1103.

Fann, M.-J. & Patterson, P. H. (1994). Neuropoietic cytokines and activin A differentially regulate the phenotype of cultured sympathetic neurons. Proc. Natl. Acad. Sci. USA 91, 43–47.

Ferguson, E. L. & Anderson, K. V. (1992). Localized enhancement and repression of the activity of the TGF-β family member, *decapentaplegic*, is necessary for dorsal-ventral pattern formation in the *Drosophila* embryo. Dev. 114, 583–597.

Forage, R. G., Ring, J. M., Brown, R. W., McInerney, B. V., Cobon, G. S., Gregson, R. P., Robertson, D. M., Morgan, F. J., Hearn, M. T. W., Findlay, J. K., Wettenhall, R. E. H., Burger, H. G., & de Kretser, D. M. (1986). Cloning and sequence analysis of cDNA species coding for the two subunits of inhibin from bovine follicular fluid. Proc. Natl. Acad. Sci., USA 83, 3091–3095.

Galway, A. B., Hsueh, A. J. W., Daneshdoost, L., Zhou, H.-H., Pavlou, S. N., & Snyder, P. J. (1990). Gonadotroph adenomas in men produce biologically active follicle-stimulating hormone. J. Clin. Endocrinol. Metab. 71, 907–912.

Gentry, L. E. & Nash, B. W. (1990). The pro domain of pre-pro-transforming growth factor β1 when independently expressed is a functional binding protein for the mature growth factor. Biochem. 29, 6851–6857.

Graham, A., Francis-West, P., Brickell, P., & Lumsden, A. (1994). The signalling molecule BMP4 mediates apoptosis in the rhombencephalic neural crest. Nature 372, 684–686.

Gray, A. M. & Mason, A. J. (1990). Requirement for activin A and transforming growth factor-β1 pro-regions in homodimer assembly. Science 247, 1328–1330.

Green, E. L. & Green, M. C. (1942). The development of three manifestations of the *short ear* gene in the mouse. J. Morphol. 70, 1–19.

Green, E. L. & Green, M. C. (1946). Effect of the short ear gene on number of ribs and presacral vertebrae in the house mouse. Am. Nat. 80, 619–625.

Green, M. C. (1951). Further morphological effects of the short ear gene in the house mouse. J. Morphol. 88, 1–21.

Green, M. C. (1958). Effects of the short ear gene in the mouse on cartilage formation in healing bone fractures. J. Exp. Zool. 137, 75–88.

Green, M. C. (1968). Mechanism of the pleiotropic effects in the short-ear mutant gene in the mouse. J. Exp. Zool. 167, 129–150.

Grüneberg, H. & Lee, A. J. (1973). The anatomy and development of brachypodism in the mouse. J. Embryol. Exp. Morphol. 30, 119–141.

Hahn, G. V., Cohen, R. B., Wozney, J. M., Levitz, C. L., Shore, E. M., Zasloff, M. A., & Kaplan, F. S. (1992). A bone morphogenetic protein subfamily: Chromosomal localization of human genes for BMP5, BMP6, and BMP7. Genomics 14, 759–762.

Hammonds, R. G., Schwall, R., Dudley, A., Berkemeier, L., Lai, C., Lee, J., Cunningham, N., Reddi, A. H., Wood, W. I., & Mason, A. J. (1991). Bone-inducing activity of mature BMP-2b produced from a hybrid BMP-2a/2b precursor. Mol. Endocrinol. 5, 149–155.

Harland, R. M. (1994). The transforming growth factor β family and induction of the vertebrate mesoderm: Bone morphogenetic proteins are ventral inducers. Proc. Natl. Acad. Sci. USA 91, 10243–10246.

Harris, S. E., Harris, M. A., Mahy, P., Wozney, J., Feng, J. Q., & Mundy, G. R. (1994). Expression of bone morphogenetic protein messenger RNAs by normal rat and human prostate and prostate cancer cells. Prostate 24, 204–211.

Hashimoto, M., Kondo, S., Sakurai, T., Yuzuru, E., Shibai, H., & Muramatsu, M. (1990). Activin/EDF as an inhibitor of neural differentiation. Biochem. Biophys. Res. Comm. 173, 193–200.

Hedger, M. P., Drummond, A. E., Robertson, D. M., Risbridger, G. P., & de Krester, D. M. (1989). Inhibin and activin regulate [^{3}H]thymidine uptake by rat thymocytes and 3T3 cells *in vitro*. Mol. Cell. Endocrinol. 61, 133–138.

Heikinheimo, K. (1994). Stage-specific expression of decapentaplegic-Vg-related genes 2, 4, and 6 (Bone morphogenetic proteins 2, 4, and 6) during human tooth morphogenesis. J. Dent. Res. 73, 590–597.

Hemmati–Brivanlou, A. & Melton, D. A. (1992). A truncated activin receptor dominantly inhibits mesoderm induction and formation of axial structures in *Xenopus* embryos. Nature 359, 609–614.

Henderson, C. E., Phillips, H. S., Pollock, R. A., Davies, A. M., Lemeulle, C., Armanini, M., Simpson, L. C., Moffet, B., Vandlen, R. A., Koliatsos, V. E., & Rosenthal, A. (1994). GDNF: A potent survival factor for motoneurons present in peripheral nerve and muscle. Science 266, 1062–1064.

Hillier, S. G. & Miró, F. (1993). Inhibin, activin, and follistatin. Ann. NY Acad. Sci. 687, 29–38.

Hüsken–Hindi, P., Tsuchida, K., Park, M., Corrigan, A. Z., Vaughan, J. M., Vale, W. W., & Fischer, W. H. (1994). Monomeric activin A retains high receptor binding affinity but exhibits low biological activity. J. Biol. Chem. 269, 19380–19384.

Iannaccone, P. M., Zhou, X., Khokha, M., Boucher, D., & Kuehn, M. R. (1992). Insertional mutation of a gene involved in growth regulation of the early mouse embryo. Dev. Dyn. 194, 198–208.

Imbeaud, S., Carré–Eusèbe, D., Rey, R., Belville, C., Josso, N., & Picard, J.–Y. (1994). Molecular genetics of the persistent Müllerian duct syndrome: A study of 19 families. Hum. Mol. Gen. 3, 125–131.

Itoh, M., Igarashi, M., Yamada, K., Hasegawa, Y., Seki, M., Eto, Y., & Shibai, H. (1990). Activin A stimulates meiotic maturation of the rat oocyte *in vitro*. Biochem. Biophys. Res. Comm. 166, 1479–1484.

Jaffe, G. J., Harrison, C. E., Lui, G. M., Roberts, W. L., Goldsmith, P. C., Mesiano, S., & Jaffe, R. B. (1994). Activin expression by cultured human retinal pigment epithelial cells. Inv. Opthamol. Vis. Sci. 35, 2924–2931.

Jones, C. M., Lyons, K. M., & Hogan, B. L. M. (1991). Involvement of bone morphogenetic protein-4 (BMP-4) and Vgr-1 in morphogenesis and neurogenesis in the mouse. Development 111, 531–542.

Jones, C. M., Simon–Chazottes, D., Guenet, J.–L., & Hogan, B. L. M. (1992). Isolation of *Vgr-2*, a novel member of the transforming growth factor-β-related gene family. Mol. Endocrinol. 6, 1961–1968.

Josso, N., Cate, R. L., Picard, J.–Y., Vigier, B., di Clemente, N., Wilson, C., Imbeaud, S., Pepinsky, R. B., Guerrier, D., Boussin, L., Legeai, L., & Carré–Eusèbe, D. (1993). Anti-Müllerian hormone: The Jost factor. Recent Prog. Horm. Res. 48, 1–59.

Jost, A. (1947). Recherches sur la différenciation sexuelle de l'embryon de lapin. Arch. Anat. Micro. Morph. Exp. 36, 271–315.

Jost, A., Vigier, B., Prepin, J., & Perchellet, J. P. (1973). Studies on sex differentiation in mammals. Recent Prog. Horm. Res. 29, 1–41.

King, J. A., Marker, P. C., Seung, K. J., & Kingsley, D. M. (1994). BMP 5 and the molecular, skeletal, and soft-tissue alterations in *short ear* mice. Development Biol. 166, 112–122.

King, T. R., Lee, B. K., Behringer, R. R., & Eicher, E. M. (1991). Mapping anti-Müllerian hormone (*Amh*) and related sequences in the mouse: Identification of a new region of homology between MMU10 and HSA19p. Genomics 11, 273–283.

Kingsley, D. M. (1994). What do BMPs do in mammals? Clues from the mouse short-ear mutation. TIGS 10, 16–21.

Kingsley, D. M., Bland, A. E., Grubber, J. M., Marker, P. C., Russell, L. B., Copeland, N. G., & Jenkins, N. A. (1992). The mouse *short ear* skeletal morphogenesis locus is associated with defects in a bone morphogenetic member of the TGF β superfamily. Cell 71, 399–410.

Knebelmann, B., Boussin, L., Guerrier, D., Legeai, L., Kahn, A., Josso, N., & Picard, J.–Y. (1991). Anti-Müllerian hormone Bruxelles: A nonsense mutation associated with the persistent Müllerian duct syndrome. Proc. Natl. Acad. Sci. USA 88, 3767–3771.

Knight, P. G., Beard, A. J., Wrathall, J. H. M., & Castillo, R. J. (1989). Evidence that the bovine ovary secretes large amounts of monomeric inhibin α subunit and its isolation from bovine follicular fluid. J. Mol. Endocrinol. 2, 189–200.

Kojima, I., Mogami, H., Kawamura, N., Yasuda, H., & Shibata, H. (1993). Modulation of growth of vascular smooth muscle cells by activin A. Exp. Cell Res. 206, 152–156.

Landauer, W. (1952). Brachypodism: A recessive mutation of house-mice. J. Hered. 43, 293–298.

Lawrence, D. A., Pircher, R., & Jullien, P. (1985). Conversion of a high molecular weight latent β-TGF from chicken embryo fibroblasts into a low molecular weight active β-TGF under acidic conditions. Biochem. Biophys. Res. Comm. 133, 1026–1034.

Lee, M. M. & Donahoe, P. K. (1993). Müllerian inhibiting substance: A gonadal hormone with multiple functions. Endocrine Rev. 14, 152–164.

Lee, S.–J. (1990). Identification of a novel member of the transforming growth factor-β superfamily. Mol. Endocrinol. 4, 1034–1040.

Lee, S.–J. (1991). Expression of growth/differentiation factor 1 in the nervous system: Conservation of a bicistronic structure. Proc. Natl. Acad. Sci., U.S.A. 88, 4250–4254.

Lefer, A. M., Tsao, P. S., Ma, X.–L., & Sampath, T. K. (1992). Anti-ischaemic and endothelial protective actions of recombinant human osteogenic protein (hOP-1). J. Mol. Cell. Cardiol. 24, 585–593.

Lillie, F. R. (1916). The theory of the free-martin. Science 43, 611–613.

Lin, L.–F. H., Doherty, D. H., Lile, J. D., Bektesh, S., & Collins, F. (1993). GDNF: A glial cell line-derived neurotrophic factor for midbrain dopaminergic neurons. Science 260, 1130–1132.

Ling, N., Ying, S.–Y., Ueno, N., Shimasaki, S., Esch, F., Hotta, M., & Guillemin, R. (1986). Pituitary FSH is released by a heterodimer of the β-subunits from the two forms of inhibin. Nature 321, 779–782.

Loeff, D. S., Imbeaud, S., Reyes, H. M., Meller, J. L., & Rosenthal, I. M. (1994). Surgical and genetic aspects of persistent Müllerian duct syndrome. J. Pediatr. Surg. 29, 61–65.

Lu, R.–Z., Matsuyama, S., Nishihara, M., & Takahashi, M. (1993). Developmental expression of activin/inhibin β_A, β_B, and α subunits, and activin receptor-IIB genes in preimplantation mouse embryos. Biol. Reprod. 49, 1163–1169.

Luyten, F. P., Cunningham, N. S., Ma, S., Muthukumaran, N., Hammonds, R. G., Nevins, W. B., Wood, W. I., & Reddi, A. H. (1989). Purification and partial amino acid sequence of osteogenin, a protein initiating bone differentiation. J. Biol. Chem. 264, 13377–13380.

Lyons, K., Graycar, J. L., Lee, A., Hashmi, S., Lindquist, P. B., Chen, E. Y., Hogan, B. L. M., & Derynck, R. (1989a). Vgr-1, a mammalian gene related to *Xenopus* Vg-1, is a member of the transforming growth factor β gene superfamily. Proc. Natl. Acad. Sci., USA 86, 4554–4558.

Lyons, K. M., Pelton, R. W., & Hogan, B. L. M. (1989b). Patterns of expression of murine Vgr-1 and BMP-2a RNA suggest that transforming growth factor- β-like genes coordinately regulate aspects of embryonic development. Genes Development 3, 1657–1668.

Lyons, K. M., Pelton, R. W., & Hogan, B. L. M. (1990). Organogenesis and pattern formation in the mouse: RNA distribution patterns suggest a role for bone morphogenetic protein-2a (BMP-2a). Development 109, 833–844.

Lyons, R. M., Keski–Oja, J., & Moses, H. L. (1988). Proteolytic activation of latent transforming growth factor-β from fibroblast-conditioned medium. J. Cell Biol. 106, 1659–1665.

Malinina, N. A., Kindiakov, B. N., & Koniukhov, B. V. (1984). Mutant gene expression in mouse aggregation chimeras. 3. The brachypodism-H gene. Ontogenez 15, 514–21.

Manova, K., Paynton, B. V., & Bachvarova, R. F. (1992). Expression of activins and TGFβ1 and β2 RNAs in early postimplantation mouse embryos and uterine decidua. Mech. Dev. 36, 141–152.

Mason, A. J., Niall, H. D., & Seeburg, P. H. (1986). Structure of two human ovarian inhibins. Biochem. Biophys. Res. Comm. 135, 957–964.

Mason, A. J. (1994). Functional analysis of the cysteine residues of activin A. Mol. Endocrinol. 8, 325–332.

Mason, A. J., Hayflick, J. S., Ling, N., Esch, F., Ueno, N., Ying, S.–Y., Guillemin, R., Niall, H., & Seeburg, P. H. (1985). Complementary DNA sequences of ovarian follicular fluid inhibin show precursor structure and homology with transforming growth factor-β. Nature 318, 659–663.

Matzuk, M. M., Finegold, M. J., Mather, J. P., Krummen, L., Lu, H., & Bradley, A. (1994). Development of cancer cachexia-like syndrome and adrenal tumors in inhibin-deficient mice. Proc. Natl. Acad. Sci., USA 91, 8817–8821.

Matzuk, M. M., Finegold, M. J., Su, J.–G. J., Hsueh, A. J. W., & Bradley, A. (1992). α-inhibin is a tumor-suppressor gene with gonadal specificity in mice. Nature 360, 313–319.

Mayo, K. E., Cerelli, G. M., Spiess, J., Rivier, J., Rosenfeld, M. G., Evans, R. M., & Vale, W. (1986). Inhibin A-subunit cDNA's from porcine ovary and human placenta. Proc. Natl. Acad. Sci., USA 83, 5849–5853.

McCarthy, S. A. & Bicknell, R. (1993). Inhibition of vascular endothelial cell growth by activin-A. J. Biol. Chem. 268, 23066–23071.

McCullagh, D. R. (1932). Dual endocrine activity of the testes. Science 76, 19–20.

McDonald, N. Q. & Hendrickson, W. A. (1993). A structural superfamily of growth factors containing a cystine knot motif. Cell 73, 421–424.

McGrath, S. A., Esquela, A. F., & Lee, S.-J. (1995). Oocyte-specific expression of growth/differentiation factor-9. Mol. Endocrinol. 9, 131–136.

McPherron, A. C. & Lee, S.–J. (1993). GDF-3 and GDF-9: Two new members of the transforming growth factor-β superfamily containing a novel pattern of cysteines. J. Biol. Chem. 268, 3444–3449.

Meunier, H., Rivier, C., Evans, R. M., & Vale, W. (1988). Gonadal and extragonadal expression of inhibin α, βA, and βB subunits in various tissues predicts diverse functions. Proc. Natl. Acad. Sci. USA 85, 247–251.

Miyazono, K., Hellman, U., Wernstedt, C., & Heldin, C.-H. (1988). Latent high molecular weight complex of transforming growth factor β1. J. Biol. Chem. 263, 6407–6415.

Moore, A., Krummen, L. A., & Mather, J. P. (1994). Inhibins, activins, their binding proteins and receptors: Interactions underlying paracrine activity in the testis. Mol. Cell. Endocrinol. 100, 81–86.

Murata, M., Eto, Y., Shibai, H., Sakai, M., & Muramatsu, M. (1988). Erythroid differentiation factor is encoded by the same mRNA as that of the inhibin βA chain. Proc. Natl. Acad. Sci., USA 85, 2434–2438.

Nakao, K., Kosaka, M., & Saito, S. (1991). Effects of erythroid differentiation factor (EDF) on proliferation and differentiation of human hematopoietic progenitors. Exp. Hematol. 19, 1090–1095.

Nakase, T., Nomura, S., Yoshikawa, H., Hashimoto, J., Hirota, S., Kitamura, Y., Oikawa, S., Ono, K., & Takaoka, K. (1994). Transient and localized expression of bone morphogenetic protein 4 messenger RNA during fracture healing. J. Bone Min. Res. 9, 651–659.

Nishi, Y., Haji, M., Tanaka, S., Yanase, T., Takayanagi, R., Etoh, Y., & Nawata, H. (1992). Human recombinant activin-A modulates the steroidogenesis of cultured bovine adrenocortical cells. J. Endocrinol. 132, R1–R4.

Niswander, L., & Martin, G. R. (1993). FGF-4 and BMP-2 have opposite effects on limb growth. Nature 361, 68–71.

O, W.-S., Robertson, D. M., & de Kretser, D. M. (1989). Inhibin as an oocyte meiotic inhibitor. Mol. Cell. Endocrinol. 62, 307–311.

Ogawa, K., Abe, K., Kurosawa, N., Kurohmaru, M., Sugino, H., Takahashi, M., & Hayashi, Y. (1993). Expression of α, βA and βB subunits of inhibin or activin and follistatin in rat pancreatic islets. FEBS Lett. 319, 217–220.

Ogawa, Y., Schmidt, D. K., Nathan, R. M., Armstrong, R. M., Miller, K. L., Sawamura, S. J., Ziman, J. M., Erickson, K. L., de Leon, E. R., Rosen, D. M., Seyedin, S. M., Glaser, C. B., Chang, R.-Y., Corrigan, A. Z., & Vale, W. (1992). Bovine bone activin enhances bone morphogenetic protein-induced ectopic bone formation. J. Biol. Chem. 267, 14233–14237.

Owens, E. M. & Solursh, M. (1982). Cell-cell interaction by mouse limb cells during *in vitro* chondrogenesis: Analysis of the brachypod mutation. Development Biol. 91, 376–388.

Özkaynak, E., Rueger, D. C., Drier, E. A., Corbett, C., Ridge, R. J., Sampath, T. K., & Oppermann, H. (1990). OP-1 cDNA encodes an osteogenic protein in the TGF-β family. EMBO J. 7, 2085–2093.

Özkaynak, E., Schnegelsberg, P. N. J., Jin, D. F., Clifford, G. M., Warren, F. D., Drier, E. A., & Opperman, H. (1992). Osteogenic protein-2, a new member of the transforming growth factor-β superfamily expressed early in embryogenesis. J. Biol. Chem. 267, 25220–25227.

Özkaynak, E., Schnegelsberg, P. N. J., & Opperman, H. (1991). Murine osteogenic protein (OP-1): High levels of mRNA in kidney. Biochem. Biophys. Res. Comm. 179, 116–123.

Padgett, R. W., St. Johnston, R. D., & Gelbart, W. M. (1987). A transcript from a *Drosophila* pattern gene predicts a protein homologous to the transforming growth factor-β family. Nature 325, 81–84.

Paralkar, V. M., Weeks, B. S., Yu, Y. M., Kleinman, H. K., & Reddi, A. H. (1992). Recombinant human bone morphogenetic protein-2B stimulates PC12 cell differentiation: Potentiation and binding to type IV collagen. J. Cell Biol. 119, 1721–1728.

Pepinsky, R. B., Sinclair, L. K., Chow, E. P., Mattaliano, R. J., Manganaro, T. F., Donahoe, P. K., & Cate, R. L. (1988). Proteolytic processing of Müllerian inhibiting substance produces a transforming growth factor-β-like fragment. J. Biol. Chem 263, 18961–18964.

Perides, G., Safran, R. M., Downing, L. A., & Charness, M. E. (1994). Regulation of neural cell adhesion molecule and L1 by the transforming growth factor-β superfamily. J. Biol. Chem. 269, 765–770.

Perides, G., Safran, R. M., Rueger, D. C., & Charness, M. E. (1992). Induction of the neural cell adhesion molecule and neuronal aggregation by osteogenic protein 1. Proc. Natl. Acad. Sci., USA 89, 10326–10330.

Petraglia, F., Sacerdote, P., Cossarizza, A., Angioni, S., Genazani, A. D., Franceschi, C., Muscettola, M., & Grasso, G. (1991). Inhibin and activin modulate human monocyte chemotaxis and human lymphocyte interferon-γ production. J. Clin. Endocrinol. Metab. 72, 496–502.

Picon, R. (1969). Action du testicule fœtal, sur le développement *in vitro* des canaux de Müller chez le rat. Arch. Anat. Micro. Morph. Exp. 58, 1–19.

Rao, V. V. N. G., Loffler, C., Wozney, J. M., & Hansmann, I. (1992). The gene for BMP 2A is localized to human chromosome 20p12 by radioactive and nonradioactive *in situ* hybridization. Hum. Genet. 90, 299.

Reddi, A. H. (1981). Cell biology and biochemistry of endochrondral bone development. Coll. Res. 1, 209–226.

Roberts, V. J., Sawchenko, P. E., & Vale, W. (1991). Expression of inhibin/activin subunit messenger ribonucleic acids during rat embryogenesis. Endocrinol. 128, 3122–3129.

Robertson, D. M., Foulds, L. M., Prisk, M., & Hedger, M. P. (1992). Inhibin/activin β-subunit monomer: Isolation and characterization. Endocrinol. 130, 1680–1687.

Robertson, D. M., Giacometti, M., Foulds, L. M., Lahnstein, J., Goss, N. H., Hearn, M. T. W., & de Krester, D. M. (1989). Isolation of inhibin α-subunit precursor proteins from bovine follicular fluid. Endocrinol. 125, 2141–2149.

Sakai, R., Eto, Y., Ohtsuka, M., Hirafuji, M., & Shinoda, H. (1993). Activin enhances osteoclast-like cell formation *in vitro*. Biochem. Biophys. Res. Comm. 195, 39–46.

Sampath, T. K., Coughlin, J. E., Whetstone, R. M., Banach, D., Corbett, C., Ridge, R. J., Özkaynak, E., Oppermann, H., & Rueger, D. C. (1990). Bovine osteogenic protein is composed of dimers of OP-1 and BMP-2A, two members of the transforming growth factor-β superfamily. J. Biol. Chem. 265, 13198–13205.

Sampath, T. K., Maliakal, J. C., Hauschka, P. V., Jones, W. K., Sasak, H., Tucker, R. F., White, K. H., Coughlin, J. E., Tucker, M. M., Pang, R. H. L., Corbett, C., Özkaynak, E., Oppermann, H., & Rueger, D. C. (1992). Recombinant human osteogenic protein-1 (hOP-1) induces new bone formation *in vivo* with a specific activity comparable with natural bovine osteogenic protein and stimulates osteoblast proliferation and differentiation *in vitro*. J. Biol. Chem. 267, 20352–20362.

Sampath, T. K. & Reddi, A. H. (1981). Dissociative extraction and reconstitution of extracellular matrix components involved in local bone differentiation. Proc. Natl. Acad. Sci., USA 78, 7599–7603.

Sawchenko, P. E., Plotsky, P. M., Pfeiffer, S. W., Cunningham, E. T., Vaughan, J., Rivier, J., & Vale, W. (1988). Inhibin β in central neural pathways involved in the control of oxytocin secretion. Nature 334, 615–617.

Schaar, D. G., Sieber, B.–A., Dreyfus, C. F., & Black, I. B. (1993). Regional and cell-specific expression of GDNF in rat brain. Exp. Neurol. 124, 368–371.

Schlunegger, M. P., & Grütter, M. G. (1992). An unusual feature revealed by the crystal structure at 2.2 Å resolution of human transforming growth factor-β2. Nature 358, 430–434.

Schubert, D., Kimura, H., LaCorbiere, M., Vaughan, J., Karr, D., & Fischer, W. H. (1990). Activin is a nerve cell survival molecule. Nature 344, 868–870.

Schwall, R. H., Robbins, K., Jardieu, P., Chang, L., Lai, C., & Terrell, T. G. (1993). Activin induces cell death in hepatocytes *in vivo* and *in vitro*. Hepatol. 18, 347–356.

Shimell, M. J., Ferguson, E. L., Childs, S. R., & O'Connor, M. B. (1991). The *Drosophila* dorsal-ventral patterning gene *tolloid* is related to human bone morphogenetic protein 1. Cell 67, 469–481.

Shiozaki, M., Sakai, R., Tabuchi, M., Eto, Y., Kosaka, M., & Shibai, H. (1989). *In vivo* treatment with erythroid differentiation factor (EDF/Activin A) increases erythroid precursors (CFU-E and BFU-E) in mice. Biochem. Biophys. Res. Comm. 165, 1155–1161.

Shiozaki, M., Sakai, R., Tabuchi, M., Nakamura, T., Sugino, K., Sugino, H., & Eto, Y. (1992). Evidence for the participation of endogenous activin A/erythroid differentiation factor in the regulation of erythropoiesis. Proc. Natl. Acad. Sci. USA 89, 1553–1556.

Smith, J. C., Price, B. M. J., van Nimmen, K., & Huylebroeck, D. (1990). Identification of a potent *Xenopus* mesoderm-inducing factor as a homologue of activin A. Nature 345, 729–731.

Spencer, S. J., Rabinovici, J., & Jaffe, R. B. (1990). Human recombinant activin-A inhibits proliferation of human fetal adrenal cells *in vitro*. J. Clin. Endocrinol. Metab. 71, 1678–1680.

Spencer, S. J., Rabinovici, J., Mesiano, S., Goldsmith, P. C., & Jaffe, R. B. (1992). Activin and inhibin in the human adrenal gland. J. Clin. Invest. 90, 142–149.

Springer, J. E., Mu, X., Bergmann, L. W., & Trojanowski, J. Q. (1994). Expression of GDNF mRNA in rat and human nervous tissue. Exp. Neurol. 127, 167–170.

Storm, E. E., Huynh, T. V., Copeland, N. G., Jenkins, N. A., Kingsley, D. M., & Lee, S.–J. (1994). Limb alterations in *brachypodism* mice due to mutations in a new member of the TGFβ-superfamily. Nature 368, 639–643.

Strömberg, I., Bjorklund, L., Johansson, M., Tomac, A., Collins, F., Olson, L., Hoffer, B., & Humpel, C. (1993). Glial cell line-derived neurotrophic factor is expressed in the developing but not adult striatum and stimulates developing dopamine neurons *in vivo*. Exp. Neurol. 124, 401–412.

Tabas, J. A., Hahn, G. V., Cohen, R. B., Seaunez, H. N., Mosi, W. S., Wozney, J. M., Zasloff, M., & Kaplan, F. S. (1993). Chromosomal assignment of the human gene for bone morphogenetic protein 4. Clin. Orthopaed. Rel. Res. 293, 310–316.

Tabas, J. A., Zasloff, M., Wasmuth, J. J., Emanuel, B. S., Altherr, M. R., McPherson, J. D., Wozney, J. M., & Kaplan, F. S. (1991). Bone morphogenetic protein: Chromosomal localization of human genes for BMP1, BMP2A, and BMP3. Genomics 9, 283–289.

Takahashi, M., Koide, S. S., & Donahoe, P. K. (1986). Müllerian inhibiting substance as oocyte meiosis inhibitor. Mol. Cell. Endocrinol. 47, 225–234.

ten Dijke, P., Hansen, P., Iwata, K. K., Pieler, C., & Foulkes, J. G. (1988). Identification of another member of the transforming growth factor type β gene family. Proc. Natl. Acad. Sci., USA 85, 4715–4719.

Thomsen, G., Woolf, T., Whitman, M., Sokol, S., Vaughan, J., Vale, W., & Melton, D. A. (1990). Activins are expressed early in *Xenopus* embryogenesis and can induce axial mesoderm and anterior structures. Cell 63, 485–493.

Thomsen, G. H. & Melton, D. A. (1993). Processed Vg1 protein is an axial mesoderm inducer in *Xenopus*. Cell 74, 433–441.

Totsuka, Y., Tabuchi, M., Kojima, I., Shibai, H., & Ogata, E. (1988). A novel action of activin A: Stimulation of insulin secretion in rat pancreatic islets. Biochem. Biophys. Res. Comm. 156, 335–339.

Urist, M. R. (1965). Bone: Formation by autoinduction. Science 150, 893–899.

Urist, M. R. (1994). In: Bone Grafts, Derivatives and Substitutes (M. R. Urist, B. T. O'Conner, & R. G. Burwell, eds.), pp. 315–362, Butterworth Heinemann, London.

Vainio, S., Karavanova, I., Jowett, A., & Thesleff, I. (1993). Identification of BMP-4 as a signal mediating secondary induction between epithelial and mesenchymal tissues during early tooth development. Cell 75, 45–58.

Vale, W., Hsueh, A., Rivier, C., & Yu, J. (1991). In: Peptide Growth Factors and Their Receptors: Handbook of Experimental Pharmacology (M. B. Sporn & A. B. Roberts, eds.), pp. 211–248, Springer-Verlag, Berlin.

Vale, W., Rivier, J., Vaughan, J., McClintock, R., Corrigan, A., Woo, W., Karr, D., & Spiess, J. (1986). Purification and characterization of an FSH releasing protein from porcine ovarian follicular fluid. Nature 321, 776–779.

van den Eijnden–van Raaij, A. J. M., Feijen, A., Lawson, K. A., & Mummery, C. L. (1992). Differential expression of inhibin subunits and follistatin, but not of activin receptor type II, during early murine embryonic development. Dev. Biol. 154, 356–365.

Vassalli, A., Matzuk, M. M., Gardner, H. A. R., & Lee, K.–F. (1994). Activin/inhibin βB subunit gene disruption leads to defects in eyelid development and female reproduction. Genes Dev. 8, 414–427.

Vigier, B., Watrin, F., Magre, S., Tran, D., & Josso, N. (1987). Purified bovine AMH induces a characteristic freemartin effect in fetal rat prospective ovaries exposed to it *in vitro*. Development 100, 43–55.

Vukicevic, S., Helder, M. N., & Luyten, F. P. (1994). Developing human lung and kidney are major sites for synthesis of bone morphogenetic protein-3 (Osteogenin). J. Histochem. Cytochem. 42, 869–875.

Wang, E. A., Rosen, V., D'Alessandro, J. S., Bauduy, M., Cordes, P., Harada, T., Israel, D. I., Hewick, R. M., Kerns, K. M., LaPan, P., Luxenberg, D. P., McQuaid, D., Moutsatsos, I. K., Nove, J., & Wozney, J. M. (1990). Recombinant human bone morphogenetic protein induces bone formation. Proc. Natl. Acad. Sci., USA 87, 2220–2224.

Wilson, C. A., di Clemente, N., Ehrenfels, C., Pepinsky, R. B., Josso, N., Vigier, B., & Cate, R. L. (1993). Müllerian inhibiting substance requires its N-terminal domain for maintenance of biological activity, a novel finding within the transforming growth factor-β superfamily. Mol. Endocrinol. 7, 247–257.

Wozney, J. M. (1992). The bone morphogenetic protein family and osteogenesis. Mol. Reprod. Dev. 32, 160–167.

Wozney, J. M. & Celeste, A. J. (1993). BMP-9 Compositions. Int. Patent Application No. PCT/US92/05374.

Wozney, J. M., Rosen, V., Celeste, A. J., Mitsock, L. M., Whitters, M. J., Kriz, R. W., Hewick, R. M., & Wang, E. A. (1988). Novel regulators of bone formation: Molecular clones and activities. Science 242, 1528–1534.

Yamashita, N., Nakajima, T., Takahashi, H., Kaneoka, H., Mizushima, Y., & Sakane, T. (1993). Effects of activin A on IgE synthesis and cytokine production by human peripheral mononuclear cells. Clin. Exp. Immunol. 94, 214–219.

Yasuda, H., Inoue, K., Shibata, H., Takeuchi, T., Eto, Y., Hasegawa, Y., Sekine, N., Totsuka, Y., Mine, T., Ogata, E., & Kojima, I. (1993a). Existence of activin-A in A- and D-cells of rat pancreatic islet. Endocrinol. 133, 624–630.

Yasuda, H., Mine, T., Shibata, H., Eto, Y., Hasegawa, Y., Takeuchi, T., Asano, S., & Kojima, I. (1993b). Activin A: An autocrine inhibitor of initiation of DNA synthesis in rat hepatocytes. J. Clin. Invest. 92, 1491–1496.

Yu, J., Shao, L.–E., Lemas, V., Yu, A. L., Vaughan, J., Rivier, J., & Vale, W. (1987). Importance of FSH-releasing protein and inhibin in erythrodifferentiation. Nature 330, 765–767.

Yu, J., Shao, L.–E., Vaughan, J., Vale, W., & Yu, A. L. (1989). Characterization of the potentiation effect of activin on human erythroid colony formation *in vitro*. Blood 73, 952–960.

Zhou, B., Watts, L. M., & Hutson, J. M. (1993a). Germ cell development in neonatal mouse testes *in vitro* requires Müllerian inhibiting substance. J. Urol. 150, 613–616.

Zhou, X., Sasaki, H., Lowe, L., Hogan, B. L. M., & Kuehn, M. R. (1993b). *Nodal* is a novel TGF-β-like gene expressed in the mouse node during gastrulation. Nature 361, 543–547.

THE TGFβ RECEPTORS AND SIGNALING PATHWAYS

Michael B. Datto, Craig H. Bassing, and Xiao-Fan Wang

I. Introduction/Overview . . . 396
II. TGFβ Ligands . . . 397
III. The TGFβ Receptors . . . 398
  A. Identification of Cell-Surface TGFβ-Binding Proteins . . . 398
  B. Cloning and the Structural Features of the TGFβ Receptors . . . 399
IV. TGFβ Signal Transduction . . . 404
  A. The Biological Effects of TGFβ Are Mediated by Multiple Signaling Pathways . . . 404
  B. The TGFβ Signaling Receptors . . . 404
  C. The Mechanism of TGFβ Receptor Activation . . . 406
  D. The Type III Receptor/Betaglycan Presents TGFβ to the Type II Receptor . . 410
  E. Identification of Receptor Interacting Proteins . . . 411
  F. Effects of TGFβ on the Cell-Cycle Machinery . . . 413
  G. TGFβ Regulates the Expression of Genes with Diverse Functions . . . 419
V. The Role of the TGFβ Signaling Pathways in Carcinogenesis . . . 420

Growth Factors and Cytokines in Health and Disease
Volume 1B, pages 395–432.

ISBN: 0-7623-0091-4

VI. Hereditary Telangiectasia Is a Result of Mutations in Endoglin 422
VII. Summary 423
Acknowledgments 424
References 424

I. INTRODUCTION/OVERVIEW

TGFβ is a group of multipotent polypeptide hormones which have diverse effects on many different cell types and tissues. TGFβ has been described as a factor which can inhibit the growth of many cell types (Roberts and Sporn, 1990; Massague, 1990). TGFβ has also been shown to stimulate proliferation of fibroblasts through its induction of PDGF (Leof et al., 1986). TGFβ is able to induce the differentiation of bronchial epithelial cells (Masui et al., 1986) and prechondrocytes (Seyedin et al., 1985), and inhibit the differentiation of myoblasts (Massague et al., 1986; Olson et al., 1986) and adipocytes (Roberts and Sporn, 1990). Besides its effects on cell growth and differentiation, TGFβ has been shown to induce the formation of extracellular matrix through its induction of many genes which encode proteins, including collagen, fibronectin, elastin, thrombospondin, plasminogen activator inhibitor, and tissue inhibitor of metalothioproteinases (Roberts and Sporn, 1990).

In addition to these cellular effects, TGFβ has been demonstrated to play important roles in regulating the functions of many complex multicellular systems. For example, TGFβ is a potent immunoregulatory factor which can enhance the function of monocytes and suppress that of lymphocytes. This is in part due to the ability of TGFβ to suppress the proliferation of lymphocytes and to antagonize the functions of several cytokines including IL-1, IL-2, IL-3, TNF, and INF-α (Roberts and Sporn, 1990; Barnard et al., 1990). TGFβ has also been implicated in wound healing and tissue repair, partly due to the effects of TGFβ on extracellular matrix formation and angiogenesis. TGFβ regulates normal wound healing but excessive activity of TGFβ may be responsible for tissue damage in cirrhosis, glomerulonephritis, scarring, and fibrosis (Broder and Ruoslahti, 1992; Broder et al., 1992). In addition to its involvement in these processes, TGFβ has multiple complicated functions in the disregulated growth seen in oncogenesis and the highly regulated process of embryogenesis (Roberts and Sporn, 1990).

Because of the broad importance of TGFβ in many physiological and pathological processes, tremendous effort has been made to understand the mechanisms of TGFβ action. This chapter summarizes many of the important findings in this fast growing field ranging from the activation of TGFβ receptors to the effects of TGFβ on cell-cycle control and regulation of specific genes. The understanding of the complex functions of TGFβ started with the understanding of TGFβ ligands, their regulation, and characterization as members of a larger superfamily of related proteins. Researchers have been able to identify several cell-surface TGFβ-binding proteins. Four of these cell-surface receptors have been cloned and further characterized. Two of these TGFβ receptors, the type I and type II receptors, are members of a larger family of transmembrane receptor serine-threonine kinases. Recently, extensive studies have led to the elucidation of the mechanism through which these receptors are activated by ligand to transmit a signal. Research at the level of the TGFβ receptors is presently focused on identifying substrates and effectors of the receptor kinases.

Significant progress has also been made in understanding events downstream of receptor activation mediating TGFβ growth inhibitory effects and transcriptional activation of specific genes. These studies have led to a greater understanding of the cell cycle control through cyclins, cyclin dependent kinases (Cdk's), and a newly emerging family of Cdk inhibitory proteins. Finally, the role of the TGFβ receptors and their signaling pathways has been studied in several disease processes. Mutations in the TGFβ receptor system and loss of the growth inhibitory effects of TGFβ have been implicated as a potential etiology for disregulated cell growth and subsequent tumorigenesis. In addition, mutations in a homologue of the type III TGFβ receptor, endoglin, has been linked to hereditary hemorrhagic telangiectasia, a disease characterized by vascular malformations.

II. TGFβ LIGANDS

The five structurally related TGFβ isoforms represent a single distinct family in the extended TGFβ superfamily. This superfamily contains many other structurally related polypeptide hormones, including activins and inhibins, Müllerian inhibiting substance (MIS), *Drosophila* morphogen decapentaplegic factor (DPP), bone morphogenic proteins, *Xenopus* genes Vg-1 and Vgr-1, growth differentiation factors GDF-1,-3 and -5 (Kingsley, 1994). Although these molecules exhibit structural similarities, they elicit diverse biological effects by interacting with their respective receptors.

The cDNAs for the TGFβ family members TGFβ1, β2, and β3 have been isolated from mammalian sources, whereas cDNAs for TGFβ4 and β5 have been isolated from chicken and *Xenopus*, respectively (Roberts and Sporn, 1990; Massague, 1990). These five isoforms of TGFβ exhibit a high degree of sequential similarity, between 64–82% identity at the amino acid level. The structural similarities of these molecules include nine invariant cysteine residues. These conserved regions and features of the TGFβ isoforms may play an important role in recognition of these ligands by their receptors (Schlunegger and Grutter, 1992; Daopin et al., 1992).

The biologically active form of these TGFβ isoforms is a 25-kDa disulfide-linked dimer. TGFβ, however, is secreted in an inactive form of a 100-kDa latent precursor. Prior to secretion, the latent TGFβ precursor is proteolytically cleaved to yield the secreted precursor molecule. This precursor consists of a TGFβ dimer noncovalently associated with an N-terminal dimer of the precursor peptide. Dissociation of the latent complex from the TGFβ dimer is a regulated process which may be important for modulating the bioavailability of TGFβ (Barnard et al., 1990).

The three mammalian isoforms of TGFβ are believed to play both distinct and overlapping roles (Roberts and Sporn, 1990). The different isoforms of TGFβ are under the control of different promoter elements and are differentially regulated (Roberts and Sporn, 1990). Thus the evolution of so many different forms of TGFβ with similar biological activities might be for differential regulation under the control of unique promoters (Roberts and Sporn, 1990). All three isoforms of TGFβ exert their effects by interacting with the same cellular receptors (Massague et al., 1990). Upon binding to their cellular receptor, the three isoforms of TGFβ initiate a wide variety of similar biological responses. There is, however, some degree of receptor specificity for the different isoforms of TGFβ. The different affinities of the TGFβ receptors for the different isoforms of TGFβ may serve some biologically significant functions.

III. THE TGFβ RECEPTORS

The diverse and complex biological effects of TGFβ are mediated through its interaction with membrane-associated receptors. One of the biggest challenges and greatest accomplishments in the TGFβ field in recent years has been the identification, cloning, and characterization of several of these receptors. These studies have advanced our understanding of the TGFβ signaling pathways and have also identified two of these receptors as prototypic members of a family of transmembrane receptor serine-threonine kinases. A better understanding of the biochemical and structural properties of these TGFβ receptors, therefore, will lead to understanding the role of these molecules in TGFβ-mediated signal transduction and also the mechanisms of action of this larger family of related receptors.

A. Identification of Cell-Surface TGFβ-Binding Proteins

A classical biochemical method has been used to identify cell-surface proteins which specifically bind to TGFβ. This assay consists of cross-linking radio-iodinated TGFβ to cell-surface proteins. The cross-linked proteins are subsequently analyzed by gel electrophoresis and characterized by molecular weight (Massague, 1990). The TGFβ type I, type II, and type III receptors were the first identified using this method (Massague, 1990). They were characterized by the range of their molecular weights of 50–60 kDa, 70–95 kDa, and 280–300 kDa, respectively, and found to be widely expressed in almost all TGFβ-responsive cell types. This assay has also been used to identify several other cell-surface TGFβ-binding proteins with a much more limited distribution. One of these receptors, the type IV TGFβ receptor, is a 70-kDa protein found only in pituitary tumor cells (Cheifetz et al., 1988). This receptor binds to TGFβ and also binds with an equally high affinity to activin and inhibin, two other members of the TGFβ superfamily. In addition, a TGFβ1 binding protein of 180 kDa and two TGFβ2 binding proteins of 60 and 140 kDa have been identified only in BHE and MG-63 human osteosarcoma cells (Cheifetz and Massague, 1991).

The only TGFβ-binding proteins not identified by this method are the type V receptor and endoglin. The type V receptor is a 400-kDa protein widely expressed in many cell types (O'Grady et al., 1991). Endoglin is a cell-surface protein composed of two 95-kDa subunits found in vascular endothelial cells, erythroblasts, macrophages and various myeloid and lymphoid leukemias (Gougos and Letarte, 1990). Endoglin can bind TGFβ1 and TGFβ3 with high affinity and it contains an RGD sequence which implicates it as a mediator of cell adhesion (Cheifetz et al., 1992).

Currently only type I, type II, type III receptors, and endoglin have been cloned and further characterized. The molecular identity of the other TGFβ-binding proteins and their roles in signal transduction are the subject of current research.

B. Cloning and the Structural Features of the TGFβ Receptors

The Type III TGFβ Receptor

The type III receptor, also known as betaglycan, is a large (280–300kDa) proteoglycan which can bind to all three isoforms of TGFβ with equally high affinity (Massague, 1990). The large and heterogeneous molecular weight of this receptor is partly contributed by its covalently linked heparin sulfate and chondroitin sulfate glycosaminoglycans. These carbohydrate moieties are not required for TGFβ binding (Lopez–Casillas et al., 1991). The type III receptor exist in both a membrane-associated and a soluble form and is widely expressed at high levels in many TGFβ-responsive cell types.

The cDNA for the type III receptor was independently isolated by expression cloning in COS cells and microsequencing of peptides derived from purified proteins (Lopez–Casillas et al., 1991; Wang et al., 1991). The cDNA encodes a 853 amino acid protein with a large cysteine-rich N-terminal extracellular domain. Contained in this extracellular domain are six consensus N-glycosylation sites, at least one consensus site for glycosaminoglycan addition, and numerous additional Ser-Gly residues that are potential sites for glycosaminoglycan (GAG) conjugation (Lopez–Casillas et al., 1991; Wang et al., 1991). Near the transmembrane segment of the molecule is a proline-rich sequence which has been hypothesized to function as a hinge at the base of the extracellular domain (Lopez–Casillas et al., 1991). Two potential sites for proteolytic cleavage are also located in this region, and it is believed that proteolytic cleavage generates the soluble form of the type III receptor (Lopez–Casillas et al., 1991). This extracellular domain is preceded by a 24 amino acid N-terminal signal sequence and followed by a 27-residue sequence with strong hydrophobicity, representing the putative transmembrane domain. The C-terminus of the type III receptor contains a relatively short, serine-rich 41 amino acid intracellular domain. Other than a single consensus of phosphorylation site for PKC, this small intracellular portion of the type III receptor contains no signaling motifs (Wang et al., 1991).

The cDNA for the type III receptor contains some sequential similarity to endoglin, another cell-surface TGFβ-binding molecule. This homology is strongest in the transmembrane and cytoplasmic regions of these molecules which share 63% amino acid identity. The function of this highly conserved region, however, remains unknown. Less sequential similarity exists in the extracellular domains of these two molecules; overall they share only 27% sequential identity. There are, however, two small highly conserved regions in the extracellular domains of the type III receptor and endoglin, with 41% and 48% sequential identity (Cheifetz et al., 1992). There are also eight conserved cysteine residues in the extracellular domains of these molecules. Deletion analysis of these highly conserved regions has shown that they may be important in the interaction of these proteins with TGFβ (Lopez–Casillas et al., 1994). The type III receptor also has homology in its extracellular domain to the sperm receptors ZP2 and ZP3, uromodulin and the zymogen granule membrane protein, GP-2. The significance of this homology to proteins of such diverse functions is unknown (Lopez–Casillas et al., 1994).

The Type II TGFβ Receptor

The type II TGFβ receptor is a member of a growing family of transmembrane receptor serine-threonine kinases. Members of this family include the type II receptors for activin and Müllerian inhibiting substance, the *C. elegans* daf-4 and daf-1 receptors, and the *Drosophila* AtrII gene (Lin and Lodish, 1993; Kingsley, 1994). All of these receptors directly interact with their ligands and contain intracellular kinase domains with a substrate specificity for serine and threonine phosphorylation.

The type II TGFβ receptor was first cloned using a COS cell transfection and ligand-binding expression cloning strategy (Lin et al., 1992). The type II receptor cDNA encodes a 565 amino acid protein. The N-terminal of this protein contains a 23 amino acid signal sequence followed by a hydrophilic extracellular domain of 136 amino acids. This extracellular region is cysteine-rich and contains three consensus N-linked glycosylation sites. The pattern and spacing of cysteines in this extracellular domain is a conserved feature of the type II serine-threonine kinase receptor family members. Following the extracellular domain is a hydrophobic domain of 30 amino acids, which represents the single putative transmembrane domain, and an intracellular domain of 376 amino acids. This intracellular domain contains sequences which identify the type II receptor as a kinase. Specifically, this region contains 18 of the 21 invariant residues, that define serine-threonine kinases, and can be further divided into the eleven kinase subdomains, that contain both invariant and highly conserved regions (Figure 1). In addition, this kinase domain contains two short kinase inserts: insert I between the subdomains VIII and IX and insert II between the subdomains X and XI. These kinase inserts are also a conserved feature of the type II receptor family members. The type II receptor contains a 56 amino acid juxtamembrane region between the kinase domain and

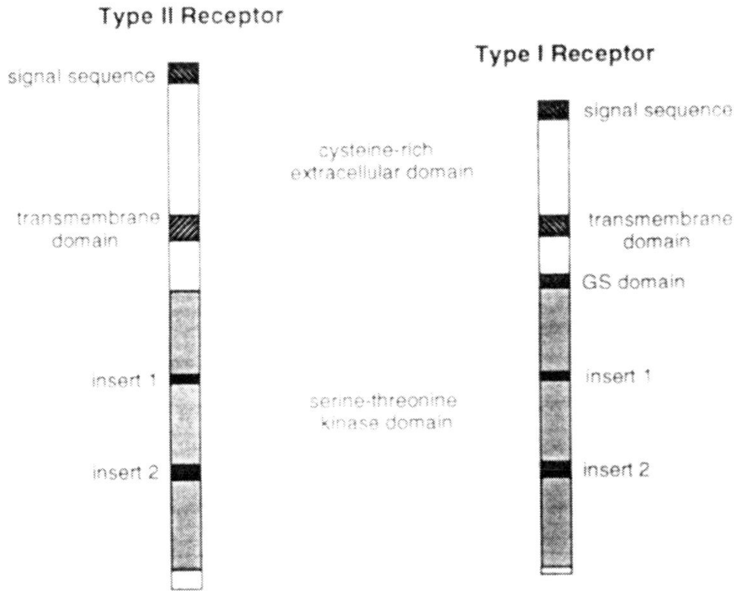

Figure 1. Schematic representation of the type I and type II TGFβ receptors.

the transmembrane domain. This region is highly conserved among different species, suggesting that it may be functionally important. Following the serine-threonine kinase domain, this cDNA encodes a small C-terminal tail of 22 amino acids.

The Type I TGFβ Receptor

Several groups have identified a number of cDNA clones encoding proteins originally classified as type I TGFβ receptors, based on their ability to bind TGFβ when overexpressed with a functional type II receptor in COS cells. These cDNAs include TSK-7L (Ebner et al., 1993), TSR-1 (Attisano et al., 1993), ALK-5 (Franzen et al., 1993), and R4 (He et al., 1993; Bassing et al., 1994a). The R4 and ALK-5 clones are homologues from rat and human, respectively, and they encode a protein which can bind to TGFβ in the presence of a type II TGFβ receptor. Functionally, this type I receptor can complement all known TGFβ signaling cascades in a mutant mink lung epithelial cell line lacking a type I TGFβ receptor, in response to all three TGFβ isoforms (Franzen et al., 1993; Bassing et al., 1994a, 1994b; ten Dijke et al., 1994; Carcamo et al., 1994). In addition, only antisera specific to R4/ALK5 were able to immunoprecipitate endogenous type I and type II TGFβ receptor complexes from a number of TGFβ-responsive cell types, suggesting that R4/ALK5 functions as the type I TGFβ receptor in a variety of cell types (ten Dijke et al., 1994).

The overall structure of the type I receptor is similar to that of the type II receptor, being also a member of the transmembrane receptor serine-threonine kinase family. Functionally, however, these two receptors are distinguished by the inability of the type I receptor to bind ligand independently. The type I receptor can only interact with TGFβ in the presence of a functional type II receptor. The cDNA for the type I receptor encodes a 54-kDa glycoprotein. The N-terminal of this protein contains a 22 amino acid signal sequence followed by a 101 amino acid hydrophilic extracellular domain. This extracellular domain contains one potential N-glycosylation site and, like the type II TGFβ receptor, is cysteine-rich, containing 10 cysteine residues. Unlike the type II subfamily of receptors, however, the positions of these cysteines are not as highly conserved among the members of the type I receptor subfamily. Overall, this extracellular domain has only 17% sequential identity to the type II TGFβ receptor. This extracellular domain is followed by a 23 amino acid hydrophobic transmembrane domain and a 355 amino acid intracellular domain. The cytoplasmic region of the type I receptor contains sequences which define it as a serine-threonine kinase. In addition, this kinase domain has two conserved kinase insert regions between kinase subdomains VIA and VIB and between X and XI, similar to those seen in the type II receptor subfamily members. Overall, the type I kinase domain has 41% amino acid similarity to the type II TGFβ receptor kinase domain. Between the kinase domain and the transmembrane region, the type I receptor contains a highly conserved serine and glycine-rich segment, termed the GS domain. This domain is a conserved feature of all type I receptor subfamily members. Following the kinase domain, the type I receptor contains a C-terminal tail of only five amino acids (Figure 1).

Although both TSK-7L and TSR-1 can bind TGFβ in affinity-labeling assays performed in COS cells, they are unable to restore the deficiency in TGFβ responsiveness in type I TGFβ receptor mutant cell lines (Attisano et al., 1993). These receptors may serve certain still undefined roles in TGFβ signal transduction by forming functionally distinct receptor complexes with the known single type II receptor. The differential interactions of these different type I receptors with the type II TGFβ receptor may provide a molecular basis for the multifunctional nature of TGFβ. Alternatively, however, these type I receptors may serve as receptors for other ligands in the TGFβ superfamily. Their binding and/or cross-linking to TGFβ observed in COS cells may have simply resulted from the overexpression of these receptors to high levels, representing nonphysiological conditions. This is possible due to the high degree of structural similarity between these related type I receptors and the different members of the TGFβ ligand superfamily.

Taken together, available evidence suggests that R4/ALK5 is probably the only physiological type I TGFβ receptor identified so far. The other isolated type I-like receptors most likely represent type I receptors for other ligands of the TGFβ superfamily. It is possible, therefore, that a single receptor complex is responsible for mediating TGFβ signals across the membrane and that the differential TGFβ effects are mediated by diverse intracellular effectors distributed in various cell types.

Endoglin

Endoglin was originally characterized as an antigen specifically recognized by a monoclonal antibody generated against a pre-B-cell leukemic cell line (Gougos and Letarte, 1990). Endoglin is expressed at high levels in vascular endothelial cells and syncytiotrophoblast and at low levels in erythroblasts, macrophages, and various myeloid and lymphoid leukemic cells (Gougos and Letarte, 1990). This molecule was shown to be a glycoprotein of two-95-kDa disulfide-linked homodimers. Endoglin was subsequently cloned and the cDNA was found to encode a 633 amino acid protein with a predicted molecular mass of 69 kDa (Gougos and Letarte, 1990). Endoglin contains a 561 amino acid extracellular domain with four potential N-linked glycosylation sites and 16 cysteine residues. In addition, this region contains an arginine-glycine-aspartate (RGD) sequence, a motif known to mediate specific interactions of extracellular matrix proteins with integrin (Gougos and Letarte, 1990). This extracellular domain is followed by a 27 amino acid hydrophobic transmembrane domain and a short 47 amino acid cytoplasmic domain. This cytoplasmic domain contains no known signaling motifs.

Based on the existence of the RGD motif, endoglin was originally thought to be a protein involved in cell–cell adhesion. After the cloning of the type III receptor, however, marked sequential similarity was noted between these two molecules. Subsequent studies demonstrated that endoglin, like the type III receptor, is able to bind TGFβ with high affinity (Cheifetz et al., 1992). Unlike the type III receptor, however, endoglin can bind only to TGFβ1 and TGFβ3 (Cheifetz et al., 1992). This difference in ligand specificity may have important implications for endoglin function in mediating TGFβ responses, because only these two isoforms of TGFβ can inhibit the proliferation of endothelial cells, a cell type which has a high expression of endoglin and low expression of the type III receptor (Cheifetz et al., 1992). It is possible that these two molecules serve a similar function in different cell types.

The Type V TGFβ Receptor

The type V TGFβ receptor is a widely distributed, high molecular weight protein originally identified in an attempt to purify the TGFβ receptors (O'Grady et al., 1991). Subsequent studies of the type V receptor have shown that it can specifically bind TGFβ and is internalized upon TGFβ treatment (O'Grady et al., 1991). Further experiments have shown that purified type V receptor can undergo autophosphorylation in the presence of TGFβ at a concentration of 4 nM (O'Grady et al., 1992). The majority of this autophosphorylation is on serine residues, as determined by one dimensional phosphoamino acid analysis. This suggests that, like the type I and type II receptors, the type V receptor may be a functional serine-threonine kinase. Confirmation for the function of the type V receptor in TGFβ-mediated signal transduction will have to come with its cloning and further characterization.

IV. TGFβ SIGNAL TRANSDUCTION

Over the past several years, significant progress has been made in understanding the mechanisms through which TGFβ mediates its biological effects. Pioneering works on the TGFβ receptors have shed light on the mechanism through which this novel family of transmembrane receptor serine-threonine kinases are activated by ligand to transmit a signal. These studies have elucidated the nature of the functional relationship between the type I, type II, and type III receptors in signal transduction. Progress has also been made in understanding events downstream of receptor activation, including the identification of one protein which interacts with the TGFβ type I receptor. In addition, studies of the growth inhibitory actions of TGFβ have led to significant advances in understanding cell-cycle control mediated by cyclins, their kinase partners, and an emerging family of cyclin-dependent kinase inhibitors.

A. The Biological Effects of TGFβ Are Mediated by Multiple Signaling Pathways

As discussed earlier in the chapter, TGFβ has many diverse biological functions. These diverse effects probably result from activation of multiple signaling pathways in response to TGFβ and the interactions of these pathways with signaling pathways initiated from other extracellular signals. In an attempt to dissect the TGFβ signaling pathways, researchers have focused on certain effects of TGFβ on cells grown in culture. One of these effects is the growth inhibitory effect of TGFβ on many cell types. Upon treatment with TGFβ, many types of cells undergo a reversible G1 cell-cycle arrest. This inhibition of cell-cycle progression is thought to be the molecular basis for many of the more global effects of TGFβ in the whole organism, including immune response modulation, developmental regulation, and cell growth homeostasis. Another principal effect of TGFβ on which research has focused is its ability to induce the expression of many genes of diverse functions. A subset of these TGFβ-inducible genes encode proteins that are essential in many physiological and pathological processes, including the formation of extracellular matrix, wound healing, and fibrosis. As described in the next several sections, the measurement of TGFβ-mediated growth inhibitory response and gene activation response have served as convenient and useful tools in dissecting and elucidating the TGFβ signal-transduction pathways in mammalian cells.

B. The TGFβ Signaling Receptors

As discussed in Section IV, the TGFβ receptors were originally classified by ligand-binding affinity labeling assays. These assays did not, however, allow the identification of those receptors responsible for mediating the biological effects of TGFβ, specifically, those most commonly studied in cultured cells, growth inhibition and induction of extracellular matrix genes. Initially, several lines of evidence suggested that the type I and type II receptors were the "signaling" receptors for TGFβ. First, the biological effects of TGFβ can be achieved by addition of concentrations of TGFβ corresponding to the known binding constants for the type I and type II receptors (Roberts and Sporn, 1990). Second, these two proteins are ubiquitously expressed in all cell types which are responsive to the known effects of TGFβ, in comparison to the other TGFβ receptors which have a more limited distribution of expression. Conversely, cell lines which are not growth inhibited by TGFβ, such as several tumor cell lines including retinoblastoma, pheochromocytoma, neuroblastoma, and breast carcinoma cells, often lack detectable expression of either one or both of these receptors.

More direct evidence for the involvement of the type I and type II TGFβ receptors in signal transduction came from a series of studies in which cell lines resistant to the growth inhibitory and gene activation effects of TGFβ were created (Boyd and Massague, 1989; Laiho et al., 1990b, 1991b). These studies were performed on mink lung epithelial (CCL64) cells, a cell line which expresses the type I, type II, and type III TGFβ receptors and whose growth is markedly inhibited by TGFβ. In these experiments, mink lung epithelial cells were first mutagenized using EMS treatment and then grown in the presence of high concentrations of TGFβ. Individual clonal cell lines were expanded and tested for responsiveness to TGFβ and expression of the cell-surface receptors. In theory, those clones which grew in the presence of TGFβ have been mutagenized to inactivate the TGFβ signaling pathway. Some of these mutants should contain mutations at the level of the signaling receptors. Presumably, those receptors which are selectively lost in the mutant cells are required for mediation of TGFβ growth inhibition.

Using this approach, three classes of TGFβ resistant mutant cell lines were generated as defined by the pattern of cell surface receptor expression: cells lacking detectable levels of the type I receptor (R mutants), cells lacking detectable levels of both the type I and type II receptors (DR mutants), and cells with normal receptor levels and composition (S mutants) (Boyd and Massague, 1989; Laiho et al., 1990b, 1991b). In addition to being resistant to the growth inhibitory action of TGFβ, all of the mutant cell lines isolated also lost responsiveness to TGFβ-mediated morphological changes and induction of fibronectin. An intriguing finding was that all of the mutant cell lines that display decreased or diminished levels of the type II receptor were also defective in the detectability of the type I receptor. The frequency of occurrence of those double receptor mutants, however, was equivalent to that for the single receptor R mutants, suggesting a peculiar relationship between the two types of receptors (Laiho et al., 1990b).

To study this phenomenon, somatic hybrids between R (containing no detectable type I receptor) and DR (containing no detectable type II or type I receptor) mutants were generated. Further analysis demonstrated that these somatic cell hybrids expressed normal levels of both type I and type II receptors and were fully responsive to TGFβ (Laiho et al., 1991b). This suggests that the DR mutants contain a single mutation in the type II TGFβ receptor, but retain a functional type I receptor undetectable by TGFβ cross-linking in the absence of a type II receptor.

From these experiments, the following conclusions were drawn: The type I and type II TGFβ receptors are "signaling receptors." The absence of either the type I or the type II receptor abolishes TGFβ-mediated signal transduction for growth inhibitory and gene response pathways. In the absence of the type II receptor, type I receptor is incapable of interacting with the ligand. Based on these findings, the research into TGFβ signal transduction began to focus on the type I and type II signaling receptors.

C. The Mechanism of TGFβ Receptor Activation

Characterization of the Functional Properties of the TGFβ Signaling Receptors

It was only after the cloning the type I and type II receptors that progress could be made in understanding the mechanism through which these receptors mediate the biological effects of TGFβ. As discussed in Section IV, the type II receptor was cloned by expression cloning and found to be a transmembrane receptor serine-threonine kinase. In an *in vitro* assay, the cytoplasmic kinase domain of the type II receptor was found to undergo autophosphorylation on serine and threonine residues (Lin et al., 1992). The functional importance of the type II receptor in TGFβ signaling was confirmed using the TGFβ resistant DR and R mutant mink lung cell lines (Wrana et al., 1992). When introduced into DR mutant cell lines, the TGFβ type II receptor cDNA restored both growth inhibitory and gene activation responses to TGFβ. Furthermore, the kinase activity of the type II receptor was demonstrated to be essential for signal transduction, because a mutant type II receptor with a lysine 277 to arginine (K277R) substitution in the ATP binding domain, which eliminates its ability to phosphorylate its targets, was unable to restore TGFβ responsiveness to the DR mutant cell lines (Wrana et al., 1992). Confirming the earlier findings from the somatic hybrids studies, reintroducing the type II receptor into DR cell lines also restores the ability of the endogenous type I receptor to interact with ligand. However, reintroduction of the type II receptor cDNA into the R mutant cell lines did not complement the defective TGFβ signaling pathways (Wrana et al., 1992). These results support the hypothesis that the type II or type I receptors alone are insufficient to mediate the biological effects of TGFβ. Furthermore, interaction of the type I receptor with ligand is dependent on the presence of the type II receptor (Wrana et al., 1992).

Subsequent studies involving mutagenesis of the type II receptor have further confirmed the importance of the receptor kinase activity and the interaction of the type I and type II receptors in TGFβ-mediated signal transduction. In these experiments, truncated mutants of the type II receptor were created which contained an intact extracellular TGFβ-binding domain, but lacked the intracellular regions responsible for the kinase activity of the receptor (Chen et al., 1993; Brand et al., 1993). These truncated versions of the type II receptor could presumably bind to TGFβ and interact with the wildtype endogenous type I receptor to form non-functional receptor complexes. In this way, these truncated type II receptors could act in a dominant negative fashion to antagonize the normal function of the endogenous type II TGFβ receptor by competitively forming complexes with the type I receptor. Interestingly, in these studies, the expression of a truncated type II receptor resulted in the reduced response of cells to the TGFβ growth inhibitory signal, but the gene activation response remained normal (Chen et al., 1993). Based on those results, it was argued that the type II receptor was responsible for mediating TGFβ signals to the growth inhibition pathway, whereas the type I receptor is responsible for the gene activation pathway. A more likely explanation, however, may be that the thresholds of initial signals generated by the receptor complex for the activation of these two distinct downstream pathways are significantly different: a much stronger signal is required for initiating the signaling cascade for growth inhibition than that for gene activation. According to this model, sufficient numbers of endogenous type I receptors may still be available to form a functional signaling complex with the endogenous type II receptor initiating the gene activation cascade, even in the presence of abundant nonfunctional receptor complexes consisting of the type I receptor and the truncated type II receptor. Therefore, both type I and type II receptors are definitely required for activating either growth inhibitory or gene activation pathways. This notion is consistent with the previously discussed phenomenon in which the type II receptor kinase null mutant failed to restore any TGFβ responses in DR mutant cells even though the interaction of ligands to the endogenous type I receptor has been fully recovered (Wrana et al., 1992).

More extensive mutagenesis of the type II receptor has shown that the cytoplasmic tail region and the kinase insert 1, a region close to the catalytic center of the kinase, are not required for signaling to the growth inhibitory or gene activation pathways (Wieser et al., 1994). This contrasts with receptor tyrosine kinases which require these conserved regions for signaling. In addition, the kinase insert 2, a region between the kinase subdomains X and XI, which contains no sites for autophosphorylation, is essential for the kinase autophosphorylation activity and the signaling properties of the type II receptor. This region can be interchanged between the activin and TGFβ type II receptors without affecting the functions of either receptor. This suggests some functional similarities between the two members of the type II receptor serine-threonine kinase family (Wieser et al., 1994). It is not known, however, whether this insert region is required for structural integrity of the kinases or is the binding domain for a common substrate/effector.

A similar approach has been taken in studying the function of the type I TGFβ receptor; a combination of functional complementation studies in cell lines lacking a type I receptor and site specific mutagenesis to identify its important functional domains. As discussed in Section IV, when reintroduced into mink lung epithelial cell R mutants (lacking a functional type I receptor), the R4/ALK5 type I TGFβ receptor cDNA is capable of restoring growth inhibition and induction of specific extracellular matrix genes in response to TGFβ (Franzen et al., 1993; Bassing et al., 1994a, 1994b; Carcamo et al., 1994; ten Dijke et al., 1994). To address the question of whether the kinase activity of the type I receptor is also required for signaling, kinase negative mutants of the type I receptor (K230R) were created. As was the case with the type II receptor kinase, the kinase negative receptor mutant failed to restore any TGFβ responses in R mutant cells (Bassing et al., 1994a, 1994b; Carcamo et al., 1994). These functional assays firmly confirm the identity of this cDNA as a type I receptor for TGFβ and formally establish the model that both type I and type II receptors are necessary for TGFβ-mediated signal transduction. In addition, these studies demonstrate that the kinase activity of the type I receptor is required for TGFβ-mediated signal transduction.

Functional Interaction between Type I and Type II Receptors

Significant progress has recently been made in understanding the nature of functional interactions between type I and type II TGFβ receptors (Wrana et al., 1992, 1994; Moustakas et al., 1993; Henis et al., 1994; Yamashita et al., 1994). The type II receptors are thought to be present on the cell surface as homo-oligomers (Henis et al., 1994). In the absence of TGFβ, the type II receptor is constitutively phosphorylated, a feature contrasting with the receptor tyrosine kinases for which autophosphorylation is regulated by ligand binding (Wrana et al., 1994). The phosphorylation of the type II receptor is partly due to the autokinase activity of the type II receptor in these oligomeric complexes. This is demonstrated by kinase negative mutant type II receptors which are incapable of autophosphorylation and do not display several phosphorylation sites present in the wild-type receptor. When tested in DR mutant cells which do not express the endogenous type II receptor, these kinase negative mutant receptors still retained several sites of phosphorylation, presumably due to phosphorylation by other unidentified cellular kinases. Furthermore, TGFβ treatment of cells does not alter this basal level of phosphorylation of the type II receptor, and its association with the type I receptor does not alter the phosphorylation status of the type II receptor (Wrana et al., 1994).

In contrast, the type I receptor is not phosphorylated in the absence of TGFβ (Wrana et al., 1994). Binding of TGFβ to the type II receptor extracellular domain results in the recruitment of type I receptor into the complex with the type II receptor, forming a ternary receptor-ligand complex (Wrana et al., 1994; Yamashita et al., 1994). Consequently, the type I receptor becomes phosphorylated on serine and threonine in a ratio of three to one, presumably by the type II receptor kinase, because this phosphorylation is dependent on the kinase activity of the type II receptor but not on the type I receptor. The phosphorylation of the type I TGFβ receptor occurs in the GS domain, a region between the kinase domain and the

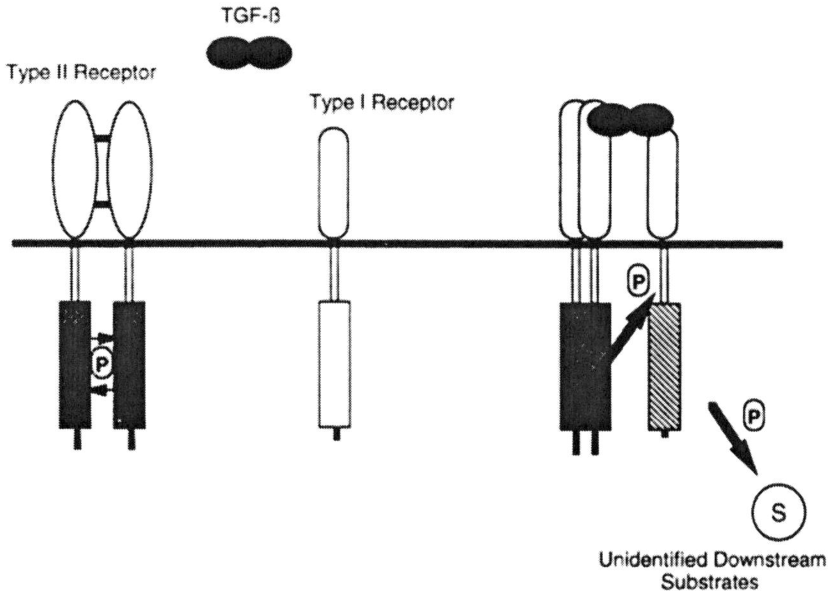

Figure 2. Proposed model for the mechanism of activation of the TGFβ receptor (Wrana et al., 1994).

transmembrane domain which is highly conserved in all type I receptor family members. Mutation of specific serine and threonine residues in this 20 amino acid region indicate that a majority, if not all, of the phosphorylation by the type II receptor occurs in this GS domain. More importantly, elimination of several phosphorylation sites in the GS domain results in a type I receptor mutant which fails to transmit TGFβ growth inhibitory and gene activation signals (Wrana et al., 1994). This suggests that phosphorylation in the GS region of the type I receptor by the type II receptor upon TGFβ treatment is a critical step in signal transduction.

Based on these results, a working model for TGFβ signal transduction across the membrane has emerged (Figure 2). In the absence of TGFβ, the type II TGFβ receptors exist as homo-oligomers on the cell surface. The type II receptor kinase is at a constitutively active state which undergoes autophosphorylation and is phosphorylated by other intracellular kinases. The phosphorylation of the type II receptor may be necessary for maintaining a proper conformation or specific interactions with other cellular components required for subsequent ligand-binding-induced complex formation. In contrast, the type I receptor kinase exists in an inactive and unphosphorylated state in the absence of TGFβ. It is not clear, however, whether the type I receptors interact with each other to form homo-oligomers or exist as single units on the cell surface under such a condition. Upon addition of

TGFβ, a complex consisting of the type I receptor, type II receptor, and the ligand is formed. After being recruited into the signaling complex, the type I receptor is phosphorylated by the type II receptor on the GS domain, activating the type I receptor kinase. The activated type I kinase can subsequently phosphorylate or interact with specific substrates/effectors, sending signals to the downstream pathways. It is still unknown whether the type II receptor could send a separate signal by phosphorylating distinct substrates once the receptor complex is formed. Alternatively, the type II kinase may act coordinately with the type I kinase to phosphorylate a common substrate.

D. The Type III Receptor/Betaglycan Presents TGFβ to the Type II Receptor

The type III TGFβ receptor, also known as betaglycan, falls into a general category of receptors which are not directly involved in signal transduction, but are unique and important in facilitating ligand-receptor interaction. This category of receptors generally does not contain cytoplasmic signaling motifs, tend to bind to ligand with low affinities, or exist in high abundance. In many cases, the presence of these receptors may not be essential for ligand-binding-initiated signal transduction, as indicated by the observations that several cell lines are fully responsive to TGFβ but lack the expression of the type III receptor. The type III receptor, however, has been demonstrated to participate in the recognition of various TGFβ isoforms by the signaling type II receptor.

The type II receptor can exist in both a high-affinity and a low-affinity TGFβ-binding state. In the absence of the type III receptor, which binds all three TGFβ isoforms with equal affinity, only 10% of the cell-surface type II receptors can interact with the TGFβ1 isoform with high affinity in the range of physiological responses (50 pM TGFβ1) (Lopez–Casillas et al., 1993). The percentage of high-affinity receptors for the TGFβ2 isoform is even lower. The ability of the type II receptor to bind to the different isoforms of TGFβ in the absence of the type III receptor correlates with the ability of cells to respond to these different isoforms (Lopez–Casillas et al., 1993). For instance, cell lines lacking the type III receptor are significantly growth-inhibited by TGFβ1 and β3 at concentrations of 50 pM but are much less affected by TGFβ2. When the type III receptor is reintroduced into these cells lacking endogenous expression, the discrepancies between the biological potency of different TGFβ isoforms are eliminated (Lopez–Casillas et al., 1993). With the presence of the type III receptor, nearly all of the type II receptors exist in a high-affinity state for both TGFβ1 and β2; and, consequently, those cells are now growth inhibited with equal potency by both TGFβ isoforms. This change in binding affinity is believed to result from the direct interaction of the type III receptor-ligand complex with the type II receptor. The formation of this ternary complex is dependent on the presence of ligand and independent of the type I TGFβ receptor. This notion is supported by the evidence that mutants of the type

III receptor, which are unable to bind TGFβ, do not augment the binding of ligand to the type II receptor (Lopez–Casillas et al., 1993).

Based on these results, a working model for the functional role of the type III receptor has been proposed: upon binding to the TGFβ ligand, the type III receptor forms a transient complex with the type II receptor, which allows a high-affinity interaction between the type II receptor and the ligand. The TGFβ bound type II receptor, then, dissociates from the type III receptor and simultaneously recruits the type I receptor to form an active signaling complex. In this mode, the equal binding affinity of the type III receptor to the three isoforms of TGFβ ligand is translated into equal potency of the three ligands for the signaling type II receptor (Lopez–Casillas et al., 1993).

In addition to its role in presenting TGFβ ligands to the signaling receptor, the type III receptor may also antagonize the effects of TGFβ (Lopez–Casillas et al., 1994). The extracellular domain of the type III receptor can be proteolytically released from the cell membrane and shed as a soluble molecule (Andres et al., 1989). This soluble form of the type III receptor may serve as a binding protein for TGFβ and block its association with the surface receptors, thus, preventing TGFβ from exerting its biological effects. Although small amounts of soluble type III receptor can be detected in serum and in the extracellular matrix (Andres et al., 1989), it must be determined if this processing of the type III receptor to a soluble form is a mechanism of regulation of TGFβ function *in vivo*.

The type III receptor may play roles other than ligand-presenting as discussed above. While the serine-rich cytoplasmic domain of the type III receptor does not contain any known catalytic motifs, it is highly conserved among different species analyzed so far. Furthermore, this domain shares a high degree of sequential similarity with the corresponding domain of another TGFβ binding protein, endoglin. It is possible, therefore, that the cytoplasmic domain could associate with still unidentified cellular components which may contain catalytic capabilities. Examples in supporting this model already exist, as in the case of integrin molecules and receptors for certain cytokines (Wang et al., 1991). Although the interactions between the type III receptor with those proteins may not be directly involved in TGFβ signal transduction, they may modulate the ultimate outcome of TGFβ effects. Alternatively, the type III receptor and its associated catalytic partners may transduce TGFβ signals to elicit certain specific biological effects other than the ones most commonly measured.

E. Identification of Receptor Interacting Proteins

Because phosphorylation of the type I TGFβ receptor by the type II receptor is essential for TGFβ signal transduction, it has been proposed that the type I receptor is responsible for transmitting signals to intracellular signaling molecules (Wrana et al., 1994). Consequently, identification of proteins, with which the type I TGFβ receptor interacts, is critical in elucidating the intracellular signaling cascades activated by TGFβ. To identify such proteins, a two-hybrid screening system in yeast was established to screen a rat heart cDNA library, using the intracellular domain of one of the receptor transmembrane serine-threonine kinases, R1/Tsk-7L, as bait (Wang et al., 1994). The cDNA for a FK506 binding protein, FKBP12, was repeatedly isolated based on the ability of the FKBP12 protein to directly bind the R1 receptor kinase domain.

FKBP12 is a small, ubiquitously expressed protein that serves as a cellular receptor for the immunosuppressive drugs FK506 and rapamycin (Schreiber, 1991; Heitman et al., 1992; Fruman et al., 1994). FKBP12 contains an intrinsic peptidyl-prolyl *cis/trans* isomerase (PPI) activity that is inhibited by the binding of either FK506 or rapamycin. The immunosuppressive effects of these drugs, however, are independent of their ability to block the PPI activity of FKBP12. The effects of rapamycin and FK506 derive from the ability of the drug-FKBP12 complexes to interfere with the signaling cascades essential for the antigen-triggered stimulation of T-lymphocyte proliferation. FK506-FKBP12 prevents the activation of quiescent T-lymphocytes by binding and inactivating calcineurin, a serine-threonine phosphatase (Liu et al., 1991). The activity of calcineurin is required to activate genes that are essential for T-lymphocytes to reenter the cell cycle. The target of rapamycin-FKBP12 is a large protein which is highly homologous to the 110-kDa catalytic subunit of PI-3 kinase (Kunz et al., 1993; Brown et al., 1994; Sabatini et al., 1994). Although the immediate effect that rapamycin-FKBP12 has upon binding this protein is unknown, rapamycin causes a G1 arrest in other mammalian cells by inactivating cyclin-dependent kinases required for the G1 to S transition (Albers et al., 1993; Jayaraman and Marks, 1993; Morice et al., 1993a, 1993b).

In the same two-hybrid screening, it was demonstrated that FKBP12 also interacts with the cytoplasmic domains of the type I receptors for activin (R2), BMP (R3), and TGFβ (R4) (Wang et al., 1994). FKBP12 failed to interact with the cytoplasmic domain of the kinase-defective R4 type I receptor, suggesting that the binding of FKBP12 to the receptor may be regulated by the phosphorylation state of the receptor. The interaction between FKBP12 and the cytoplasmic domain of the type I TGFβ receptor could be disrupted by the addition of FK506 and rapamycin (Wang et al., 1994; Bassing et al., unpublished results). A FKBP12 mutant incapable of binding FK506 also failed to bind to the type I TGFβ receptor (Wang et al., 1994). In addition, FKBP12 mutants, containing point mutations in residues located on the surface of FKBP12 proximal to the active site, all show reduced abilities to bind the cytoplasmic domain of the type I TGFβ receptor. A PPI-defective mutant of FKBP12 also exhibits decreased binding of the type I TGFβ receptor (Bassing et al., unpublished results). Taken together, these results suggest that the cytoplasmic domain of the type I TGFβ receptor binds to FKBP12, specifically, within the active site and along the proximal surface of the molecule.

It has been suggested that endogenous ligands may exist which bind FKBP12 and mimic the effects of FK506 and rapamycin. Because the type I TGFβ receptor specifically interacts with FKBP12 and appears to bind along the drug binding pocket of FKBP12, it was proposed that the type I TGFβ receptor may be an endogenous ligand of FKBP12 (Wang et al., 1994). An *in vivo* interaction between FKBP12 and the type I TGFβ receptor has been observed in mink lung epithelial cells (J. Wrana and J. Massague, personal communication). The exact role of FKBP-12 in TGFβ receptor complex mediated signaling remains unclear. FKBP12 may serve a common function in its interaction with members of the type I serine-threonine kinase receptor family. Because the isomerase activity of FKBP12 appears required for this interaction, however, FKBP12 may not be directly involved in signal transduction but rather may act to aid in folding these receptors into their proper conformation. Alternatively, FKBP12 may serve a function similar to that recently demonstrated for the 14-3-3 protein family by bringing relevant intracellular components into close association or by stabilizing existing complexes consisting of multiple components for signaling. Therefore, current research in this area is focused on the possibility that the interaction between FKBP12 and the type I TGFβ receptor may result in the functional association of other molecules, such as calcineurin, with the type I receptor.

Although the yeast two-hybrid system has yielded only limited success in identifying receptor-interacting proteins, it remains until now the only successful approach. There are, however, several shortcomings of this approach. The most significant is the difficulty in identifying interacting proteins truly involved in mediating signals derived from the TGFβ receptor complex. It is very difficult to recapitulate, in yeast, the physiological condition under which the type I receptor kinase is activated through its physical interaction with and covalent modification by the type II receptor. It is possible that the effectors/substrates for TGFβ receptor complex can associate only with the type I receptor under these conditions. Therefore, the development of a system for identifying downstream targets of the TGFβ signaling receptor complex will have to take into account these complex considerations.

F. Effects of TGFβ on the Cell-Cycle Machinery

One of the most important effects of TGFβ is its inhibitory effect on cell proliferation. The anti-proliferative effects of TGFβ are thought to be the molecular basis of many of its broader biological effects in tissues, organs, and whole animals. Many cell types undergo a reversible G1 cell-cycle arrest when treated with TGFβ (Lyons and Moses, 1990; Massague, 1990; Roberts and Sporn, 1990). It has been demonstrated that cell-cycle progression through the G1 phase is dependent on the sequential formation, activation, and subsequent inactivation of the G1 cyclins and cyclin-dependent kinases, cyclins D1, D2, D3, and E and their kinase partners Cdk4, Cdk6, and Cdk2 (Xiong et al., 1991; Sherr, 1993; Ohtsubo and Roberts, 1993). The phosphorylation of cellular targets of these G1 cyclin-Cdk complexes is thought to be essential for entry into S phase (Sherr, 1993). One of these targets is the retinoblastoma susceptibility gene product Rb, which binds to and inactivates the transcription factor E2F when it is hypophosphorylated (Lin et al., 1991; Lees et al., 1991; Hinds et al., 1992; Kato et al., 1993; Dowdy et al., 1993; Hamel et al.,

1993). Preceding the G1 to S-transition, Rb becomes hyperphosphorylated and releases E2F (Buchkovich et al., 1989; Chen et al., 1989; Decapria et al., 1989; Mihara et al., 1989). E2F then activates the transcription of a number of genes including DNA polymerase-alpha, dihydrofolate reductase, thymidine kinase, and many others essential for entry into S phase (Hiebert et al., 1992; Zamanian and La Thangue, 1993; Hiebert, 1993). This is obviously a highly simplified model, because Rb is probably not the only target of the G1 cyclin/Cdk complexes and E2F is probably not the only target of Rb. Furthermore, both Rb and E2F are the prototype of each respective family which consists of multiple members sharing functional and structural similarities. This model does, however, set a conceptual framework for understanding the rationale for and the implications of the experiments that have defined the role of TGFβ in regulating cell cycle progression.

One of the first observed molecular effects of TGFβ on cell-cycle control was that, in TGFβ treated cells, Rb is held in a hypophosphorylated state, thus, preventing activation of the transcription factor, E2F (Laiho et al., 1990a). Several hours before the G1-S transition, Rb becomes hyperphosphorylated and TGFβ can no longer have an effect on cell cycle progression. The phosphorylation of Rb, therefore, represents a commitment point for cell entry into S phase, after which TGFβ cannot halt cell cycle progression (Laiho et al., 1990a). This notion is further supported by the observation that Rb-inactivating viral oncoproteins, which bind to Rb and disrupt Rb-E2F interaction, could prevent TGFβ from inducing cell-cycle arrest (Laiho et al., 1990a). Recent evidence suggests that, if E2F is overexpressed to a level which alleviates its regulation by Rb, TGFβ can no longer block cell-cycle progression (Schwarz et al., 1995). Based on these results, it is believed that the regulation of Rb and, consequently, the transcription factor E2F by the TGFβ signal is critical in TGFβ-induced growth inhibition.

The inhibition of Rb phosphorylation by TGFβ is thought to be mediated through its regulatory effects on the G1 cyclin-Cdk complexes. Several groups have shown that TGFβ can decrease the levels of several G1 cyclins and Cdks (Ewen et al., 1993; Geng and Weinberg, 1993; Koff et al., 1993). It has been reported that TGFβ can cause moderate decreases in the levels of cyclin E and Cdk2 in human keratinocytes. This is, however, dependent on the specific cell lines studied and the conditions of TGFβ treatment (Koff et al., 1993; Geng and Weinberg, 1993). In mink lung epithelial cells, TGFβ causes a marked decrease in the levels of Cdk4 (Ewen et al., 1993). Overexpression of Cdk4 in these cells blocks the ability of TGFβ to cause cell-cycle arrest, which correlates with the loss of TGFβ's effect on the phosphorylation state of Rb (Ewen et al., 1993). These results suggest that the growth inhibitory effects of TGFβ may be partly mediated by a downregulation of the Cdk4 level. The specific mechanism through which TGFβ decreases the level of Cdk4 remains unknown. It is unlikely that TGFβ regulates Cdk-4 expression at the transcriptional level, because several groups have reported that only minor decreases in the level of Cdk-4 mRNA upon TGFβ treatment could be observed (Ewen et al., 1993; Geng and Weinberg, 1993). It is possible that TGFβ acts posttranscriptionally to affect either the translational rate or the stability of the Cdk-4 protein.

Although TGFβ does not always alter the levels of the other G1 cyclins and Cdks, it is capable of inhibiting the activities of those G1 cyclin-Cdk complexes. Specifically, it has been shown that TGFβ causes a significant decrease in the ability of cyclin E-Cdk2 complexes to phosphorylate histone H1 in mink lung epithelial cells (Koff et al., 1993). Overexpression of Cdk2, however, does not elevate the activity of Cdk2 kinase or block TGFβ-mediated growth arrest (Ewen et al., 1993). Phosphorylation of Cdks on critical threonine residues has been shown to be essential for their activation and association with their cyclin partners. Cdk-2 is normally phosphorylated on threonine-160 by the cyclin-activating kinases in late G1 (Fisher and Morgan, 1994; Makela et al., 1994). Upon TGFβ treatment, the phosphorylation of Cdk2 is inhibited and, consequently, it is no longer able to associate with cyclin E (Koff et al., 1993). Therefore, TGFβ may render Cdk2 inactive by preventing its phosphorylation (Polyak et al., 1994a, 1994b). It is not known whether this inhibition of Cdk2 phosphorylation results from either direct or indirect downregulation of the activities of the cyclin-activating kinase, MO15/Cdk7 and its partner, cyclin H.

Recently, research on the mechanism of TGFβ regulation of cell cycle progression has focused on the ability of TGFβ to induce several proteins which can directly bind to and inhibit the activities of cyclin-Cdk complexes. These cyclin-dependent kinase inhibitors, CdkIs, consist of a family of low molecular weight proteins which include p21 (WAF1/Cip1), p27 (Kip1), p16 (INK4/MTS1), and p15 (INK4B/MTS2) (Hunter, 1993; Peter and Herskowitz, 1994; Peters, 1994; Marx, 1994; Nasmyth and Hunt, 1994). All of these proteins contain conserved domains that are involved in their interactions with and inhibition of Cdk complexes.

p21 was the first of these inhibitors identified as a component of the quaternary cyclin D, Cdk, PCNA complex (Xiong et al., 1992). Subsequent cloning and characterization of p21 indicated that it is under the transcriptional control of the tumor suppressor gene, p53 (El–Deiry et al., 1993), is induced by γ-irradiation (Dulic et al., 1994), and acts as an inhibitor of multiple cyclin-cdk complexes, including the G1 complexes cyclin E-Cdk2 and cyclin D-Cdk4 (Harper et al., 1993; El–Deiry et al., 1993; Xiong et al., 1993; Gu et al., 1993). The p21 levels were found high in cells that are senescent, and overexpression of p21 leads to a decreased cell entry into S phase (Hunter, 1993). In addition to inhibiting the Cdk kinase activity, p21 is able to directly inhibit DNA synthesis through its interaction with PCNA, a component of the DNA replication machinery (Waga et al., 1994). In the context of DNA damage-induced, cell-cycle arrest, the level of p21 is significantly elevated, resulting in the increased association of p21 with cyclin E-Cdk2 complexes and inhibition of kinase activity (Dulic et al., 1994; El–Deiry et al., 1994). This induction of p21 can be blocked by the papilloma virus E6 oncoprotein which binds to and inactivates p53. Correlated with this blockage of p21 induction, E6 expression inhibits the DNA damage-induced cell-cycle arrest, suggesting that p21 induction is important for the mediation of cell-cycle arrest initiated by DNA damage (Dulic et al., 1994).

The functional characteristics of p21 as a potent inhibitor of cell-cycle progression led to studies to determine if it could serve as a downstream effector for the TGFβ growth inhibitory signal. The study of TGFβ's effect on p21 was performed in HaCaT, a cell line of spontaneously immortalized human keratinocytes which can be growth-arrested by TGFβ in G1 (Baukamp et al., 1988; Geng and Weinberg, 1993). Upon TGFβ treatment, p21 mRNA steady-state levels are increased approximately sixfold (Datto et al., 1994). This increase in p21 mRNA is accompanied by a similar increase in p21 protein level, as well as an association of p21 with at least two of its targets in G1, cyclin D1 and Cdk2. Using a luciferase reporter gene under the transcriptional control of the p21 promoter region, it was determined that TGFβ induces p21 expression by activating transcription from the p21 promoter. It has previously been shown that p53 can also transcriptionally activate the p21 promoter. However, the elements in the promoter responsible for the transcriptional activation by p53 and TGFβ are distinct. A further characterization of the p21 promoter through the identification of the promoter elements and the *trans* acting factors responsible for TGFβ-mediated transcriptional activation should lead to elucidation of the mechanism by which the induction of p21 occurs.

Similar to p21, p15 is also a member of the Cdk inhibitor family whose activity is regulated by TGFβ (Hannon and Beach, 1994). p15 was originally identified as a homologue of p16, another member of the CdkI family, by its high degree of sequential similarity in the analysis of the genomic locus of p16. p16 was found to be homozygously deleted in a high frequency in a wide variety of human tumor-derived cell lines (Kamb et al., 1994a; Nobori et al., 1994) and is also mutated or deleted in several specific types of primary tumors (Caldas et al., 1994; Hussussian et al., 1994; Kamb et al., 1994b; Mori et al., 1994; Spruck et al., 1994). p15 was subsequently cloned from a cDNA library of TGFβ treated HaCaT human keratinocytes (Hannon and Beach, 1994). TGFβ treatment of HaCaT cells results in a thirtyfold increase in the levels of p15 mRNA. Similarly to the p21 induction, this increase in p15 mRNA leads to an increase in p15 protein and an association of p15 with cyclin D-Cdk6 and cyclin D-Cdk4 complexes (Hannon and Beach, 1994). *In vitro*, p15 can inhibit the kinase activity of these cyclin-Cdk complexes. It remains to be determined if this induction of p15 is sufficient to block the kinase activity of these complexes *in vivo*. It is likely that the induction of both p21 and p15 will contribute to the inhibition of these complexes in TGFβ-mediated cell-cycle arrest. Although the mechanism through which TGFβ causes an increase in p15 mRNA remains unknown, it is possible that p15, like p21, is transcriptionally activated from its promoter. These two genes may be activated by a common factor in response to TGFβ through a common element in their respective promoters. Alternatively, TGFβ may activate the two genes through two distinct pathways by modulating the activities of different transcription factors.

p27 was the first member of the CdkI family shown to play a role in TGFβ-mediated cell-cycle arrest by its ability to associate with and block the activity of cyclin E-Cdk2 complexes in mink lung epithelial cells (Koff et al., 1993; Polyak et al., 1994a). p27 has since been cloned by using the two-hybrid system with cyclin D1 as bait and by microsequencing peptides derived from purified protein (Toyoshima and Hunter, 1994; Polyak et al., 1994b). Under *in vitro* assaying conditions, p27 is able to associate with cyclins D1, D2, D3, and more weakly to cyclin E, as well as with Cdk2, Cdk4 and Cdk6 (Polyak et al., 1994b). It can inhibit the kinase activities of multiple cyclin-Cdk complexes, including cyclin D1-Cdk4, cyclin A-Cdk2, cyclin E-Cdk2, and cyclin B1-Cdc2 (Toyoshima and Hunter, 1994; Polyak et al., 1994b). In addition, p27 is capable of blocking the activation of Cdk2 in *in vitro* kinase activation assays (Toyoshima and Hunter, 1994; Kato et al., 1994). When overexpressed, p27 can inhibit entry into the S-phase (Toyoshima and Hunter, 1994). Unlike p21 and p15, however, TGFβ does not affect the activity of p27 through transcriptional regulation (Polyak et al., 1994a, 1994b). In fact, there is little fluctuation in p27 level throughout the cell cycle (Polyak et al., 1994b).

Based on these studies, a model for the mechanism of p27 activation by TGFβ has been proposed (Peters, 1994; Polyak et al., 1994b). In untreated mink lung epithelial cells, p27 is largely associated with cyclin D-Cdk4 complexes. Upon treatment of TGFβ, the level of Cdk4 decreases leading to a reduced number of cyclin D1-Cdk4 complexes. This, in turn, results in an increase in the amount of "free" intracellular p27 which can subsequently interact with cyclin E-Cdk2 complexes. This interaction is thought to block the kinase activity of this complex and inhibit the phosphorylation of Cdk-2 by cyclin-activating kinases. Thus, the regulation of cyclin E-Cdk2 complex activity by p27 in response to TGFβ is achieved through subtle changes in the level of "free" p27 by decreasing the levels of other interacting proteins, especially, cyclin D1-Cdk4 complexes. Although this model may explain the TGFβ-induced activation of p27 in mink lung epithelial cells, it is not supported by evidence derived from studies in HaCaT cells. As shown in the studies of TGFβ-mediated induction of p15, no significant changes in cyclin D-Cdk4 complex levels were observed upon TGFβ treatment (Hannon and Beach, 1994). It is possible that the level of "free" p27 is modulated by fluctuations in the levels of other interacting components in response to TGFβ. Alternatively, the activity of p27 could be regulated by TGFβ through other mechanisms, such as posttranslational modifications.

The evidence discussed above clearly supports the notion that TGFβ affects cell-cycle progression mainly by inducing the activities of multiple cyclin-dependent kinase inhibitors through at least two different mechanisms. These inhibitors, p21, p27, and p15, may serve both unique and overlapping functions in inhibiting the activities or the activation of the G1 cyclin-Cdk complexes in response to TGFβ. Specific inhibitors may function preferentially to inhibit specific complexes. Thus, the combined action of multiple inhibitors may be necessary to completely inhibit the progression of cells into the S-phase. Conversely, each inhibitor alone may be sufficient to block the G1 cyclin-Cdk complexes, causing cell-cycle arrest. In this case, simultaneous induction of several inhibitors by TGFβ may have evolved evolutionarily to serve as a functional safeguard insuring a proper response to a growth inhibitory signal. It is also possible that different individual inhibitors act in different cell types or under different growth conditions to mediate the growth inhibitory effects of TGFβ. This notion is supported by the evidence that no induction of p21 is observed in mink lung epithelial cells, whereas no activation of p27 could be detected in HaCaT cells in response to TGFβ (Datto et al., unpublished results).

Previous studies on the mechanism of TGFβ-mediated growth inhibition have shown that TGFβ could downregulate the expression of a proto-oncogene, c-myc (Pietenpol et al., 1990; Munger et al., 1992). c-myc is a helix-loop-helix transcription factor which causes transformation when overexpressed in cooperation with other oncogenes (Land et al., 1983). In addition to its transforming activities, c-myc is believed essential for the normal cell-cycle progression through G1 into S-phase (Lewin, 1991; Hunter, 1991; 1993; Evan and Littlewood, 1993). In experimental systems in which the expression of c-myc is blocked by an antisense oligonucleotide, cells undergo a G1 cell-cycle arrest (Heikkila et al., 1987; Prochownik et al., 1988). Ectopic induction of c-myc activity could overcome the deficiency in mitogenic stimuli and drive the cells into proliferation (Eilers et al., 1991). These activities of c-myc are believed mediated through its association with its partner protein, max, and the subsequent binding of the myc/max complex in a sequence-specific manner to its target elements in various promoters (Blackwood et al., 1991; Ayer et al., 1993; Zervos et al., 1993; Amati et al., 1993). Although the consensus binding sequence for c-myc has been defined, specific target genes of the c-myc activity have yet to be identified.

TGFβ treatment of human keratinocytes results in a significant decrease in c-myc transcription and growth inhibition (Murphy et al., 1991; Munger et al., 1992). When transformed by the human papilloma virus -16 and 18 or the SV40 virus, cells become nonresponsive to the growth inhibitory effects of TGFβ and no longer have a decrease in the expression of c-myc (Pietenpol et al., 1990). Further studies indicate that expression of the viral oncoproteins E7, E1A, and SV40 large T-antigen blocks TGFβ-mediated growth inhibition and c-myc down-regulation (Pietenpol et al., 1990). Because the ability of TGFβ to cause growth inhibition is correlated with its ability to inhibit the expression of c-myc, it is believed that the c-myc down-regulation is important in TGFβ-mediated growth arrest. This effect is not observed when these viral oncoproteins are mutated in their Rb binding domains. These results suggest that Rb may play a role in this TGFβ-mediated inhibition of c-myc expression (Pietenpol et al., 1990).

More direct evidence for the involvement of Rb in the TGFβ-mediated c-myc down-regulation derived from the characterization of the c-myc promoter region. The elements in the c-myc promoter responsible for the TGFβ inhibitory effect were found to be identical to those mediating the Rb inhibitory effect on c-myc expression. Although several proteins have been shown to bind to this TGFβ–Rb-responsive element, the molecular identities of these factors remain unknown (Pietenpol et al., 1994). The potential involvement of Rb in the regulation of c-myc may become a focus of further study. As discussed earlier in the review, one of the functions of Rb is to sequester and inactivate the E2F family of transcription factors. Accumulating evidence suggests that the expression of c-myc may be controlled by a member of the E2F transcription family, as apparent by the presence of a E2F consensus-binding site in the c-myc promoter.

G. TGFβ Regulates the Expression of Genes with Diverse Functions

Parallel to the regulatory role of TGFβ on cell-cycle progression, TGFβ can induce the expression of many genes with diverse functions (Barnard et al., 1990). These include the genes for type I, type II, and type III collagens (Ignotz and Massague, 1986; Centrella et al., 1987; Fine and Goldstein, 1987), fibronectin (Ignotz and Massague, 1986; Laiho et al., 1991), thrombospondin (Laiho et al., 1991), elastin (Marigo et al., 1993), PDGF (Leof et al., 1986), plasminogen activator inhibitor (Keeton et al., 1991), tissue inhibitor of metalloproteinases (Edwards et al., 1987), IG-alpha constant gene (Lin and Stavnezer, 1992), TGFβ1 (Kim et al., 1990), junB (Laiho et al., 1991), and c-jun (Pertovaara et al., 1989). The TGFβ regulation of these multiple genes undoubtedly reflects the diverse effects of TGFβ on multiple signaling pathways involving a variety of transcription factors. The promoter regions of several of these genes have been isolated and their TGFβ-responsive elements have been identified. Some of these elements represent consensus sites for transcription factors which have been previously defined. Other TGFβ-responsive elements represent novel sites which are not consensus sites for previously described transcription factors.

The effect of TGFβ on several promoters has been partly attributed to its induction of the AP-1 activity. The AP-1 transcription complex is composed of two families of leucine-zipper transcription factors, c-fos and c-jun (Hunter, 1991). These two proteins are thought to play important roles in the growth response to mitogenic stimuli and in transformation under certain conditions, making them unusual candidates for TGFβ-inducible genes. Members of these two families of proteins can form heterodimers, and for some of them homodimers, which bind to and activate transcription from a consensus AP-1 site TGAg/cTCA (Lee et al., 1987). Furthermore, the binding specificity of this AP-1 complex can be significantly altered by its interactions with other transcription factors, as demonstrated by the role of AP-1 in regulating c-AMP-responsive promoter elements in association with NF-κB (Lewin, 1991). TGFβ has been found to increase the expression levels of junB, c-jun and c-fos (Pertovaara et al., 1989). As would be expected from the conflicting effects of TGFβ and AP-1 on cell proliferation, the activation event of JunB by TGFβ is clearly distinct from the ability of TGFβ to cause cell-cycle arrest. Cell lines which are transformed with the viral oncoprotein, SV40 large T-antigen, lose their responsiveness to the TGFβ growth inhibitory signal, but retain the response of JunB induction by TGFβ (Laiho et al., 1991). The TGFβ-mediated induction of the components of the AP-1 complex leads to an increase in AP-1 activity, which, in turn, could lead to the activation of several TGFβ target genes containing AP-1 sites in the 5′-region of their promoters. The genes that contain nearly consensus AP-1 sites include the TGFβ1 gene itself (Kim et al., 1990), the tissue inhibitor of metalloproteinase gene (Campbell et al., 1991), and the plasminogen activator inhibitor (PAI-1) gene (Keeton et al., 1991). These AP-1 consensus sites are required for their induction by TGFβ (Keeton et al., 1991). The role of AP-1 activity in TGFβ-mediated gene responses is also supported by the finding that introduction of an AP-1 site in a TGFβ-nonresponsive promoter can confer TGFβ responsiveness (Wrana et al., 1992). Therefore, even though the activation of AP-1 is probably not essential in TGFβ-mediated growth arrest, it may be important in regulating a diverse set of genes involved in the mediation of TGFβ effects other than cell-cycle control.

Several promoters have been identified that are responsive to TGFβ but do not contain AP-1-responsive elements. The promoter region of the alpha (I) collagen gene contains an NF-1 like binding site and a novel TGFβ activating element (Ritzenthaler et al., 1991, 1993). The factors interacting with this novel sequence have not been identified. Like the alpha 1 collagen promoter, the promoter of the germline Ig alpha constant region gene is also activated through a novel TGFβ-responsive element (Lin and Stavnezer, 1992). Further studies of these TGFβ regulatory elements will lead to identification of factors and pathways that mediate the diverse biological effects of TGFβ.

V. THE ROLE OF THE TGFβ SIGNALING PATHWAYS IN CARCINOGENESIS

Cell proliferation is normally controlled by the opposing actions of growth stimulatory and growth inhibitory signals. An alternation in the balance between these signals can lead to proliferation or growth arrest (Merlino, 1994). Oncogenesis results, at least partly, from a tipping of the balance between these opposing signals toward proliferation. Often this is due to an alternation or upregulation of multiple components of growth stimulatory signaling pathways. These normal cellular factors which are commonly altered or upregulated in oncogenesis are referred to as proto-oncogenes. Conversely, cells can be stimulated to grow by losing components of growth inhibitory pathways (Bishop, 1991). These factors which are commonly lost in oncogenesis are called tumor-suppressor genes. Because of the well-characterized roles that TGFβ may play in inhibiting normal cell proliferation, it is conceivable that the TGFβ ligand and components of the TGFβ signaling pathway may act as tumor-suppressor genes to regulate the proliferation of cells *in vivo*. Loss of TGFβ responsiveness, therefore, could lead to uncontrolled growth and may be important in tumor initiation or progression.

This hypothesis is supported by numerous studies which have shown a loss of TGFβ-mediated growth inhibition in many primary tumor cells and tumor cell lines, including T-cells (Kadin et al., 1994), squamous carcinoma cells (Shipley et al., 1986), retinoblastoma cells (Kimchi et al., 1988), breast cancer cells (Arteaga et al., 1988), colon carcinoma cells (Hoosein et al., 1989), and prostate adenocarcinoma cells (Murphy et al., 1991). It is important to point out, however, that many tumor cells still retain their responsiveness to the TGFβ growth inhibitory signal (Roberts and Sporn, 1990). This suggests that, although losing the ability to be growth inhibited by TGFβ may provide some advantage to proliferation *in vivo*, it is not required in tumorigenesis. Therefore, the observed alterations in TGFβ responsiveness in those tumor cell lines may not represent an initiating event in tumor formation, but a later event occurring during tumor progression and expansion. This hypothesis is supported by studies of colon carcinoma cell lines which showed that, as these cells progress to a more advanced tumorigenic stage, both *in vitro* and *in vivo*, they have a reduced response to the inhibitory effects of TGFβ (Hoosein et al., 1989; Manning et al., 1991; Hsu et al., 1994). This loss of TGFβ responsiveness in more advanced tumors has also been reported with T-cell malignancies (Kadin et al., 1994).

The loss of TGFβ responsiveness in many tumor cell lines has been, in part, attributed to a decrease in the expression of TGFβ receptor levels (Shipley et al., 1986; Arteaga et al., 1988; Kimchi et al., 1988; Norgaard et al., 1993; Kadin et al., 1994). In human gastric cancer cells, this decrease in the levels of TGFβ receptors results from mutations in the TGFβ receptor genes (Park et al., 1994). The cell lines examined in this study contain either normal, truncated, or no detectable type II TGFβ receptor mRNA (Park et al., 1994). Although the cell lines which do not express the type II receptor mRNA or express a truncated version of the type II receptor are nonresponsive to the growth inhibitory effects of TGFβ, those cell lines expressing a functional type II receptor respond normally to TGFβ. These results suggest that inactivation of the TGFβ signaling pathway can occur in tumor cells through mutations of the type II TGFβ receptor. Furthermore, three of the four cell lines examined in this study, which express truncated mutants of the type II receptor, were established from malignant ascites. This suggests that mutations in the receptor gene may have occurred late in tumor progression.

The role that this loss of the type II TGFβ receptor plays in tumorigenesis was studied by reintroducing a functional type II receptor into a type II receptor-deficient, breast cancer cell line MCF-7 and studying the tumorigenic properties of these cells in a nude mouse model system (Sun et al., 1994). The parental MCF-7 cells are not growth inhibited by TGFβ and express no detectable type I or type II TGFβ receptor. MCF-7 cells stably expressing the type II TGFβ receptor were generated by the reintroduction of a functional type II receptor cDNA. Both TGFβ binding to the type II receptor and to the endogenous type I receptor and the responsiveness to the TGFβ growth inhibitory effects were restored. These stable cell lines showed decreased ability to form colonies in soft agar and were growth arrested in a monolayer culture at a lower density in comparison to the parental MCF-7 cells. The stable cell lines also showed a significant reduction in tumorigenicity as well as a longer delay in tumor formation when introduced into ovariectomized, estrogen-supplemented nude mice. In addition, tumor growth in these mice was associated with the denovo loss of the exogenous type II receptor. These results confirm the hypothesis that a functional TGFβ growth inhibitory signaling pathway is important in regulating the growth of tumor cells *in vivo*. When this pathway is disrupted due to mutations in the TGFβ receptor, cells acquire a growth advantage. Restoring the TGFβ growth inhibitory responsiveness could dramatically reduce tumorigenicity, probably by the partial restoration of a delicate balance between the growth inhibitory and the growth stimulatory signal input.

Besides its role as a growth inhibitory factor, TGFβ could play much more complicated roles in oncogenesis. In fact, TGFβ was first described as a factor which could stimulate cell growth in soft agar, a property indicative of a transformed cell type. This effect is probably due to the ability of TGFβ to induce extracellular matrix deposition, which may provide a substrate on which cells could adhere and grow in these soft agar assays. *In vivo*, TGFβ has also been shown to promote tumor growth, possibly through its positive effect on the production of specific proteolytic enzymes important in metastasis and its negative effect on the function of the host's immune system (Welch et al., 1990; Arteaga et al., 1993). Thus, the paracrine action of TGFβ on extracellular matrix formation and immune response may aid in tumor progression and metastasis.

VI. HEREDITARY TELANGIECTASIA IS A RESULT OF MUTATIONS IN ENDOGLIN

The most direct link between the dysfunction of the TGFβ receptor system and a disease state has been recently discovered in the study of an inherited disorder, hereditary hemorrhagic telangiectasia (HHT) (McAllister et al., 1994). HHT is an autosomal dominant disorder characterized by multisystemic vascular dysplasia and recurrent hemorrhage. Patients with this disorder suffer from recurrent hemorrhage from vascular lesions, especially in the nasal mucosa and gastrointestinal tract. Additionally, this disease is characterized by telangiectases, dysplastic capillary beds consisting of dilated channels lined by a single layer of endothelium in the dermus, nasal mucosa, and gastrointestinal tract.

Genetic linkage for HHT has been established to markers on chromosome 9q33-q34 (McDonald et al., 1994; Shovlin et al., 1994). Because the chromosomal location of the TGFβ-binding protein endoglin is located in this region and endoglin expression is restricted in a few cell types, including endothelial cells, endoglin was studied as a candidate gene for HHT (McAllister et al., 1994). Subsequently, it was found that three individuals affected with HHT harbored the same mutated allele of the endoglin gene. All of these patients have mutations in the endoglin gene which would lead to a premature termination and the synthesis of a truncated form of endoglin protein.

The molecular mechanism underlying the development of HTT from a single germline mutant allele of the endoglin gene is still unknown. It is possible that having only a single copy of the endoglin gene decreases the ability of endothelial cells to respond appropriately to TGFβ, which could lead to abnormal vascular remodeling in response to injury, a characteristic of HHT. It is also possible that, although HHT is an autosomal dominant disorder, on the cellular level the single allele mutation in endoglin behaves as a recessive mutation. In this model which is similar to that for the tumor-suppressor genes, cells carrying a single functional copy of the endoglin gene behave normally. However, cells in the discrete lesions found in patients with HHT have reduced the heterozygosity in the endoglin locus to a homozygous state, often resulting from a second mutation or deletion in the single functional allele of endoglin gene. This hypothesis is supported by the findings that outside of the discrete regions of affected vessels, there is no evidence of abnormality in blood vessel structure (McAllister et al., 1994).

Not all patients with HHT, however, have a mutated allele of endoglin. It has been proposed that HHT may result from the loss of endoglin function and the dysfunction of other components of the TGFβ signaling pathway, especially the type III receptor/betaglycan which share structural similarities with endoglin (McAllister et al., 1994). Genetic linkage studies of families with normal endoglin genotype have identified an additional chromosomal locus linked to HHT. One of these families was linked to 3p22, a chromosomal region containing the type II TGFβ receptor (McAllister et al., 1994). Therefore, HHT may also develop from mutations in at least one other component of the TGFβ signaling pathway, the type II receptor.

VII. SUMMARY

Currently, research in the TGFβ field is at a very exciting stage. Within the past four years, the TGFβ signaling receptors have been cloned and characterized. Subsequent studies have begun to demonstrate how these transmembrane receptor serine-threonine kinases are activated by ligand to transduce a signal. Researchers have also made significant progress in understanding the molecular mechanisms through which TGFβ inhibits cell-cycle progression and activates the expression of a diverse set of TGFβ-inducible genes. In addition, researchers have begun to understand the roles of TGFβ and the TGFβ signaling receptors in the pathological processes of several diseases, including oncogenesis and HHT.

With all the progress discussed in this review, many important questions remain unanswered. The main challenge is to determine the molecular identities of substrates/effectors of the TGFβ receptor complex. Although one type I receptor associated protein, FKBP-12, has been identified, the functional consequences of this interaction remains uncertain. Currently, no known substrates of the type I or type II receptor kinases (other than the receptors themselves) have been identified as responsible for directly transmitting a signal from the receptor complex to the down-stream pathways. A better understanding of the TGFβ signaling pathways undoubtedly will come as more immediate targets of the TGFβ receptor kinases are characterized and the molecular links to the regulators of cell-cycle control and TGFβ-inducible genes are elucidated.

ACKNOWLEDGMENTS

We thank H. Symmonds for technical assistance and helpful discussions of the manuscript. This work was supported by Grant DK45746 from the National Institutes of Health and Grant 3613 from the Council for Tobacco Research to X.–F. Wang. M.B. Datto was supported by an M.S.T.P. grant from the National Institutes of Health, and C.H. Bassing was supported by a National Science Foundation Graduate Research Fellowship. X.–F. Wang is a Leukemia Society Scholar.

REFERENCES

Albers, M. W., Williams, R. T., Brown, E. J., Tanaka, A., Hall, F. L., & Schreiber, S. L. (1993). FKSP-rapamycin inhibits a cyclin-dependent kinase activity and cyclin D1-cdk association in early G1 of an osteosarcoma cell line. J. Biol. Chem. 268, 22825–22829.

Amati, B., Brooks, M. W., Levy, N., Littlewood, T. D., Evan, G. I., & Land, H. (1993). Oncogenic activity of the c-Myc protein requires dimerization with Max. Cell 72, 233–245.

Andres, J. L., Stanley, K., Cheifetz, S., & Massague, J. (1989). Membrane-anchored and soluble forms of betaglycan, a polymorphic proteoglycan that binds transforming growth factorβ. J. Cell Biol. 109, 3137–3145.

Arteaga, C. L., Carty–Dugger, T., Moses, H. L., Hurd, S. D., & Pietenpol, J. A. (1993). Transforming growth factor β1 can induce estrogen-independent tumorigenicity of human breast cancer cells in athymic mice. Cell Growth Differentiation 4, 193–201.

Arteaga, C. L., Tandon, A. L., Von Hoff, D. D., & Osborn, C. K. (1988). Transforming growth factor β: Potential autocrine growth inhibitor of estrogen receptor-negative human breast cancer cells. Cancer Res. 48, 3898–3904.

Attisano, L., Carcamo, J., Ventura, F., Weis, F. M. B., Massague, J., & Wrana, J. L. (1993). Identification of human activin and TGFβ type I receptors that form heteromeric kinase complexes with type II receptors. Cell 75, 671–680.

Ayer, D. E., Kretzner, L., & Eisenman, R. N. (1993). Mad: A heterodimeric partner for Max that antagonizes Myc transcriptional activity. Cell 72, 211–222.

Barnard, J. A., Lyons, R. M., and Moses, H. L. (1990). The cell biology of transforming growth factor β. Biochem. Biophys. Acta 1032, 79–87.

Bassing, C. H., Howe, D. J., Segarini, P. R., Donahoe, P. K., & Wang, X–F. (1994b). A single heteromeric receptor complex is sufficient to mediate biological effects of transforming growth factor β ligands. J. Biol. Chem. 269, 1–4.

Bassing, C. H., Yingling, J. M., Howe, D. J., Wang, T., He, W. W., Gustafson, M. L., Shah, P., Donahoe, P. K., & Wang, X–F. (1994a). A transforming growth factor β type I receptor that signals to activate gene expression. Science 263, 87–89.

Bishop, J. M. (1991). Molecular themes in oncogenesis. Cell 64, 235–248.

Blackwood, E. M., Luscher, B., Kretzner, L., & Eisenman, R. N. (1991). The Myc: Max protein complex and cell growth regulation. Cold Spr. Harb. Symp. Quant. Biol. 56, 109–117.

Boyd, F. & Massague, J. (1989). Transforming growth factor β inhibition of epithelial cell proliferation linked to the expression of a 53-kDa membrane receptor. J. Biol. Chem. 264, 2272–2278.

Brand, T., Maclellan, W. R., and Schneider, M. D. (1993). A dominant-negative receptor for type β transforming growth factors created by deletion of the kinase domain. J. Biol. Chem. 268, 11500–11503.

Broder, W. A. & Ruoslahti, E. (1992). Transforming growth factor β in disease: The dark side of tissue repair. J. Clin. Invest. 90, 1–7.

Broder, W. A., Noble, N. A., Yamamoto, T., Harper, J. R., Yamaguchi, Y., Pierschbacher, M. D., and Ruoslahti, E. (1992). Natural inhibitor of transforming growth factor β protects against scarring in experimental kidney disease. Nature 360, 361–364.

Brown, E. J., Albers, M. W., Shin, T. B., Ichikawa, K., Keith, C. T., Lane, W. S., & Schreiber, S. L. (1994). A mammalian protein targeted by G1-arresting rapamycin-receptor complex. Nature 369, 756–758.

Buchkovich, K., Duffy, L. A., & Harlow, E. (1989). The retinoblastoma protein is phosphorylated during specific phases of the cell cycle. Cell 58, 1097–1105.

Caldas, C., Hahn, S. A., da Costa, L. T., Redston, M. S., Schutte, M., Seymour, A. B., Weinstein, C. L., Hruban, R. H., Yeo, C. J., & Kern, S. E. (1994). Frequent somatic mutations and homozygous deletions of the p16 (MTS1) gene in pancreatic adenocarcinoma. Nature Genet. 8, 27–32.

Campbell, C. E., Flenniken, A. M., Skup, D., & Williams, B. R. G. (1991). Identification of a serum- and phorbal ester-responsive element in the murine tissue inhibitor of metalloproteinase gene. J. Biol. Chem. 266, 7199–7206.

Carcamo, J., Weis, F. M. B., Ventura, F., Wieser, R., Wrana, J. L., Attisano, L., & Massague, J. (1994). Type I receptors specify growth-inhibitory and transcriptional responses to transforming growth factor β and activin. Mol. Cell Biol. 14, 3810–3821.

Centrella, M., McCarthy, T. L., & Canalis, E. (1987). Transforming growth factor β is a bifunctional regulator of replication and collagen synthesis in osteoblast-enriched cell cultures from fetal rat bone. J. Biol. Chem. 262, 2869–2874.

Cheifetz, S. & Massague, J. (1991). Isoform-specific transforming growth factor β binding proteins with membrane attachments sensitive to phosphatidylinositol-specific phospholipase C. J. Biol. Chem. 266, 20767–20772.

Cheifetz, S., Bellon, T., Cales, C., Vera, S., Bernabeu, C., Massague, J., & Letarte, M. (1992). Endoglin is a component of the transforming growth factor β receptor system in human endothelial cells. J. Biol. Chem. 267, 19027–19030.

Cheifetz, S., Ling, N., Guillemin, R., & Massague, J. (1988). A surface component on GH3 pituitary cells that recognizes transforming growth factor β, activin, and inhibin. J. Biol. Chem. 263, 17225–17228.

Chen, P.–L., Scully, P., Shew, J.–Y., Wang, J. Y., & Lee, W.–H. (1989). Phosphorylation of the retinoblastoma gene product is modulated during the cell-cycle and cellular differentiation. Cell 58, 1193–1198.

Chen, R–H., Ebner, R., & Derynck, R. (1993). Inactivation of the type II receptor reveals two receptor pathways for the diverse TGFβ activities. Science 260, 1335–1338.

Daopin, S., Piez, K. A., Ogawa, Y., & Davies, D.R. (1992). Crystal structure of transforming growth factor β2: An unusual fold for the superfamily. Science 257, 369–373.

Datto, M. B., Li, Y., Panus, J., Howe, D. J., Xiong, Y., & Wang, X–Y. (1995). Transforming growth factor β induces the cyclin-dependent kinase inhibitor p21 through a p53-independent mechanism. Proc. Natl. Acad. Sci. 92, 5545–5549.

Decapria, J. A., Ludlow, J. W., Lynch, D., Furukawa, Y., Griffin, J., Piwnica–Worms, H., Huang, C.–M., & Livingston, D. M. (1989). The product of the retinoblastoma susceptibility gene has properties of a cell cycle regulatory element. Cell 58, 1085–1095.

Dowdy, S. F., Hinds, P. W., Louie, K., Reed, S. I., Arnold, A., & Weinberg, R. A. (1993). Physical interaction of the retinoblastoma protein with human D cyclins. Cell 73, 499–511.

Dulic, V., Kaufmann, W. K., Wilson, S. J., Tlsty, T. D., Lees, E., Harper, J. W., Elledge, S. J., & Reed, S. I. (1994). p53-Dependent inhibition of cyclin-dependent kinase activities in human fibroblasts during radiation-induced G1 arrest. Cell 76, 1013–1023.

Ebner, R., Chen, R–H., Shum, L., Lawler, S., Zioncheck, T. F., Lee, A., Lopez, A. R., & Derynck, R. (1993). Cloning of a type I TGFβ receptor and its effect on TGFβ binding to the type II receptor. Science 260, 1344–1348.

Edwards, D. R., Murphy, G., Reynolds, J. J., Whitham, S. E., Docherty, A. J. P., Angel, P., & Heath, J. K. (1987). Transforming growth factor β modulates the expression of collagenase and metalloproteinase inhibitor. EMBO J. 6, 1899–1904.

Eilers, M., Picard, D., Yamamoto, K. R., & Bishop, J. M. (1991). The Myc protein activates transcription of the α-prothymosin gene. EMBO J. 10, 133–141.

El–Deiry, W. S., Harper, J. W., O'Connor, P. M., Velculescu, V. E., Canman, C. E., Jackman, J., Pietenpol, J. A., Burrell, M., Hill, D. E., Wang, Y., Wiman, K. G., Mercer, W. E., Kastan, M. B., Kohn, K. W., Elledge, S. J., Kinzler, K. W., & Vogelstein, B. (1994). WAF1/CIP1 is induced in p53-mediated G1 arrest and apoptosis. Cancer Res. 54, 1169–1174.

El–Deiry, W. S., Tokino, T., Velculescu, V. E., Levy, D. B., Parsons, R., Trent, J. M., Lin, D., Mercer, W. E., Kinzler, K. W., & Vogelstein, B. (1993). WAF1, a potential mediator of p53 tumor suppression. Cell 75, 817–825.

Evan, G. I. & Littlewood, T. D. (1993). The role of c-myc in cell growth. Curr. Opin. Genet. Dev. 3, 44–49.

Ewen, M. E., Sluss, H. K., Whitehouse, L. L., & Livingston, D. M. (1993). TGFβ inhibition of Cdk4 synthesis is linked to cell cycle arrest. Cell 74, 1009–1020.

Fine, A. & Goldstein, R.H. (1987). The effect of transforming growth factor β on cell proliferation and collagen formation by lung fibroblasts. J. Biol. Chem. 262, 3897–3902.

Fisher, R. P. & Morgan, D.O. (1994). A novel cyclin associates with MO15/CDK7 to form the CDK-activating kinase. Cell 78, 713–724.

Franzen, P., ten Dijke, P., Ichijo, H., Yamashita, H., Schulz, P., Heldin, C–H., & Miyazono, K. (1993). Cloning of a TGFβ type I receptor that forms a heteromeric complex with the TGFβ type II receptor. Cell 75, 681–692.

Fruman, D. A., Burakoff, S. L., & Bierer, D. E. (1994). Immunophilins in protein folding and immunosuppression. FASEB 8, 391–400.

Geng, Y. & Weinberg, R. A. (1993). Transforming growth factor β effects on expression of G1 cyclins and cyclin-dependent protein kinases. Proc. Natl. Acad. Sci. USA 90, 10315–10319.

Gougos, A. & Letarte, M. (1990). Primary structure of endoglin, an RGD-containing glycoprotein of human endothelial cells. J. Biol. Chem. 265, 8361–8364.

Gu, Y., Turck, C. W., & Morgan, D.O. (1993). Inhibition of CDK2 activity *in vivo* by an associated 20Kd regulatory subunit. Nature 336, 707–710.

Hamel, P. A., Phillips, R. A., Muncaster, M., & Gallie, B. L. (1993). Speculations on the roles of RG1 in tissue-specific differentiation, tumor initiation, and tumor progression. FASEB 7, 846–854.

Hannon, G. J. & Beach, D. (1994). p15I^{NK4B} is a potential effector of TGFβ-induced cell-cycle arrest. Nature 371, 257–261.

Harper, J. W., Adami, G. R., Wei, N., Keyomarsi, K., & Elledge, S. J. (1993). The p21 Cdk-interacting protein Cip1 is a potent inhibitor of G1 cyclin-dependent kinases. Cell 75, 805–816.

He, W. W., Gustafson, M. L., Hirobe, S., & Donahoe, P. K. (1993). Developmental expression of four novel serine/threonine kinase receptors homologous to the activin/transforming growth factor β type II receptor family. Dev. Dyn. 196, 133–142.

Heikkila, R., Schwab, G., Wickstrom, E., Loke, S. L., Pluznik, D. H., Watt, R., & Neckers, L. M. (1987). A c-myc antisense oligodeoxy-nucleotide inhibits entry into S phase but not progress from G0 to G1. Nature 328, 445–449.

Heitman, J., Mocca, N. R., & Hall, M. H. (1992). Proline isomerases at the crossroads of protein folding, signal transduction, and immunosuppression. The New Biologist 4, 448–460.

Henis, Y. I., Moustakas, A., Lin, H. Y., & Lodish, H. F. (1994). Type II and type III transforming growth factor β receptors form homo-oligomers. J. Cell Biol. 126, 139–154.

Hiebert, S. W. (1993). Regions of the retinoblastoma gene product required for its interaction with the E2F transcription factor are necessary for E2 promoter repression and pRb-mediated growth suppression. Mol. Cell Biol. 13, 3384–3391.

Hiebert, S. W., Chellappan, S. P., Horowitz, J. M., & Nevins, J.R. (1992). The interaction of RB with E2F coincides with an inhibition of the transcriptional activity of E2F. Genes Dev. 6, 117–185.

Hinds, P. W., Mittnacht, S., Dulic, V., Arnold, A., Reed, S. I., & Weinberg, R.A. (1992). Regulation of retinoblastoma protein functions by ectopic expression of human cyclins. Cell 70, 993–1006.

Hoosein, N. M., McKnight, M. K., Levine, A. E., Mulder, K. M., Childress, K. E., Brattain, D. E., & Brattain, M. G. (1989). Differential sensitivity of subclasses of human colon carcinoma cell lines to the growth inhibitory effects of transforming growth factor β1. Exp. Cell Res. 181, 442–453.

Hsu, S., Huang, F., Hafez, M., Winawer, S., & Friedman, E. (1994). Colon carcinoma cells switch their response to transforming growth factor β1 with tumor progression. Cell Growth Differentiation 5, 267–275.

Hunter, T. (1991). Cooperation between oncogenes. Cell 64, 249–270.

Hunter, T. (1993). Braking the cell cycle. Cell 75, 839–841.

Ignotz, R. A. & Massague, J. (1986). Transforming growth factor β stimulates the expression of fibronectin and collagen and their incorporation into the extracellular matrix. J. Biol. Chem. 261, 4337–4345.

Jayaraman, T. & Marks, A. R. (1993). Rapamycin-FKBP12 blocks proliferation, induces differentiation, and inhibits cdc2 kinase activity in a myogenic cell line. J. Biol. Chem. 268, 25385–25388.

Kadin, M. E., Cavaille–Coll, M. W., Gertz, R., Massague, J., Cheifetz, S., & George, D. (1994). Loss of receptors for transforming growth factor β in human T-cell malignancies. Proc. Natl. Acad. Sci. USA 91, 6002–6006.

Kamb, A., Gruis, N. A., Weaver–Feldhaus, J., Liu, Q., Harshman, K., Tavitgian, S. V., Stockert, E., Day, R. S., Johnson, B. E., & Skolnick, M. H. (1994a). A cell cycle regulator potentially involved in genesis of many tumor types. Science 264, 436–440.

Kamb, A., Shattuck–Eidens, D., Eeles, R., Liu, Q., Gruis, N. A., et al., Meyer, L. J., Youl, P., Zone, J. J., Skolnick, M. H., & Cannon–Albright, L. A. (1994b). Analysis of the p16 gene (CDKN2) as a candidate for the chromosome 9p melanoma susceptibility locus. Nature Genet. 8, 22–26.

Kato, J–Y., Matsushime, H., Hiebert, S. W., Ewen, M. E., & Sherr, C. J. (1993). Direct binding of cyclin D to the retinoblastoma gene product (pRb) and pRb phosphorylation by the cyclin D-dependent kinase CDK4. Genes Dev. 7, 331–342.

Kato, J.–Y., Matsuoka, M., Polyak, K., Massague, J., & Sherr, C. J. (1994). Cyclin AMP-induced G1 phase arrest mediated by an inhibitor (p27kip1) of cyclin-dependent kinase 4 activation. Cell 79, 487–496.

Keeton, M. R., Curriden, S. A., van Zonneveld, A.–J., & Loskutoff, D. J. (1991). Identification of regulatory sequences in the type 1 plasminogen activator inhibitor gene responsive to transforming growth factor β. J. Biol. Chem. 266, 23048–23052.

Kim, S.–J., Angel, P., Lafyatis, R., Hattori, K., Kim, K. Y., Sporn, M. B., Karin, M., & Roberts, A. B. (1990). Autoinduction of transforming growth +factor β1 is mediated by the AP-1 complex. Mol. Cell Biol. 10, 1492–1497.

Kimchi, A., Wang, X.–F., Weinberg, R. A., Cheifetz, S., & Massague, J. (1988). Absence of TGFβ receptors and growth inhibitory responses in retinoblastoma cells. Science 240, 196–199.

Kingsley, D. M. (1994). The TGFβ superfamily: New members, new receptors, and new genetic tests of function in different organisms. Genes Dev. 8, 133–146.

Koff, A., Ohtsuki, M., Polyak, K., Roberts, J. M., & Massague, J. (1993). Negative regulation of G1 in mammalian cells: Inhibition of cyclin E-dependent kinase by TGFβ. Science 260, 536–539.

Kunz, J., Henriquez, R., Schneider, U., Deuter–Reinhard, M., Movva, N. R., & Hall, M. H. (1993). Target of rapamycin in yeast, TOR2, is an essential phosphatidylinositol kinase homolog required for G1 progression. Cell 73, 585–596.

Laiho, M., DeCaprio, J. A., Ludlow, J. W., Livingston, D. M., & Massague, J. (1990a). Growth inhibition by TGFβ linked to suppression of retinoblastoma protein phosphorylation. Cell 62, 175–185.

Laiho, M., Ronnstrand, L., Heino, J., Decapria, J. A., Ludlow, J. W., Livingston, D. M., & Massague, J. (1991a). Control of JunB and extracellular matrix protein expression by transforming growth factor β1 is independent of simian virus 40 T antigen-sensitive growth-inhibitory events. Mol. Cell Biol. 11, 972–978.

Laiho, M., Weis, F. M. B., & Massague, J. (1990b). Concomitant loss of transforming growth factor (TGF)-β receptor type I and II in TGFβ-resistant cell mutants implicates both receptor types in signal transduction. J. Biol. Chem. 265, 18518–18524.

Laiho, M., Weis, F. M. B., Boyd, F. T., Ignotz, R. A., & Massague, J. (1991b). Responsiveness to transforming growth factor β (TGFβ) restored by genetic complementation between cells defective in TGFβ receptors I and II. J. Biol. Chem. 266, 9108–9112.

Land, H., Farada, L. F., & Weinberg, R. A. (1983). Tumorigenic conversion of primary embryo fibroblasts requires at least two cooperating oncogenes. Nature 304, 596–602.

Lee, W., Mitchell, P., & Tjian, R. (1987). Purified transcription factor AP-1 interacts with TPA-inducible enhancer elements. Cell 49, 741–752.

Lees, J. A., Buchkovich, K. J., Marshak, D. R., Anderson, C. W., & Harlow, E. (1991). The retinoblastoma protein is phosphorylated on multiple sites by human cdc2. EMBO J. 10, 4279–4290.

Leof, E. B., Proper, J. A., Goustin, A. S., Shipley, G. D., DiCorleto, P. E., & Moses, H. L. (1986). Induciton of c-sis mRNA and activity similar to platelet-derived growth factor by transforming growth factor β: A proposed model for indirect mitogenesis involving autocrine activity. Proc. Natl. Acad. Sci. USA 83, 2453–2457.

Lewin, B. (1991). Oncogenic conversion by regulatory changes in transcription factors. Cell 64, 303–312.

Lin, H. Y. & Lodish, H. F. (1993). Receptors for the TGFβ superfamily: Multiple polypeptides and serine/threonine kinases. Trends Cell Biol. 3, 14–19.

Lin, H. Y., Wang, X.–F., Ng–Eaton, E., Weinberg, R. A., & Lodish, H. F. (1992). Expression cloning of the TGFβ type II receptor, a functional transmembrane serine/threonine kinase. Cell 68, 775–785.

Lin, Y.–C. A. & Stavnezer, J. (1992). Regulation of transcription of the germ-line Ig-α constant region gene by an ATG element and by novel transforming growth factor-β1-responsive elements. J. Immunology 149, 2914–2925.

Liu, J., Farmer, J. D., Lane, W. S., Friedman, J., Weissman, I., & Schreiber, S. L. (1991). Calcineurin is a common target of cyclophilin-cyclosporin A and FKBP-FK506 complexes. Cell 68, 807–815.

Lopez–Casillas, F., Cheifetz, S., Doody, J., Andres, J. L., Lane, W. S., & Massague, J. (1991). Structure and expression of the membrane proteoglycan betaglycan, a component of the TGFβ receptor system. Cell 67, 785–795.

Lopez–Casillas, F., Payne, H. M., Andres, J. L., & Massague, J. (1994). Betaglycan can act as a dual modulator of TGFβ access to signaling receptors: Mapping of ligand binding and GAG attachment sites. J. Cell Biol. 124, 557–568.

Lopez–Casillas, F., Wrana, J. L., & Massague, J. (1993). Betaglycan presents ligand to the TGFβ signaling receptor. Cell 73, 1435–1444.

Lyons, R. M. & Moses, H. L. (1990). Transforming growth factors and the regulation of cell proliferation. Eur. J. Biochem. 187, 467–473.

Makela, T. P., Tassan, J–P., Nigg, E. A., Frutiger, S., Hughes, G. J., & Weinberg, R. A. (1994). A cyclin associated with the CDK-activating kinase MO15. Nature 371, 254–256.

Manning, A. M., Williams, A. C., Game, S. M., & Paraskeva, C. (1991). Differential sensitivity of human colonic adenoma and carcinoma cells to transforming growth factor β (TGFβ): Conversion of an adenoma cell line to a tumorgenic phenotype is accompanied by a reduced response to the inhibitory effects of TGFβ. Oncogene 6, 1471–1476.

Marigo, V., Volpin, D., & Bressan, G. M. (1993). Regulation of the human elastin promoter in chick embryo cells. Tissue-specific effect of TGFβ. Biochim. Biophys. Acta 1172, 31–36.

Marx, J. (1994). New tumor supressor may rival p53. Science 264, 4–5.

Massague, J. (1990). The transforming growth factor β family. Annu. Rev. Cell Biol. 6, 597–641.

Massague, J., Cheifetz, S., Boyd, F. T., & Andres, J. L. (1990). TGFβ receptors and TGFβ binding proteoglycans: Recent progress in identifying their functional properties. Ann. NY Acad. Sci. 59–72.

Massague, J., Cheifetz, S., Endo, T., & Nadal–Ginard, B. (1986). Type β transforming growth factor is an inhibitor of myogenic differentiation. Proc. Natl. Acad. Sci. USA 83, 8206–8210.

Masui, T., Wakefield, L. M., Lechner, J. F., LaVeck, M. A., Sporn, M. B., & Harris, C. C. (1986). Type β transforming growth factor is the primary differentiation-inducing serum factor for normal human bronchial epithelial cells. Proc. Natl. Acad. Sci. USA 83, 2438–2442.

McAllister, K. A., Grogg, K. M., Johnson, D. W., Gallione, C. J., Baldwin, M. A., Jackson, C. E., Helmbold, E. A., Markel, D. S., Mckinnon, W. C., Murrell, J., McCormick, M. K., Pericak–Vance, M. A., Heutink, P., Oostra, B. A., Haitjema, T., Westerman, C. J. J., Parteous, M. E., Guttmacher, A. E., Letarte, M., & Marchuk, D. A. (1994). Endoglin, a TGFβ-binding protein of endothelial cells, is the gene for hereditary hemorrhagic telangiectasia type 1. Nature Genetics 8, 345–351.

McDonald, M. T., Papenberg, K. A., Ghosh, S., Glatfelter, A. A., Biesecker, B. B., Helmbold, E. A., Markel, D. S., Zolotor, A., McKinnon, W. C., & Vanderstoep, J. L. (1994). A disease locus for hereditary hemorrhagic telangiectasia maps to chromosome 9q33-34. Nature Genetics 6, 197–204.

Merlino, G. (1994). Regulatory imbalances in cell proliferation and cell death during oncogenesis in transgenic mice. Seminars in Cancer Biol. 5, 13–20.

Mihara, K., Cau, X. R., Yen, A., Chandler, S., Driscoll, B., Murphree, A. L., T'Ang, A., & Fung, Y. K. (1989). Cell cycle-dependent regulation of phosphorylation of the human retinoblastoma gene product. Science 246, 1300–1303.

Mori, T., Miura, K., Aoki, T., Nishihira, T., Shozo, N., & Nakamura, Y. (1994). Frequent somatic mutation of the MTS1/CDK4I gene in esophageal squamous cell carcinoma. Cancer Res. 54, 3396–3397.

Morice, W. G., Brunn, G. J., Wiederrech, G., Siekierka, J. J., & Abraham, R. T. (1993a). Rapamycin-induced inhibition of p34 cdc2 kinase activation is associated with G1/S-phase growth arrest in T-lymphocytes. J. Biol. Chem. 268, 3734–3738.

Morice, W. G., Wiederrech, G., Brunn, G. J., Siekierka, J. J., & Abraham, R. T. (1993b). Rapamycin inhibition of interleukin-2-dependent p33 cdk2 and p34 cdc2 kinase activation in T-lymphocytes. J. Biol. Chem. 268, 22737–22745.

Moustakas, A., Lin, H. Y., Henis, Y. I., Plamondon, J., O'Connor–McCourt, M. D., & Lodish, H. (1993). The transforming growth factor β receptors types I, II, and III form hetero-oligomeric complexes in the presence of ligand. J. Biol. Chem. 268, 22215–22218.

Munger, K., Pietenpol, J. A., Pittelkow, M. R., Holt, J. T., & Moses, H. L. (1992). Transforming growth factor β1 regulation of c-myc expression, pRB phosphorylation, and cell-cycle progression in keratinocytes. Cell Growth Differentiation 3, 291–298.

Murphy, C. S., Pietenpol, J. A., Munger, K., Howley, P. M., & Moses, H. L. (1991). c-myc and pRB: Role in TGFβ1 inhibition of keratinocyte proliferation. Cold Spr. Harb. Symp. Quant. Biol. 56, 129–135.

Nasmyth, K. & Hunt, T. (1994). Dams and sluices. Nature 336, 634–635.

Nobori, T., Mlura, K., Wu, D. J., Lois, A., Takabayashi, K., & Carson, D. A. (1994). Deletion of the cyclin-dependent kinase 4 inhibitor gene in multiple human cancers. Nature 368, 753–756.

Norgaard, P., Damstrup, L., Rygaard, K., Spang–Thomsen, M. S., & Poulsen, H. S. (1993). Growth suppression by transforming growth factor β1 of human small-cell lung cancer cell lines is associated with expression of the type II receptor. British J. Cancer 69, 802–808.

O'Grady, P., Kuo, M.–D., Baldassare, J. J., Huang, S. S., & Huang, J. S. (1991). Purification of a new type high molecular weight receptor (type V receptor) for transforming growth factor β (TGFβ) from bovine liver. J. Biol. Chem. 226, 8583–8589.

O'Grady, P., Liu, Q., Huang, S. S., & Huang, J. S. (1992). Transforming growth factor β (TGFβ) type V receptor has a TGFβ stimulated serine/threonine-specific autophosphorylation activity. J. Biol. Chem. 267, 21033–21037.

Ohtsubo, M. & Roberts, J. M. (1993). Cyclin-dependent regulation of G1 in mammalian fibroblasts. Science 259, 1908–1912.

Olson, E. N., Sternberg, E., Hu, J. S., Spizz, G., & Wilcox, C. (1986). Regulation of myogenic differentiation by type β transforming growth factor. J. Cell Biol. 103, 1799–1805.

Park, K., Kim, S.–J., Bang, Y.–J., Park, J.–G., Kim, N. K., Roberts, A. B., & Sporn, M. B. (1994). Genetic changes in the transforming growth factor β (TGFβ) type II receptor gene in human gastric cancer cells: Correlation with sensitivity to growth inhibition by TGFβ. Proc. Natl. Acad. Sci. USA 91, 8772–8776.

Pertovaara, L., Sistonen, L., Bos, T. J., Vogt, P. K., Keski–Oja, J., & Alitalo, K. (1989). Enhanced jun gene expression is an early genomic response to transforming growth factor β stimulation. Mol. Cell Biol. 9, 1255–1262.

Peter, M. & Herskowitz, I. (1994). Joining the complex: Cyclin-dependent kinase inhibitory proteins and the cell cycle. Cell 79, 181–184.

Peters, G. (1994). Stifled by inhibitions. Nature 371, 204–205.

Pietenpol, J. A., Munger, K., Howley, P. M., Stein, R. W., & Moses, H. L. (1994). Factor-binding element in the human c-myc promoter involved in transcriptional regulation by transforming growth factor β1 and by the retinoblastoma gene product. Proc. Natl. Acad. Sci. USA 88, 10227–10231.

Pietenpol, J. A., Stein, R. W., Moran, E., Yaciuk, P., Schlegel, R., Lyons, R. M., Pittelkow, M. R., Munger, K., Howley, P. M., & Moses, H. (1990). TGFβ1 inhibition of c-myc transcription and growth in keratinocytes is abrogated by viral transforming proteins with pRB binding domains. Cell 61, 777–785.

Polyak, K., Kato, J., Solomon, M. J., Sherr, C. J., Massague, J., Roberts, J. M., & Koff, A. (1994a). p27Kip1, a cyclin-cdk inhibitor, links transforming growth factor β and contact inhibition to cell cycle arrest. Genes Dev. 8, 9–22.

Polyak, K., Lee, M.–H., Erdjument–Bromage, H., Koff, A., Roberts, J. M., Tempst, P., & Massague, J. (1994b). Cloning of p27Kip1, a cyclin-dependent kinase inhibitor and a potential mediator of extracellular antimitogenic signals. Cell 78, 59–66.

Prochownik, E. V., Kukowska, J., & Rodgers, C. (1988). c-myc antisense transcripts accelerate differentiation and inhibit G1 progression in murine erythroleukemia cells. Mol. Cell. Biol. 8, 3683–3695.

Ritzenthaler, J. D., Goldstein, R. H., Fine, A., & Smither, B. D. (1993). Regulation of the α1(I) collagen promoter via a transforming growth factor β activation element. J. Biol. Chem. 268, 13625–13631.

Ritzenthaler, J. D., Goldstein, R. H., Fine, A., Lichtler, A., Rowe, D. W., & Smith, B. D. (1991). Transforming growth factor β activation elements in the distal promoter regions of the rat α1 type I collagen gene. Biochem. J. 280, 157–162.

Roberts, A. B. & Sporn, M. B. (1990). The transforming growth factor β. In: Handbook of Experimental Pharmacology, Peptide Growth Factors and Their Receptors. (Sporn, M. B. & Roberts, A. B., eds.), Springer, Heidelberg, pp. 419–431.

Sabatini, D. M., Erdjument–Bromage, H., Lui, M., Tempst, P., & Snyder, S. H. (1994). RAFT1: A mammalian protein that binds to FKBP12 in a rapamycin-dependent fashion and is a homolog to yeast TORs. Cell 78, 35–43.

Schlunegger, M. P. & Grutter, M. G. (1992). An unusual feature revealed by the crystal structure of 2.2 Å resolution of human transforming growth factor β2. Nature 358, 430–434.

Schreiber, S. L. (1991). Chemistry and biology of the immunophilins and their immunosuppressive ligands. Science 251, 283–287.

Schwarz, J. K., Bassing, C. H., Kovesdi, I., Datto, M. B., Blazing, M., George, S., Wang, X.–F., & Nevins, J. R. (1995). Expression of the E2F1 transcription factor overcomes type β transforming growth factor-mediated growth suppression. Proc. Natl. Acad. Sci. USA 92, 483–487.

Seyedin, S. M., Thomas, T. C., Thompson, A. Y., Rosen, D. M., & Piez, K. A. (1985). Purification and characterization of two cartilage-inducing factors from bovine demineralized bone. Proc. Natl. Acad. Sci. USA 82, 2267–2271.

Sherr, C. J. (1993). Mammalian G1 cyclins. Cell 73, 1059–1065.

Shipley, G. D., Pittelkow, M. R., Wille, J. J., Scott, R. E., & Moses, H. L. (1986). Reversible inhibition of normal human prokeratinocyte proliferation by type β transforming growth factor-growth inhibitor in serum-free medium. Cancer Res. 46, 2068–2071

Shovlin, C. L., Hughes, J. M., Tuddenham, E. G., Temperley, I., Perembelon, Y. F., Scott, J., Seidman, C. E., & Seidman, J. G. (1994). A gene for hereditary hemorrhagic telangiectasia maps to chromosome 9q3. Nature Genet. 6, 205–209.

Spruck, C. H. III, Gonzalez–Zulueta, M., Shibata, A, Simoneau, A. R., Lin, M.–F., Gozales, F., Tsai, Y. C., & Jones, P. A. (1994). p16 gene in uncultured tumors. Nature 370, 183–184.

Sun, L., Wu, G., Willson, J. K. V., Zborowska, E., Yang, J., Rajkarunanayake, I., Wang, J., Gentry, L. E., Wang, X.–F., & Brattain, M. G. (1994). Expression of transforming growth factor β type II receptor leads to reduced malignancy in human breast cancer MCF-7 cells. J. Biol. Chem. 269, 26449–26455.

ten Dijke, P., Iohijo, H., Franzen, P., Schults, P., Saras, J., Toyoshima, H., Heldin, C.–H., Miyazono, K. (1993). Activin receptor-like kinases: A novel subclass of cell surface receptors with predicted serine/threonine kinase activity. Oncogene 8, 2879–2887.

Toyoshima, H. & Hunter, T. (1994). p27, a novel inhibitor of G1 cyclin-cdk protein kinase activity, is related to p21. Cell 78, 67–74.

Waga, S., Hannon, G. J., Beach, D., & Stillman, B. (1994). The p21 inhibitor of cyclin-dependent kinases controls DNA replication by interaction with PCNA. Nature 369, 574–578.

Wang, T., Donahoe, P. K., & Zervos, A. (1994). Specific interaction of type I receptors of the TGFβ family with the immunophilin FKBP-12. Science 265, 674–676.

Wang, X.–F., Lin, W. Y., Ng–Eaton, E., Downward, J., Lodish, H. F., & Weinberg, R. A. (1991). Expression cloning and characterization of the TGFβ type III receptor. Cell 67, 797–805.

Welch, D. R., Fabra, A., & Nakajima, M. (1990). Transforming growth factor β stimulates mammary adenocarcinoma cell invasion and metastatic potential. Proc. Natl. Acad. Sci. USA 87, 7678–7682.

Wieser, R., Attisano, L., Wrana, J. L., & Massague, J. (1994). Signaling activity of transforming growth factor β type II receptors lacking specific domains in the cytoplasmic region. Mol. Cell Biol. 13, 7239–7247.

Wrana, J. L., Attisano, L., Carcamo, J., Zentella, A., Doody, J., Laiho, M., Wang, X.–F., & Massague, J. (1992). TGFβ signals through a heteromeric protein kinase receptor complex. Cell 71, 1003–1014.

Wrana, J. L., Attisano, L., Wieser, R., Ventura, F., & Massague, J. (1994). Mechanism of activation of the TGFβ receptor. Nature 370, 341–347.

Xiong, Y., Connelly, T., Futcher, B., & Beach, D. (1991). Human D-type cyclin. Cell 65, 691–699.

Xiong, Y., Hannon, G. J., Zhang, H., Casso, D., Kobayashi, R., & Beach, D. (1993). p21 is a universal inhibitor of cyclin kinases. Nature 366, 701–704.

Xiong, Y., Zhang, H., & Beach, D. (1992). D-Type cyclins associate with multiple protein kinases and the DNA replication and repair factor PCNA. Cell 71, 505–514.

Yamashita, H., Dijke, P. T., Franzen, P., Miyazono, K., & Heldin, C.–H. (1994). Formation of hetero-oligomeric complexes of type I and type II receptors for transforming growth factor β. J. Biol. Chem. 269, 20172–20178.

Zamanian, M. & La Thangue, N. B. (1993). Transcriptional repression by the Rb-related protein p107. Mol. Biol. Cell 4, 389–396.

Zervos, A. S., Gyuris, J., & Brent, R. (1993). Mxi1, a protein that specifically interacts with Max to bind Myc-Max recognition sites. Cell 72, 223–232.

THE VASCULAR ENDOTHELIAL CELL GROWTH FACTOR FAMILY AND ITS RECEPTORS:

MOLECULAR AND BIOLOGICAL PROPERTIES

Denis Gospodarowicz

I. Introduction . . . 434

II. Structural Characterization of the Various VEGF Isoforms and Related Growth Factors . . . 435

III. VEGF/VPF Receptors . . . 437

A. The VEGF/VPF Binding Sites . . . 437

B. Cloning of the VEGF/VPF Receptors . . . 438

C. Modulation of VEGF/VPF Binding to Its Receptors . . . 440

IV. Induction of VEGF/VPF Expression . . . 440

A. Differentiating Agents . . . 441

B. Hormonal Agents . . . 441

C. Hypoxia . . . 442

V. Expression of VEGF/VPF and Its Receptors in Embryonic Development . . . 443

Growth Factors and Cytokines in Health and Disease
Volume 1B, pages 433–464.

ISBN: 0-7623-0091-4

VI. Expression of VEGF/VPF and Its Receptors in the Female Reproductive System . . . 444
VII. VEGF/VPF and Therapeutic Angiogenesis . . . 446
VIII. VEGF/VPF and the Disease State . . . 447
A. Ocular Diseases . . . 447
B. VEGF/VPF and Tumor Angiogenesis . . . 448
C. VEGF/VPF in Psoriasis and Rheumatoid Arthritis . . . 450
IX. Target Cells and Tissues Other Than the Vascular Endothelium Responsive to VEGF/VPF . . . 451
X. Conclusion . . . 453
Note Added in Proofs . . . 454
References . . . 454

I. INTRODUCTION

Angiogenesis is a multistep phenomenon which involves capillary endothelial cell proliferation, migration, and tissue infiltration (Folkman, 1986). It plays a central role in a variety of physiological and pathological processes, such as embryonic development, wound healing, atherosclerosis, and tumor growth (Folkman, 1986; Gospodarowicz et al., 1987). Several factors that induce angiogenesis have recently been isolated and characterized. Among these are the acidic and basic form of fibroblast growth factor (FGF), both capable of stimulating capillary endothelial cell growth *in vitro* and being chemotactic for that cell type (Gospodarowicz et al., 1987). In addition, both FGF isoforms stimulate extracellular proteolysis which is believed to be an essential component of the angiogenic process (reviewed by Mullins and Rifkind, 1988). Numerous angiogenesis factors have been identified and characterized in the last ten years (reviewed by Klagsburn and D'Amore, 1991). The list includes tumor necrosis factor alpha (TNFα), transforming growth factor beta (TGFβ), transforming growth factor alpha (TGFα), epidermal growth factor (EGF) and interleukin-8 which are also angiogenic *in vivo* (reviewed in Gospodarowicz, 1990; Koch et al., 1992). However, with the exception of TGFα and EGF at high concentrations, these growth factors are not mitogenic for capillary endothelial cells. Their action on the angiogenic process is therefore probably indirect, resulting from such activities as attracting macrophages (Leibovich et al., 1987; Wahl et al., 1987) which in turn release direct angiogenic factor(s), one of which could be basic FGF (Baird et al., 1985).

Despite the evidence that FGF is angiogenic, some of its properties force us to question its suitability as a physiological or pathological regulator of angiogenesis. First, FGF lacks the hydrophobic signal sequences that govern secretion (Abraham et al., 1986; Jaye et al., 1986), and, although it reached the cell surface, it stays closely associated with it (Florkiewicz et al., 1991). For acceptance as an angiogenic factor, any putative mediator should be a diffusible substance which induces new capillary formation from a microcirculatory bed. Second, FGF is produced by endothelial cells themselves (Schweigerer et al., 1987; Vlodavsky et al., 1987). If FGF is present in and around endothelial cells and yet the cells are quiescent, other factors must come into play to trigger angiogenesis. Finally, there is no definitive evidence for a temporal or spatial correlation between the expression of the various FGF isoforms and its receptors and angiogenesis *in vivo* (Heuer et al., 1990; Wanaka et al., 1991; Peters et al., 1992).

There is, however, mounting evidence that a newly identified class of growth factors, specific for endothelial cells, the vascular endothelial cell growth factors (VEGF), also named vascular permeability factors (VPF), could be a prime regulator of physiological as well as pathological angiogenesis. VEGF/VPF are secreted growth factors which are mitogenic specifically for endothelial cells in culture. They are angiogenic *in vivo*, and they induce vascular permeability that leads to vascular stroma production (Connelly, 1991; Ferrara et al., 1992; Dvorak et al., 1991); the expression of VEGF/VPF and its receptors correlates strongly with periods of vasculogenesis and angiogenesis during embryonic development (Millauer et al., 1993; Jakeman et al., 1993; Peters et al., 1993; Breier et al., 1992) and with programmed neovascularization that occurs in the female reproductive cycle (Shweiki et al., 1993); finally, in tumors, VEGF/VPF is expressed in tumor cells, whereas the VEGF/VPF receptor is expressed exclusively in adjacent small blood vessels. These blood vessels bind VEGF/VPF protein and are hyperpermeable (Dvorak et al., 1991; Shweiki et al., 1992; Plate et al., 1992). Therefore, VEGF/VPF best fits all of the criteria defining an angiogenic factor.

II. STRUCTURAL CHARACTERIZATION OF THE VARIOUS VEGF ISOFORMS AND RELATED GROWTH FACTORS

VEGF and VPF were purified independently using two separate lines of investigation (reviewed in Connolly, 1991; Ferrara et al., 1992). A specific vascular endothelial cell mitogenic activity was found in the conditioned medium of pituitary folliculostellate cells (Gospodarowicz et al., 1989; Ferrara and Henzel, 1989), AtT-20 pituitary cells (Plouet et al., 1989) and some tumor cell lines (reviewed by Ferrara et al., 1992)—hence, the name VEGF—and the factor causing this activity was purified from these sources. At about the same time, an activity that enhances the permeability of blood vessels was found in the conditioned medium of guinea pig tumor cells (Senger et al., 1983, 1986) and human histiocytic lymphoma U-937 cells (Connolly et al., 1989)—hence, the name VPF—using the Miles assay, a method used to measure extravasation of Evans blue dye. When both genes were cloned, VEGF and VPF were found to be identical (Keck et al., 1989; Leung et al., 1989; Tischer et al., 1989; Conn et al., 1990). VEGF/VPF is a 46-kDa glycoprotein composed of two identical disulfide-linked subunits. It has a limited sequence similarity to PDGF (18–20%), particularly, in the distribution of cysteine residues. An interesting feature of VEGF/VPF is that it exists in four forms composed of 121, 165, 189 and 206 amino acid residues long which are the result of alternative splicing (Tischer et al., 1991). Of these, $VEGF_{165}$ is the most commonly expressed by various tissues (Garrido et al., 1993; Tischer et al., 1991). All four forms appear to be mitogenic for endothelial cells and have vascular permeability activity (Houck et al., 1991; Charnock–Jones et al., 1993). All of the forms are synthesized with signal peptides, yet $VEGF/VPF_{121}$ and $VEGF/VPF_{165}$ are secreted into the medium, whereas $VEGF/VPF_{189}$ and $VEGF/VPF_{206}$ are mostly cell-associated (Houck et al., 1992). The two larger VEGF/VPF forms have a very strong affinity for heparin and may be sequestered by cell-surface and matrix heparan sulfate proteoglycans (Houck et al., 1992).

Related to VEGF/VPF is the placenta growth factor (PIGF) first isolated from a term placenta cDNA library (Maglione et al., 1991). PIGF is a dimeric glycosylated 149 amino acid long protein. The main feature common to both PIGF and VEGF/VPF is the high amino acid sequence similarity in the PDGF-like region. In fact, the PIGF protein presents a similarity of ≈71% to the PDGF-like domain of human VPF if both the identical and conservative amino acid changes are considered. As in VEGF/VPF, two isoforms of PIGF have been characterized. Sequential comparison of the polypeptide coded by the two different isolated cDNA indicates that they are identical except for the insertion of a highly basic 21AA stretch at the end of the molecule ($PIGF_{170}$; Maglione et al., 1993).

Structural characterization of the human VEGF/VPF gene shows that it comprises eight exons (Tischer et al., 1991). Mature $VEGF_{189}$ retains coding sequence from exons 2–8 (exon 1 representing the majority of the secretion signal portion of the primary translation product), whereas $VEGF_{165}$ lacks the amino acids encoded by exon 6, and $VEGF_{121}$ lacks the residues encoded by exons 6 and 7. There is no intron between the coding sequences for the 24 amino acid insertion in $VEGF_{189}$ and the additional 17 amino acid insertion in $VEGF_{206}$. The 5′-end of the 51 base pair insertion of $VEGF_{206}$ begins with GT, the consensus sequence for the 5′-splice donor necessary for mRNA processing. Thus, definition of the 5′-splice donor site for removal of this 1-kilobase intron appears to be variable (Houck et al., 1991). Analysis of the VEGF gene-promoter region revealed a single major transcription start, which lies near a cluster of potential Sp 1 factor binding sites (Tischer et al., 1991). The promoter region also contains several potential binding sites for the transcription factors AP-1 and AP-2 (Tischer et al., 1991). Consistent with the presence of these sites, Northern blot analysis has demonstrated that the level of VEGF transcripts is elevated in cultured vascular smooth muscle cells after treatment with the phorbol ester 12-*O*-tetradecanoyl-phorbol-13-acetate (TPA) (Tischer et al., 1991).

Consistent with the homologous primary structure and their evolutionary origin, the intron-exon structure of the PIGF and VEGF/VPF genes have been shown to be very similar (Tischer et al., 1991; Maglione et al., 1993). In particular, the eight cysteines, characteristic of the PDGF-like domain are encoded by exons 3 and 4 (six and two cysteine residues, respectively) in both genes. Furthermore, as in the VEGF/VPF gene, exon 1 contains all the 5′-untranslated region and the putative signal peptide of PIGF, whereas exon 7 encodes the last amino acids of PIGF and contains all the 3′-untranslated region.

In contrast to other growth factors which have a wide range of target-cell specificity, the main biological activity of VEGF/VPF seem to be restricted to vascular endothelial cells (reviewed by Ferrara et al., 1992, and Klagsbrun and Soken, 1993) and monocytes (Shen et al., 1993; Clauss et al., 1990). Based on the observation that VEGF/VPF perturbed blood vessel barrier function, studies have focused on its interaction with vascular endothelial cells (reviewed in Shen et al., 1993). VEGF/VPF binds to vascular endothelial cells with high affinity, induces an increase in cytosolic calcium (Criscuolo et al., 1989), stimulates phosphoinositide metabolism (Shen et al., 1993) and glucose transport (Pekala et al., 1990), and enhances vascular endothelial proliferation (reviewed in Neufeld et al., 1994). It also increased uPA and tPA activity and their mRNA expression. These are two keys enzymes involved in extracellular proteolysis (Pepper et al., 1991). It has also been reported to increase collagenase activity (Unemori et al., 1992). A potent synergy between VEGF/VPF and basic FGF in the induction of angiogenesis *in vitro* has been reported and it has been suggested that it could be important in controlling angiogenesis *in vivo* (Pepper et al., 1992; Goto et al., 1993). *In vivo* injection of VEGF/VPF results in a rapid peak (within 5–10 min) and reversal (by 20 min) in vascular permeability without causing mast cell degranulation or endothelial cell damage, but, rather, by causing separations at endothelial cell junctions. These kinetics and effects are similar to those of the histamine, bradykinin, with VEGF being 50,000 times more potent than histamine on a weight basis (Senger et al., 1986; Connolly, 1991). The effect of VEGF/VPF on increased vascular permeability may reflect its ability to rapidly induce disorganization of actin filaments accompanied by drastic morphological changes which are characterized by decreased cell–cell and cell–substrate adhesion *in vitro*. Those effects are related to VEGF's ability to induce tyrosine phosphorylation of several cytoplasmic proteins (Enomoto et al., 1994).

VEGF/VPF has also been shown to induce monocytes chemotaxis and activation as evidenced by expression of tissue factor (Clauss et al., 1990). Together, these observations indicate that VEGF/VPF can modulate the coagulant properties of endothelium and monocytes and can promote monocyte migration into the tumor bed. This suggests one mechanism through which tumor-derived mediators can alter vessel wall properties.

III. VEGF/VPF RECEPTORS

A. The VEGF/VPF Binding Sites

The mitogenic activity of VEGF/VPF seems to be restricted to vascular endothelial cells (Gospodarowicz et al., 1989; Plouet et al., 1989; Keck et al., 1989; Leung et al., 1989; Connolly et al., 1989), and initial characterization of VEGF receptors was therefore carried out using these cells. Binding experiments identified specific high-affinity and low-affinity binding sites for ^{125}I-VEGF$_{165}$ on the cell surface of various types of vascular endothelial cells (Vaisman et al., 1990; Plouet and Moukadiri, 1990; Olander et al., 1991). The dissociation constant of the high-affinity binding sites ranges from 10^{-12} (Vaisman et al., 1990) to 9×10^{-12} M (Olander et al., 1991), and the dissociation constant of the low-affinity binding sites ranges from 10^{-11} (Vaisman et al., 1990) to 1.5×10^{-9} M (Olander et al., 1991). The density of the high-affinity binding sites ranges from 3×10^3 (Vaisman et al., 1990) to 25×10^3 receptors per cell (Olander et al., 1991), and the density of the low-affinity sites ranges from 3×10^3 (Plouet and Moukadiri, 1990) to 180×10^3 sites per cell (Olander et al., 1991). The reason for these large discrepancies, especially with regard to the low-affinity binding sites, is unclear but could reflect the ability of VEGF/VPF to bind to heparan sulfate proteoglycans. VEGF is internalized following its binding to the VEGF receptors of human umbilical vein derived endothelial cells, and the internalized VEGF is subsequently degraded (Bikfalvi et al., 1991). Three high molecular weight complexes of 225, 195 and 175 kDa are seen following covalent cross-linking of ^{125}I-VEGF$_{165}$ to bovine aortic arch derived vascular endothelial cells (Vaisman et al., 1990). It was, therefore, predicted that the molecular mass of the receptors would be 180, 150 and 130 kDa, assuming that one VEGF dimer binds to a single receptor. The binding of VEGF$_{165}$ to these binding sites appears to be highly specific because several different growth factors did not inhibit the binding of VEGF (Vaisman et al., 1990; Plouet and Moukadiri, 1990). The expected size of the largest of these receptors is similar to the size of a 190-kDa protein that is phosphorylated on tyrosine residues following stimulation by VEGF (Myoken et al., 1991).

VEGF receptors can also detect on cell types, such as corneal endothelial cells, HeLa cells, and on several lines of human melanoma cells. However, these cell types do not respond to VEGF by proliferation (Gitay–Goren et al., 1992, 1993). Interestingly, no VEGF receptors could be detected on the cell surface of normal human melanocytes (Gitay–Goren et al., 1993). In most of these cells, except for the corneal endothelial cells, the 150- and 130-kDa receptors are the main receptor species and the larger receptor type cannot be detected or is seen very weakly (Gitay–Goren et al., 1992, 1993; Ferrara et al., 1992).

B. Cloning of the VEGF/VPF Receptors

Because of the structural homology between VEGF and PDGF, it was reasonable to assume that the VEGF receptors would be similar to the PDGF receptors (Yarden et al., 1986; Ullrich and Schlessinger, 1990). That type of receptor belongs to the class III receptor tyrosine kinase, which is characterized by a long, extracellular domain organized into 5 Ig-like loops and an intracellular catalytic domain interrupted by a kinase insert domain. Screening of a cDNA library from human placenta led to the identification of a clone which was almost identical to the published sequence of the fims-like tyrosine kinase (flt) (Shibuya et al., 1990). It had 7 Ig-like domains in its extracellular region and a tyrosine kinase sequence interrupted by a kinase insert. The flt-encoded receptor binds $VEGF_{165}$ and is expressed in vascular endothelial cells (DeVries et al., 1992). Although autophosphorylation of flt following VEGF binding has not yet been demonstrated, expression of flt mRNA in oocytes results in calcium efflux following VEGF exposure (DeVries et al., 1992).

Another RTK, which functions as a VEGF receptor, is encoded by the KDR-flk-1 gene (Terman et al., 1991, 1992). KDR was identified from a human endothelial cell cDNA library by the polymerase chain reaction using degenerate primers homologous to the kinase domain of known type III receptor tyrosine kinases (RTK). Northern blot analysis demonstrated expression of mRNA in endothelial cells but not in fibroblasts or smooth muscle cells. The amino acid sequence deduced indicates that the extracellular domain of KDR has seven Ig-like domains and that its cytoplasmic domain contains a kinase insert domain. Independently, Matthews et al. (1991) have cloned the mouse analogue of KDR called fetal liver kinase I (flk-1) from mouse cell populations enriched for hematopoietic stem and progenitor cells. Sequential analysis of this clone reveals strong homology to the c-Kit subfamily of receptor kinases, and in particular to the flt gene product. flk-1 mRNA is expressed in primitive and more mature hematopoietic cells and also in a wide variety of nonhematopoietic tissues because most tissues, with a few exceptions, such as cartilage, are vascularized. This is consistent with flk-1 expression in vascular endothelial cells (Terman et al., 1991, 1992; Millauer et al., 1993; Peters et al., 1994).

KDR, when expressed in CMT-3 cells (cells which do not contain receptors for VEGF), allows for saturable ^{125}I-VEGF binding with high affinity. Affinity cross-linking of ^{125}I-VEGF to KDR-transfected CMT-3 cells results in specific labeling of two proteins of MW equalling 195 and 235 kDa (Terman et al., 1992). Specific binding of VEGF to flk-1 was demonstrated by cross-linking and competition binding experiments. VEGF stimulation of flk-1-expressing cells led to a significant induction of tyrosine phosphorylation of the 180-kDa flk-1 receptor (Millauer et al., 1993). When expressed in *xenopus* oocytes, KDR-flk-1 binds VEGF with high affinity, undergoes autophosphorylation, and mediates VEGF-dependent Ca^{2+} efflux (Quinn et al., 1993).

KDR-flk-1 and the fms-like tyrosine kinase flt, which was previously shown to act as a VEGF receptor, share many structural features. They are type III RTK with kinase insert domains of similar length, each contains seven extracellular immunoglobulin-like domains, and there is a high degree of amino acid sequential homology between the two receptors in each structural domain. These similarities are reminiscent of those between the α- and β-chains of the PDGF receptor (Hart and Bowen–Pope, 1990; Heldin, 1995). Because VEGF is structurally similar to PDGF, it should not be surprising that the structural properties of receptors for VEGF are similar to those of PDGF receptors.

Related to flt and KDR-flk-1 is the flt4 receptor tyrosine kinase which contains 7 Ig-like loops and is expressed in multiple human tissues and cell lines. This receptor was cloned from a human HEL erythroleukemia cell library. No evidence, however, has been obtained for the expression of flt-4 in the endothelial cells of blood vessels (Pajusola et al., 1992).

C. Modulation of VEGF/VPF Binding to Its Receptors

The binding of VEGF to its receptors is inhibited by suramin, a highly sulfated drug which inhibits the biological activity of several heparin-binding growth factors (Vaisman et al., 1990; Plouet and Moukadiri, 1990; Middaugh et al., 1992; Kasayama et al., 1993). Protamine, a basic antiangiogenic protein which binds to heparin and inhibits the activity of several heparin-binding growth factors also inhibits the binding of VEGF to receptors on vascular endothelial cells (Vaisman et al., 1990). These observations suggest that heparin may be important in the interaction of VEGF with its cell-surface receptors. The binding of VEGF to the VEGF receptors of vascular endothelial cells is potentiated in the presence of low concentrations of heparin (Gitay–Goren et al., 1992). High concentrations of heparin, on the other hand, inhibit the binding (Gitay–Goren et al., 1992) and the mitogenic activity of VEGF/VPF. When heparin-like molecules are depleted from the cell surface of vascular endothelial cells by heparinase treatment, the binding of VEGF/VPF to receptors is inhibited, but can be restored by adding exogenous heparin (Gitay–Goren et al., 1992). This indicates that the binding of VEGF to its receptors is dependent on cell-surface-associated heparin-like molecules, and is reminiscent of the interaction of basic fibroblast growth factor (bFGF) with its cell surface receptors (Gitay–Goren et al., 1992; Rapraeger et al., 1991; Yayon et al., 1991). Recently, it was reported that low concentrations of heparin strongly enhance the binding of VEGF to a soluble extracellular domain of the KDR-flk-1-encoded VEGF receptor (Tessler et al., 1994).

Studies done by Terman et al. (1994) also indicate that VEGF binding to cells is differentially modulated by heparin, depending on whether the KDR receptor or the flt receptor is involved. The bimodal effect of heparin reported by others (Gitay–Goren et al., 1992) could be explained by the fact that low concentrations of heparin inhibit VEGF binding to flt while higher concentrations augment binding to KDR-flk-1 (Terman et al., 1994).

The specificity of flt and KDR-flk-1 receptor ligand-binding domains for PIGF and VEGF has been examined (Kendall et al., 1994; Terman et al., 1994). flt-1 can bind both VEGF and PIGF-1, whereas KDR-flk-1 binds only VEGF. This situation is very reminiscent of that of the structurally related gene families of the PDGF ligands and their receptors, where it is known that PDGF-AA and PDGF-BB homodimers and the PDGF-AB heterodimers show selectivity in interacting with the two PDGF-receptors (Hart and Bowen–Pope, 1990; Heldin, 1995).

IV. INDUCTION OF VEGF/VPF EXPRESSION

Increased VEGF/VPF expression can be observed in response to various agents, including agents inducing differentiation, such as TPA. VEGF/VPF expression can also be under hormonal control and, most importantly, it can be controlled by hypoxia.

A. Differentiating Agents

The four potential AP-1-binding sites implied that the VEGF gene should be inducible by serum or by the phorbol ester TPA (Curran and Franza, 1988; Mitchell and Tjian, 1989). In addition, two potential AP-2-binding sites also indicated that VEGF gene expression should be inducible by TPA and by cAMP (Imagawa et al., 1987; Mitchell and Tjian, 1989). Consistent with these predictions, experiments with fetal human vascular smooth muscle cells and granulosa cells indicate that TPA exposure indeed results in an increase in the level of VEGF mRNA (Tischer et al., 1991). In granulosa cells, TPA induces VEGF transcription in a time- and dose-dependent fashion. Induction apparently requires de novo protein synthesis because it does not occur when translation is inhibited by cycloheximide (Garrido et al., 1993). Forskolin, a naturally occurring diterpene that activates adenylylcyclase, also increases VEGF mRNA content time dependently. Induction does not require de novo protein synthesis and, in contrast to TPA, is strongly potentiated by cycloheximide (Garrido et al., 1993). Likewise, experiments performed on differentiating fat, muscle, and PC12 cells suggest that there are highly regulated signal-transduction pathways involved in activating VEGF expression (Claffey et al., 1992). VEGF mRNA can be induced by a variety of second messenger effectors. Activation of VEGF expression by phorbol ester, calcium ionophore, and cAMP analogues indicates the involvement of protein kinase C and cAMP-dependent protein kinase pathways. Activation of protein kinase C generally involves translocation of the enzyme to a membrane surface (Kikkawa et al., 1992). Thus, the actual levels of membrane-associated protein kinase C observed in transformed cells may be the determining factor regulating VEGF expression. This suggests a general mechanism for linking tumorigenesis and angiogenesis.

B. Hormonal Agents

Both estrogens and luteinizing hormone (LH) have been shown to induce VEGF expression. In the uterus, estrogen causes a rapid increase in microvascular permeability, followed later by growth of the endometrium, including the richly vascular stroma. Treatment of immature rats with estrogen rapidly stimulates VEGF mRNA expression (Cullinan–Bove et al., 1993). Cycloheximide does not prevent the estrogen-induced increase in VEGF/VPF mRNA, indicating that it does not require *de novo* protein synthesis. Thus, VEGF/VPF appears to be a primary response gene product (Herschman, 1991), at least in the uterus. Consistent with those results, it has been observed that, in endometrial carcinoma cell lines, estradiol increased steady-state VEGFmRNA in a dose- and time-dependent manner (Charnock–Jones et al., 1993).

In granulosa cells, estrogens also increased VEGF mRNA expression (Garrido et al., 1993). Confirming the effect of cAMP on VEGF gene expression, it has been reported that LH, a hormone which activates adenylylcyclase, was also able to induce the accumulation of VEGF mRNA (Garrido et al., 1993). This is of interest because a surge of LH is required for ovulation (reviewed in Yen, 1991). This coincides with the start of the angiogenic events taking place within the theca interna (Bassett, 1943; Gospodarowicz and Thakral, 1978). Finally, in fibroblastic and epithelial cells, VEGF mRNA expression can be stimulated by TGFβ (Pertovaara et al., 1994), suggesting a model for the angiogenic effects of TGFβ, and, in psoriasis, TGFα has been implicated as a modulator (Detmar et al., 1994).

C. Hypoxia

Hypoxia has been shown to be a potent stimulus for the inducting VEGF mRNA expression (Shweiki et al., 1992; Plate et al., 1992; Minchenko et al., 1994). This observation in itself, is important for the physiological control of VEGF expression. Inefficient vascular supply and the resultant reduction in tissue oxygen tension often lead to neovascularization to satisfy the needs of the tissue (Alder et al., 1990). It has also been pointed out that the initial stimulus attracting blood vessels is an increased demand for O_2 arising from increased metabolic activity (Gospodarowicz, 1978). Hypoxia can be the cause of various diseases characterized by vascular defects, in particular hyperpermeability of blood vessels. This is best seen in ocular diseases (Gospodarowicz, 1978; Pe'er et al., 1995).

Multiple similarities exist between the oxygen-sensing mechanisms regulating the expression of VEGF/VPF and erythropoietin, a major factor controlling erythropoiesis and therefore blood O_2 tension (Goldberg and Schneider, 1994). The expression of both mRNAs is significantly upregulated by hypoxia and cobalt chloride, and the half-life of both mRNAs is markedly prolonged by cycloheximide. In addition, hypoxic induction of both erythropoietin and VEGF is inhibited by carbon monoxide. VEGF/VPF and erythropoietin appear to be genes that probably use a common, widespread oxygen-sending mechanism. If VEGF and erythropoietin share a common oxygen-sensing pathway, it is likely that they may also share common *cis* regulatory elements. Computer analysis has revealed two 10-base pair sequences 5′ to the VEGF transcription start site which are 90% homologous to a region within the human Epo 3′ hypoxia-responsive enhancer (Beck et al., 1993; Blanchard et al., 1992; Semenza and Wang, 1992; Madan and Curtin, 1993). In addition, there is a 9-base pair sequence in the VEGF 3′-untranslated region which is identical to a sequence within the 3′-untranslated region of Epo mRNA that is known to specifically bind a protein in many tissue types (Rondon et al., 1991). The mechanism of oxygen sensing and the response of the VEGF gene to hypoxia are still unclear, but could involve a heme-containing protein sensor, such as the one described in the activation of the erythropoietin gene (Goldberg et al., 1988). This hemeprotein seems to be present in most cells and it is likely to be part of a generalized mechanism of oxygen sensing (Beck et al., 1993; Maxwell et al., 1993; Wang and Semenza, 1993).

V. EXPRESSION OF VEGF/VPF AND ITS RECEPTORS IN EMBRYONIC DEVELOPMENT

The growth of blood vessels during embryogenesis and in the adult organism is tightly controlled. Some developing organs, like brain or kidney, are vascularized by sprouts branching from preexisting blood vessels (Bär, 1980; Ekblom et al., 1982). This process, termed angiogenesis, is thought to be regulated by the action of soluble factors.

During mouse brain development, VEGF transcript levels are abundant in the ventricular neuroectoderm of embryonic and postnatal brain when endothelial cells proliferate rapidly, but are reduced in the adult when endothelial cell proliferation has ceased (Breier et al., 1992). The temporal and spatial expression of VEGF is consistent with the hypothesis that VEGF is synthesized and released by the ventricular neuroectoderm and may induce the ingrowth of capillaries from the perineural vascular plexus. In addition to the transient expression during brain development, a persistent expression of VEGF was observed in epithelial cells adjacent to fenestrated endothelium, for example, in choroid plexus and in kidney glomeruli (Breier et al., 1992).

flt-1, one of the VEGF receptors, has also been shown to be expressed in mouse embryonic tissue, confined to the embryonic endothelium in two embryonic structures that are known sites of endothelial differentiation, the yolk sac at E7.5 and the lung bud at E10.5 (Peters et al., 1993). flt expression was detected in cells, perhaps angioblasts, that were not associated with blood vessels (Noden, 1989; Coffin et al., 1991; Pardanaud et al., 1989). Later in the E9.5 yolk sac, flt was detected only in the endothelium (Peters et al., 1993).

During fetal development (E14.5-E16.5), a time when endothelial differentiation might be completed but blood vessel growth is ongoing, flt expression markedly decreased. This could mean that factors other than VEGF drive blood vessel growth during fetal development or that KDR-flk-1, the other type of VEGF receptor, is, then, expressed in a majority. In fact, studies on the developmental expression of KDR-flk-1 done by Millauer et al. (1993) and Quinn et al. (1993) suggest that this could be the case. Northern blot analysis of RNA, prepared from mouse embryos at different stages (E9.5 to E.18), revealed that KDR-flk-1 was expressed at high levels in all embryonal stages examined. *In situ* hybridization with sections of mouse brain and whole mouse embryos suggested a strong correlation of KDR-flk-1 gene expression with the histological distribution pattern expected for endothelial cells and their progenitors. Additional evidence for the identification of the KDR-flk-1 positive cells as endothelial cells was provided by the demonstration that capillaries of postnatal day 4 brain were highly enriched for KDR-flk-1 mRNA. Thus, in contrast with the FGF receptors bek and flg, which are expressed in endothelial cells *in vitro* but not *in vivo* (Heuer et al., 1990; Wanaka et al., 1991; Peters et al., 1992), KDR-flk-1 is expressed during physiological processes that are characterized by endothelial cell proliferation. The temporal and spatial expression patterns found in the embryonic brain correlate precisely with the development of the neural vascular system as described by Bär (1980). Vascular sprouts originating in the perineural plexus grow radially into the neuroectoderm and branch there. These sprouts express high amounts of KDR-flk-1 mRNA. In the early postnatal stages, endothelial cell proliferation is still evident and KDR-flk-1 is expressed, whereas, after completion of the vascularization process in the adult organism, the turnover of endothelial cells and the expression of KDR-flk-1 is very low (Engermann et al., 1967; Robertson et al., 1985; Millauer et al., 1993). KDR-flk-1 mRNA has also been found in the cells of the yolk sac mesodermal layer and in the peripherally located endothelial cell precursors of the blood islands, suggesting an involvement of this receptor in the early segregation of angioblasts from a committed mesodermal stem cell that may differentiate into hematopoietic precursor cells and endothelial cells. In quail and chick embryos, the homologue of KDR-flk-1 and flt named Quek 1 and Quek 2 are expressed essentially in endothelial cells. Quek 1 is expressed in the mesoderm from the onset of gastrulation and Quek 2 is first expressed on endothelial cells and white blood cells. The expression pattern of Quek1 suggests that it could identify the putative precursor of both endothelial and hematopoietic lineages, the hemangioblast. Quek 1 and Quek 2 are not expressed in all endothelial cells throughout life. At E9, after the initial phase of vasculogenesis, these genes are switched off in various compartments of the vascular network (Eichmann et al., 1993). The report by Matthews et al. (1991) of KDR-flk-1 mRNA expression in fetal liver cells that were enriched for hematopoietic stem cells supports the view that KDR-flk-1 may play a role in common progenitors of hematopoietic and endothelial cells.

VI. EXPRESSION OF VEGF/VPF AND ITS RECEPTORS IN THE FEMALE REPRODUCTIVE SYSTEM

Because the turnover rates of vascular endothelial cells are extremely low, angiogenesis is generally a quiescent process in the healthy adult organism. A marked exception is the female reproductive system, where the need for renewed microvascularization is constantly imposed by the cyclic evolution of ovarian transient structures and by the cyclic repair of damaged uterine tissues. The reproductive system provides an attractive model to critically evaluate the physiological relevance of various angiogenic factors to naturally occurring angiogenesis. It has the advantages that angiogenesis can be readily induced and synchronized by gonadotropins, and that it changes over a relatively short time with a well-characterized schedule. Specifically, in every estrous cycle, the sequential development of ovarian follicles and corpora lutea is accompanied by a concomitant development of elaborate capillary networks. The latter are formed in the periphery of each growing follicle or in the vicinity of luteal cells, respectively (Findlay, 1986; Gospodarowicz, 1980). Similarly, in the uterus, blood vessels of the functional endometrium undergo cyclic extensions and repair, compensating for vessels that have been damaged or sloughed off (Christianes et al., 1982). Extensive neovascularization of the endometrium also takes place upon decidual transformation in response to embryonic implantation. The increase in decidual mass is accompanied by induction of neovascularization and formation of an anastomosing network traversing the maternal decidua and converging on the embryo (Motro et al., 1990). This vascular network supplies the maternal contribution of the future placenta. These four processes represent independent examples of programmed self-limiting angiogenesis within the female reproductive system.

In all four processes, VEGF is expressed in space and time close to the forming vasculature. Specifically, during follicular neovascularization, VEGF was produced by cells in the theca interna and theca externa (Shweiki et al., 1993). During the formation of the vasculature of the corpus luteum (Shweiki et al., 1993; Phillips et al., 1990), VEGF production is predominantly detected in luteal cells. During neovascularization of the endometrium, VEGF is detected throughout the endometrial stroma, and during vascularization of the decidua, VEGF is expressed in the decidua basalis. These sites of VEGF expression, colocalize with the sites where new vasculature has developed. These observations suggest that, during these processes, VEGF/VPF acts primarily on nearby cells. It appears that the locally released growth factor determines the area of endothelial cell growth by creating a microenvironment supporting the proliferation of nearby endothelial cells (Shweiki et al., 1993). Furthermore, uneven distribution of extracellular VEGF in the neovascularized tissue may target capillary growth toward cells producing the highest levels of VEGF/VPF (Sweiki et al., 1993).

A principal complication of ovulation induction therapy is the ovarian hyperstimulation syndrome (OHSS). The main problem in OHSS is the shift of body fluids from the vascular bed onto the extravascular space. Because this therapy involves the use of follicle stimulating hormone combined with an unphysiological dose of human chorionic gonadotropin (LH) and because LH is known to increase VEGF mRNA (Garrido et al., 1993), it raises the possibility that overexpression of VEGF might cause OHSS. Recent studies by McClure et al. (1994) have demonstrated the presence of VEGF/VPF in OHSS ascites and liver ascites fluid.

The importance of VEGF/VPF in ovarian vascularization has also been studied transplanting juvenile rat ovaries to an ectopic site (Dissen et al., 1994). This results in a prompt and massive revascularization of the gland. VEGF mRNA levels in transplanted ovaries increase 60-fold over the control from 24 to 48 hours after transplantation (Dissen et al., 1994). The increase in mRNA abundance detected in the thecal compartment of preovulatory follicles and the appearance of the message in granulosa cells of the cumulus oophorus of such follicles suggests (Ravindranath et al., 1992; Phillips et al., 1990; Shweiki et al., 1993) that expression of the VEGF gene in the ovary is regulated by gonadotropins. Consonant with this view, treatment with a LHRH antagonist inhibits the posttransplantation increase in ovarian VEGF mRNA levels. This finding also agrees with the recent observation that treating nonhuman primates with a LHRH antagonist during the luteal phase of the menstrual cycle reduces VEGF mRNA levels in the corpus luteum (Ravindranath et al., 1992).

VEGF/VPF is one of the most potent stimulators of microvascular permeability known (Senger et al., 1986; Connolly, 1991). An early uterine response to estrogen is a marked increase in vascular permeability resulting in uterine edema (Clark and Markaverich, 1988; Astwood, 1938). The increase in VEGF mRNA in the uterus (Cullinan-Bove and Koos, 1993) immediately precedes the increase in water retention, which begins within 1–2 hours after estrogen administration and is maximal at 3–6 hours (Ham et al., 1970; Astwood, 1938), doubling the uterine wet weight by 6 hours. The edema then decreases (Ham et al., 1970; Astwood, 1938; Hechta et al., 1941; Friederici, 1967), consistent with the preceding decline in VEGF/VPF mRNA levels. Estrogen- induced water imbibition in the uterus is inhibited by puromycin, which blocks protein synthesis, and actinomycin-D, which inhibits transcription (Ui and Mueller, 1963; Mueller et al., 1961). This suggests that imbibition is due to increased synthesis of a protein factor. The possibility that this factor could be VEGF/VPF is further supported by the observation that VEGF/VPF expression is also high in the corpus luteum of the ovary (Phillips et al., 1990; Cullinan–Bove and Koos, 1993), a tissue with perhaps the most permeable blood vessels in the body (Morris and Sass, 1966). The increase in uterine vascular permeability is believed due to the formation of gaps between capillary endothelial cells (Ham et al., 1970; Friederici, 1967). Interestingly, most mediators of increased permeability, such as histamine and serotonin, act predominantly on venules rather than capillaries (Ham et al., 1970). Capillaries are probably a site of VEGF/VPF action, however, because capillary endothelial cells in the lung, kidney, and other tissues have binding sites for VEGF/VPF (Ferrara et al., 1992).

VII. VEGF/VPF AND THERAPEUTIC ANGIOGENESIS

Many of the failures of vascular reconstructions, endarterectomies, and angioplasties can be attributed to the inability of the vascular endothelium to adequately regenerate and cover the denuded area at the side of arterial disruption (Callow and O'Donnell, 1986; Ross, 1986; Lange et al., 1991). An approach toward enhancing reendothelialization lies in the use of growth factors to increase the speed and completeness of endothelial repaving. VEGF/VPF administration following balloon angioplasty (Callow et al., 1994) results in reendothelization without vascular smooth muscle proliferation, a side effect often encountered when reendothelization is induced by bFGF (Lindner et al., 1990, 1991). Therefore, VEGF would be preferred to induce endothelium repair after denudation.

In blood vessels, an endogenous production of VEGF/VPF could take place in fetal human aortic smooth muscle cells which express and release VEGF/VPF (Tischer et al., 1991). Adult bovine aortic SMCs also express the VEGF transcript and secrete VEGF (Ferrara et al., 1991). The localization of VEGF/VPF in aortic smooth muscle cells could, therefore, have significant implications for maintaining the integrity of the endothelial cells in the intima.

In another model, the hind limb ischemia model, which simulates ischemia characteristic of patients with severe lower extremity arterial occlusive disease, VEGF/VPF administration results in increased angiogenesis (Takeshita et al., 1994). This response to VEGF/VPF administration was observed at two levels. Autopsy documented an increase in vascularity at the capillary level, consistent with the classical definition of angiogenesis proposed by Klagsbrun and Folkman (1990). In addition, systematic quantification of angiographic recordings performed serially at 10, 20, and 40 days postoperation demonstrated that development of angiographically visible collateral arteries in the VEGF-treated animals exceeded that observed in the control group, both at days 20 and 40. From an anatomic standpoint, angiographic improvement in the treatment group appeared restricted to the 10 days after administration of VEGF, after which a plateau was observed for the subsequent 20-day period.

Likewise, Banai et al. (1994) have reported that intracoronary VEGF enhances the development of small coronary arteries supplying ischemic myocardium, resulting in marked augmentation of maximal collateral blood flow delivery. These results demonstrate the feasibility of pharmacologically enhancing collateral growth and suggest a new therapeutic approach for treating myocardial ischemia.

VIII. VEGF/VPF AND THE DISEASE STATE

VEGF has been implicated in a number of diseases sharing, in common, increased angiogenesis, increased vascular permeability, and hypoxic conditions. Those diseases range from ocular diseases to tumor formation and inflammatory diseases of joints and the skin.

A. Ocular Diseases

Intraocular neovascularization occurs in numerous ischemic retinal disorders, including diabetic retinopathy, ischemic retinal vein occlusion, and retinopathy of prematurity (Ashton, 1957; Michaelson, 1948). This proliferation often results in vitreous hemorrhage, retinal detachment, and neovascular glaucoma, with subsequent visual loss. Although numerous growth factors stimulate angiogenesis *in vivo*, none have yet been shown to be causally related to ocular diseases (D'Amore et al., 1981; Swalingam et al., 1990; Senger et al., 1993; Dvorak et al., 1979).

Proliferation of vascular elements in proliferative diabetic retinopathy and neovascularization of the retina and/or iris secondary to central retinal vein occlusion, retinal detachment, and intraocular tumors are always accompanied by induction of retinal VEGF expression (Pe'er et al., 1995). Furthermore, in each case, expression of VEGF is induced only in a particular layer of the retina, matching the zones affected by impaired perfusion. In a rabbit model simulating retinal vein occlusion, elevated levels of VEGF mRNA can be detected within a few days of experimental induction of retinal ischemia, exclusively in the ischemic region (Pe'er et al., 1995).

Using radioimmunoassays and radioreceptor assays, Aiello et al. (1994) have determined the concentrations of VEGF in ocular fluid of patients with diabetic retinopathy and other retinal disorders. VEGF was detected in 69 of 136 ocular fluid samples from patients with diabetic retinopathy, 29 of 38 samples from patients with neovascularization of the iris, and 3 of 4 samples from patients with ischemic occlusion of the central retinal vein, as compared with 2 of 31 samples from patients with no neovascular disorders. The concentration of VEGF was greater in the vitreous than in the aqueous humor consistent with the establishment of a gradient. It declined after successful laser photocoagulation suggesting that VEGF played a major role in mediating active intraocular neovasculization (Aiello et al., 1994).

Similar results were reported by Miller et al. (1994) following the production of retinal ischemia by laser occlusion of all branch retinal veins. VEGF was increased in the aqueous fluid, and the aqueous VEGF levels changed synchronously and proportionally with the severity of iridic neovascularization. Northern analysis and *in situ* hybridization revealed that VEGF messenger RNA was upregulated in the ischemic retina. These observations support the hypothesis that ocular neovascularization is regulated by a diffusible factor and identify VEGF as a likely candidate for a retina-derived vascular permeability and angiogenesis factor *in vivo*.

One of the earliest changes in the retina in diabetic patients is the breakdown of the blood-retinal barrier in the retinal capillaries and excessive leakiness of these vessels. Macular edema, in fact, is probably the greatest single cause of visual impairment in diabetic patients. Likewise, new blood vessels, that develop on the iris, leak significantly more than normal iridic vessels. The identity of the factor causing vascular permeabilization is unknown. While it is possible that intraocular neovascularization, per se, and abnormal leakiness of the newly formed vessels are caused by different factors, it is equally possible that these two phenotypes are caused by the same factor, because VEGF/VPF is active as an endothelial cell-specific mitogen and a potent vascular permeabilization factor (Senger et al., 1986; Connolly, 1991). Excessive production of VEGF/VPF during the progression of diabetic retinopathy and rubeosis iridis may also account for the increased vascular permeability that characterizes these diseases.

B. VEGF/VPF and Tumor Angiogenesis

Malignant brain tumors show increased vascularity and are usually embedded in edematous brain tissue. Vascular permeability activity has been demonstrated earlier in connection with many tumor studies, and it was also identified in the conditioned media of brain tumor cells (Criscuolo et al., 1988, 1989; Senger et al., 1986; Conn et al., 1990). More recent studies showed that capillary permeability-enhancing activity is present in the conditioned media of glioblastoma multiforme cells and also can be detected in brain cyst fluids, in which the activity can be blocked by neutralizing antibodies against VEGF/VPF (Berkman et al., 1993). These studies also showed that there is a significant elevation of VEGF/VPF gene expression in highly vascularized and edema-associated brain tumors when compared with brain tumors with less neovascularization and little or no edema (Berkman et al., 1993). To assess and quantify the expression of the VEGF/VPF gene and of the receptor gene, 26 surgical specimens of brain tumor tissue from 24 patients were analyzed. In most malignant gliomas, the expression level of the VEGF/VPF gene was elevated and can be up to 20 to 50-fold higher compared with low-grade tumors (Weindel et al., 1994) where significant expression of PIGF was observed. Using an intracerebral rat glioma model, Plate et al. (1993) analyzed VEGF/VPF regulation *in vitro* and expression of VEGF and its high-affinity tyrosine kinase receptors, flt-1 and KDR-flk-1, *in vivo*. *In situ* hybridization showed that VEGF is expressed *in vivo* in rat glioma cells which reside along necrotic areas, and, therefore, it closely mimics the expression pattern of VEGF observed in human glioblastoma. flt-1 and KDR-flk-1 are specifically expressed in endothelial cells in the tumor and at the border between tumor and normal brain but are absent from endothelial cells in the normal brain proper. The action of VEGF may, therefore, be restricted to tumor endothelium.

The localized upregulation of VEGF expression may be explained in two different ways. First, VEGF expression has been shown to be inducible by hypoxia (Shweiki et al., 1992; Plate et al., 1992). The rapid cell proliferation in the center of a tumor can lead to increased interstitial pressure, which may lead to compression closure of capillaries and consecutive tissue necrosis (Folkman and Klagsbrun, 1987). On the way to necrosis, the increasing hypoxia may upregulate VEGF/VPF. Alternatively, upregulation of VEGF/VPF could be attributed to factors released from necrotic tumor cells.

Two other types of tumors, renal cell carcinomas (Takahashi et al., 1994) and bladder carcinomas (Brown et al., 1993) have been analyzed for VEGF/VPF expression. Northern blot analysis showed that 26 of 27 hypervascular renal cell carcinoma tissues (96%) exhibited a markedly elevated level (3–13-fold) of VEGF/VPF mRNA compared to the adjacent normal kidney tissues. Even tumors of small size, whenever they were hypervascular, overexpressed VEGF mRNA. mRNA for PIGF was also expressed in 21 of 23 hypervascular renal cell carcinoma tissues but was not detected in the adjacent normal kidney tissues. Two hypovascular carcinoma tissues neither overexpressed VEGF mRNA nor had PIGF mRNA. In another study (Brown et al., 1993), VEGF/VPF mRNA and protein were shown to be strongly expressed in renal cell carcinomas and transitional cell carcinomas. Moreover, in contrast to endothelial cells in the normal kidney, endothelial cells lining tumor blood vessels strongly expressed mRNAs for the VPF receptors flt-1 and KDR-flk-1 and stained strongly for the VEGF/VPF ligand.

Direct proof for a role of VEGF in tumor angiogenesis, which requires the inhibition of VEGF-receptor activation *in vivo* has been provided by two different lines of investigation. Treatment with a monoclonal antibody specific for VEGF inhibited the growth of human rhabdomyosarcoma, glioblastoma multiforme, or leiomyosarcoma cell lines into nude mice. They had no effect on the growth rate of the tumor cells *in vitro*. The density of vessels was decreased in the antibody-treated tumors (Jin Kim et al., 1993). These findings demonstrate that inhibition of the action of VEGF/VPF spontaneously produced by tumor cells may suppress tumor growth *in vivo*. The biological relevance of the KDR-flk-1 receptor/ligand system for angiogenesis has been studied by Millauer et al. (1994). Use was made of a retrovirus encoding a dominant-negative mutant of the KDR-flk-1/VEGF receptor to infect endothelial target cells *in vivo*. To test this model, C6 glioblastoma cells were implanted subcutaneously in nude mice, either alone or with different relative amounts of cells producing comparable titers of recombinant retrovirus. The group coimplanted with the control virus-producing cell line showed the same growth behavior as the implant of tumor cells alone. This was also true for a group coimplanted with cells expressing wild-type KDR-flk-1. In contrast, the group inoculated with retrovirus-producing cells inducing the expression of the KDR-flk-1 dominant-negative deletion mutant had a tumor growth rate dramatically decreased. The ability to block VEGF-mediated signal transduction with the KDR-flk-1 dominant-negative mutant emphasizes the central role of the KDR-flk-1/VEGF system in tumor angiogenesis and further supports the contention that this system is the major regulator of angiogenesis and vasculogenesis in the whole organism.

C. VEGF/VPF in Psoriasis and Rheumatoid Arthritis

Psoriasis is a common, chronic skin disease characterized by recurrent erythematous skin plaques that exhibit epidermal hyperplasia, a variable inflammatory cell infiltrate, and abnormalities of the papillary dermal vasculature (Weinstein et al., 1985; Fry, 1988; Christophers et al., 1973; Ryan, 1980). Microvessels in the papillary dermis of psoriatic plaques are elongated, dilated, and hyperpermeable (Braverman and Keh Yen, 1986; Bull et al., 1992) and more closely resemble postcapillary venules than the capillary loops of normal skin (Braverman and Sibley, 1982; Jungkunz et al., 1992). Detmar et al. (1994), looking at the expression of VEGF/VPF and its receptors in psoriasis, have reported increased expression of VEGF/VPF mRNA by the hyperplastic epidermis of lesional psoriatic skin and increased expression of the two VEGF/VPF receptors in psoriatic dermal microvessels. *In vitro* investigations showed that VEGF/VPF expression and secretion of bioactive VEGF by epidermal keratinocytes were induced by TGFα, a cytokine known to be overexpressed in psoriatic epidermis (Gottlieb et al., 1988; Elder et al., 1989). Therefore, besides its mitogenic effect on epidermal keratinocytes, TGFα probably regulates VEGF/VPF expression in psoriatic epidermis by an autocrine mechanism, leading to increased vascular permeability and angiogenesis.

Angiogenesis is also important in the proliferation of inflammatory synovial tissue because development of the destructive synovial pannus in rheumatoid arthritis (RA) is associated with changes in vascular permeability (synovial fluid accumulation) and angiogenesis. Because VEGF/VPF has been implicated in both events, its expression in RA synovial tissues has been examined. *In situ* hybridization studies performed on joint tissues from patients with active RA revealed that synovial lining macrophages strongly expressed VEGF/VPF mRNA and that microvascular endothelial cells of nearby blood vessels strongly expressed mRNA for the VEGF/VPF receptors, flt-1 and KDR-flk-1. Immunohistochemistry performed on inflamed rheumatoid synovial tissue revealed that the VEGF/VPF peptide was localized to macrophages within the inflamed synovium and to microvascular endothelium, its putative target in the tissue (Fava et al., 1994). Together, these findings indicate that VEGF/VPF may be important in the pathogenesis of RA.

Koch et al. (1994) have also examined the potential role of VEGF in the RA joint using synovial tissue and synovial fluid samples from 55 patients. Antigenic VEGF was increased in synovial fluids from patients with RA compared with synovial fluids from patients with osteoarthritis or other diseases. Moreover, RA synovial fluid was chemotactic for human endothelial cells *in vitro* and anti-VEGF neutralized more than half of the chemotactic activity for endothelial cells found in these fluids. Similarly, the mitogenic activity for microvascular endothelial cells found in RA synovial tissue explant cultures was suppressed by 28% upon incubation of these conditioned media with anti-VEGF antibodies, suggesting that VEGF/VPF could be a contributing factor in that disease.

IX. TARGET CELLS AND TISSUES OTHER THAN THE VASCULAR ENDOTHELIUM RESPONSIVE TO VEGF/VPF

Although it was first thought that there was a unique cellular and tissue specificity for VEGF, this paradigm was only valid within the range of tissues analyzed for their response or expression of VEGF/VPF. While it is correct to assume that the vascular endothelium is one of the primary targets for VEGF, other tissues have also been shown to be potential targets.

As pointed out earlier, VEGF/VPF binding sites can be detected on cell types, such as corneal endothelial cells, Hela cells, and on several lines of human melanoma cells, but not on melanocytes (reviewed by Neufeld et al., 1994). Although the significance of those sites cannot yet be assessed, in view of the lack of biological effect, the possibility exists that VEGF could exert an effect on those cells which would differ from the mitogenic and increased permeability seen with vascular endothelial cells. Among cell types for which binding of VEGF has been correlated with a biological effect are Balb/c 3T3, monocytes, osteoblasts, and β cells of the pancreas.

In the case of Balb/c3T3, the VEGF induced drastic morphological changes (Enomoto et al., 1994). Upon its addition, VEGF induced the disorganization of actin-stress fiber accompanied by drastic cell rounding. Since this effect could be inhibited by a tyrosine kinase inhibitor, genistein, it was concluded that protein phosphorylation may be central in these events (Enomoto et al., 1994). In the case of monocytes, VEGF has been shown to induce monocyte migration and also monocytic procoagulant activity due to enhanced synthesis and expression of tissue factor, a cell-surface procoagulant factor (Clauss et al., 1990). This observation agrees with the ability of VEGF/VPF to modulate the procoagulant activity of the vascular endothelium and to synergize with TNFα (Clauss et al., 1990). VEGF/VPF also binds to osteoblasts and, although it does not stimulate their proliferation, it induces migration, PTH-dependent cAMP accumulation, and alkaline phosphatase increase when added to osteoblast. The maximal effects reach levels comparable to those obtained with bone morphogenetic protein 2 (BMP-2), although the VEGF/VPF concentrations required are at least 100-fold lower (Midy and Plouet, 1994). Those observations are particularly relevant to bone metabolism.

Prostaglandins (PGs) are produced in bone in response to inflammation, injury, and mechanical stress (Murray and Rushton, 1990) and have been implicated in the local regulation of bone metabolism (Raisz and Martin, 1994). Increasing evidence indicates that PGs stimulate bone formation *in vivo*.

One of the most pronounced effects of PGs *in vivo* is on the vascular system. PGE_2 and PGE_1 can cause vasodilation and stimulate angiogenesis (Form and Auerbach, 1983; Ziche et al., 1982; Spismi et al., 1992). However, PGs do not stimulate endothelial cell growth, suggesting that the angiogenic effect may be mediated by the paracrine action of angiogenic factors released from other cells. VEGF mRNA is expressed in rat calvaria-derived osteoblast-enriched cells, in clonal osteoblastic cells, and in bone tissue. PGE_2, which is a potent stimulator of osteogenesis *in vivo*, causes rapid induction of VEGF mRNA and VEGF production in osteoblastic cells (Harada et al., 1994). The rapid short-lived induction of VEGF mRNA in osteoblastic cells by PGE_2 is consistent with a tight regulation of angiogenesis in time and space. A role for VEGF in bone metabolism is further supported by the suppression of PGE_2-induced VEGF mRNA by glucocorticoids. Long-term glucocorticoid treatment inhibits bone formation and causes bone loss (Baylink, 1983). Furthermore, glucocorticoid treatment can induce avascular necrosis of bone (Cruess, 1981). Suppression of VEGF expression by glucocorticoid could thus be involved in the pathogenesis of glucocorticoid-induced bone disorders.

In pancreatic tissue, KDR-flk-1 is expressed and this correlates with the ability of VEGF/VPF to stimulate both the insulin content/islet DNA ratio and the accumulation of insulin in the culture medium without affecting the rates of β-cell replication (Öberg et al., 1994). To investigate the localization of expression of the flk-1 receptor in the pancreas, serial sections of fetal pancreata were immunostained for flk-1 and insulin. Expression of flk-1 was detected in endothelial-like cells and cells lining pancreatic ducts. The latter are considered to contain precursor cells for the endocrine pancreas (Pictet and Rutter, 1972). It suggests that flk-1 positive cells could be the precursor cells for β-cells and that flk-1 could play an important role for the maturation of β cells from precursor cells (Öberg et al., 1994).

X. CONCLUSION

VEGF/VPF is probably unique as a growth factor in its ability to cause major changes in the vascular endothelium. Its various effects on that tissue can vary from increasing vascular permeability, to stimulating tissue factor expression, and inducing cell proliferation. Since the structural characterization of VEGF/VPF, evidence has been rapidly accumulating that it plays a major role in regulating angiogenesis in normal development and in a variety of diseases where defects in neovascularization are prominent. These include ocular diseases, such as diabetic retinopathy, and other diseases, such as psoriasis or tumor growth. In the case of tumor angiogenesis, VEGF/VPF seems to be the best candidate for the long sought tumor angiogenesis factor. Its presence in tumor correlates with the fact that tumor blood vessels are leaky, a known effect of VEGF/VPF, but VEGF/VPF expression can also be induced by hypoxia. This condition is characteristic of rapidly growing tumors and is reflected in necrosis of the central areas. The identification of VEGF/VPF as a tumor angiogenic factor can lead to new approaches to tumor therapy using antiangiogenesis strategy based on inhibiting either the growth factor or inhibiting the growth factor receptor interaction. Such strategies which could involve the use of antisense agents, specific antibodies, or small inhibitors of growth factors/receptor interactions could also be used in other diseases, such as diabetic retinopathy, where VEGF/VPF has been implicated in a major role. Because estrogens have been reported to induce VEGF expression, it will also be of interest to determine to what extent their growth of estrogen-dependent tumors depend on VEGF.

The relative importance of VEGF/VPF in controlling angiogenesis is not limited to the disease state alone. Its expression in ovarian tissue has also helped to explain the cyclic renewal of the ovarian microvascularization. In particular, the control of VEGF/VPF expression by both LH and estrogens has made it of major importance in ovarian physiology, in particular, where ovulation and corpus luteum development are concerned. As in the case of tumor, blood vessels of the corpus luteum are extremely leaky and their hyperpermeability could be controlled by VEGF/VPF. The best model for increased permeability and edema unrelated to diseases is provided by the transient edema induced in the uterus by estrogens. Here again, VEGF/VPF appears to play a major role and could become a key intermediary in estrogen action.

Although rapid progress has been made in our understanding of the *in vivo* role of VEGF/VPF, numerous questions remain unanswered. Little is known of the signal-transduction mechanisms mediating the various effects of VEGF/VPF. Also, we do not know which receptors mediate the vascular permeability effect of VEGF/VPF and which mediate its mitogenic activity. The effect of VEGF/VPF in very early embryonic development, a step where other growth factors have been shown to play a major role in determining tissue and pattern formation (Klein and Melton, 1994), has also not been examined. Unresolved is whether the vascular endothelium will remain a preferential target for VEGF/VPF or whether this factor could have effects on other tissues as well. Among those could be the immune system because VEGF/VPF has already been shown to act on monocytes. Whether or not VEGF/VPF could play a role, directly or indirectly, in granuloporesis has not yet been examined. Likewise, the role of VEGF/VPF in bone metabolism, *in vivo*, and in bone marrow formation remained to be assessed.

NOTE ADDED IN PROOFS

The ligand for Flt-4 has been identified (Joukov et al., 1996). It does interact with both Flt-4 and KDR. Therefore, it can both affect lymphatic endothelia which express Flt-4 as well as vascular endothelia expressing KDR. Following the report that Flk-1 is an early marker for endothelial cell precursors (Yamaguchi et al., 1993), null mutation of the Flt-1 and KDR receptor in mice has further defined their pivotal role in endothelial development. Mice deficient in Flk-1 die *in utero* between 8.5 and 9.5 days post-coitum. Yolk sac blood islands were absent at 7.5 days, organized blood vessels could not be observed in the embryos or yolk sac at any stage and hematopoietic progenitors were severely reduced (Shalaby et al., 1995). Mice deficient in Flt-1 had disorganized blood islands as well as a disorganized vascular phenotype indicating that Flt-1 is essential for the organization of the embryonic vasculature but is not essential for vascular differentiation (Fong et al., 1995). In regard to OHSS further studies (Neulen et al., 1995) strongly suggest that it could be the result of an overproduction of VEGF. Finally, suppression of retinal neovascularization *in vivo* using soluble VEGF receptor chimeric proteins has been reported (Aiello et al., 1995).

REFERENCES

Abraham, J. A., Mergia, A., Whang, J. L., Tumolo, A., Friedman, J., Hjerrild, K. A., Gospodarowicz, D., & Fiddes, J. C. (1986). Nucleotide sequence of a bovine clone encoding the angiogenic protein, basic fibroblast growth factor. Science 233, 545–547.

Aiello, L. P., Avery, R. L., Arrigg, P. G., Kerpt, B. A., Jampel, H. D., Shah, S. T., Pasquale, L. R., Thieme, H., Kwamoto, M. A., Park, J. E., Nguyen, H. V., Aiello, L. M., Ferrara, N., & King, G. L. (1994). VEGF in ocular fluid of patients with diabetic retinopathy and others retinol disorders. N. Engl. J. Med 331, 1480–1487.

Aiello, L. P., Pierce, E. A., Foley, E. D., Takagi, H., Chen, H., Riddle, L., Ferrara, N., King, G. L., & Smith, L. E. H. (1995). Suppression of retinal neovascularization *in vivo* by inhibition of vascular endothelial growth factor (VEGF) using soluble VEGF-receptor chimeric proteins. Proc. Natl. Acad. Sci. USA 92, 10457–10461.

Alder, V. A., Ben-Nun, J., & Cringle, S. J. (1990). PO_2 profiles and oxygen consumption in cat retina with an occluded retinal circulation. Invest. Opthalmol. Vis. Sci. 31, 1029–1034.

Ashton, N. (1957). Retinal vascularization in health and disease. Am. J. Ophthalmol. 44, 7–24.

Astwood, E. B. (1938). A six-h assay for the quantitative determination of estrogen. Endocrinology 23, 25–31.

Baird, A., Mormede, P., & Bohlen, P. (1985). Biochem. Biophys. Res. Commun. 126, 358–363.

Banai, S., Jaklitsch, M. T., Shou, M., Lazarous, D. F., Scheinowitz, M., Buo, S., Epstein, S. E., & Unger, E. F. (1994). Angiogenesis-induced enhancement of collateral blood flow to ischemic myocardium by VEGF in dogs. Circulation 89, 2183–2189.

Bär, T. (1980). The vascular system of the cerebral cortex. Adv. Anat. Embryol. Cell. Biol. 59, 1–62.

Bassett, D. L. (1943). The changes in the vascular pattern of the ovary of the albino rat during the estrous cycle. Am. J. Anat. 73, 251–291.

Baylink, D. J. (1983). Glucocorticoid-induced osteoporosis. N. Engl. J. Med. 309, 306–308.

Beck, I., Weinmann, R., & Caro, J. (1993). Characterization of hypoxia responsive enhancer in the human erythropoietin gene reveals presence of hypoxia inducible 120-kDa nuclear DNA-binding protein in erythropoietin producing and nonproducing cells. Blood 82, 704–711.

Berkman, R. A., Merill, M. J., Reinhold, W. C., Tonacci, W. T., Saxena, A., Clark, W. C., Robertson, I. T., Ali, I. U., & Oldfield, E. H. (1993). Expression of the vascular permeability/vascular endothelial growth factor gene in central nervous system neoplasms. J. Clin. Invest. 91, 153–159.

Bikfalvi, A., Sauzeau, C., Moukadiri, H., Maclouf, J., Busso, N., Bryckaert, M., Plouet, J., & Tobelem, G. (1991). Interaction of vasculotropin/vascular endothelial cell growth factor with human umbilical vein endothelial cells: Binding, internalization, degradation, and biological effects. J. Cell Physiol. 149, 50–59.

Blanchard, K. L., Acquaviva, A. M., Galson, D. L., & Bunn, H. F. (1992). Hypoxic induction of the human erythropoietin gene: Cooperation between the promoter and enhancer, each of which contains steroid receptor response elements. Mol. Cell. Biol. 12, 5373–5385.

Braverman, I. M. & Sibley, J. (1982). Role of the microcirculation in the treatment and pathogenesis of psoriasis. J. Invest. Dermatol. 78, 12–22.

Braverman, I. M. & Keh–Yen, A. (1986). Three-dimensional reconstruction of endothelial cell gaps in psoriatic vessels and their morphologic identity with gaps produced by the intradermal injection of histamine. J. Invest. Dermatol. 86, 577.

Breier, G., Albrecht, U., Sterrer, S., & Risau, W. (1992). Expression of vascular endothelial growth factor during embryonic angiogenesis and endothelial cell differentiation. Development 114, 521–532.

Brock, T. A., Dvorak, H. F., & Senger, D. R. (1991). Tumor-secreted vascular permeability factor increases cytosolic Ca^{2+} and von Willebrand factor release in human endothelial cells. Am. J. Pathol. 138, 213–221.

Brown, L. F., Berse, B., Jackman, R. W., Tognazzi, K., Manseau, E. J., Dvorak, H. F., & Senger, D. R. (1993). Increased expression of VEGF/VPF and its receptors in kidney and bladder carcinoma. Am. J. Pathol. 143, 1255–1262.

Bull, R. H., Bates, D. L., & Mortimer, P. S. (1992). Intravital videocapillaroscopy for the study of microcirculation in psoriasis. Br. J. Dermatol. 126, 436.

Callow, A. D. & O'Donnell, T. F., Jr. (1986). Recurrent carotid stenosis: Frequency, clinical implications, and some suggestions concerning etiology. In: Reoperative Arterial Surgery (Bergan, J. J. & Yao, J. S. T., eds,) pp. 513–535. Grune and Stratton, New York.

Callow, A. D., Choi, E. T., Trachtenberg, J. D., Stevens, S. L., Connolly, D. T., Rodi, C., & Ryan, U. S. (1994). Vascular permeability factor accelerates endothelial regrowth following balloon angioplasty. Growth Factors 10, 223–228.

Charnock–Jones, D. S., Sharkey, A. M., Rajput–Williams, J., Burch, D., Schofield, J. P., Fountain, S. A., Boocock, C. A., & Smith, S. K. (1993). Identification and localization of alternately spliced mRNAs for vascular endothelial growth factor in human uterus and estrogen regulation in endometrial carcinoma cell lines. Biol. Reprod. 48, 1120–1128.

Christianes, G. C. M. L., Sixma, J. J., & Haspeis, A. A. (1982). Hemostasis in menstrual endometrium: A review. Obstet. Gynecol. Survey 37, 281–303.

Christophers, E., Parzefall, R., & Braun–Falco, O. (1973). Initial events in psoriasis: Quantitative assessment. Br. J. Dermatol. 89, 327.

Claffey, K. P., Wilkinson, W. O., & Spiegelman, B. M. (1992). Vascular endothelial cell growth factor: Regulation by cell differentiation and activated second messenger pathways. J. Biol. Chem. 267, 16317–16322.

Clark, J. H. & Markaverich, B. M. (1988). Actions of ovarian steroids. In: The Physiology of Reproduction (Knobil, E. & Neill, J., eds.) pp. 675–724, Raven Press, New York.

Clauss, M., Gerlach, M., Gerlach, H., Brett, J., Wang, F., Familetti, P. C., Pan, Y. C., Olander, J. V., Connolly, D. T., & Stern, D. (1990). Vascular permeability factor: A tumor-derived polypeptide that induces endothelial cell and monocyte procoagulant activity and promotes monocyte migration. J. Exp. Med. 172, 1535–1545.

Coffin, J., Harrison, J., Schwartz, S., & Heimark, R. (1991). Angioblast differentiation and morphogenesis of the vascular endothelium in the mouse embryo. Dev. Biol. 148, 51–62.

Conn, G., Bayne, M. L., Soderman, D. D., Kwok, P. W., Sullivan, K. A., Palisi, T. M., Hope, D. A., & Thomas, K. A. (1990). Amino acid and cDNA sequences of a vascular endothelial cell mitogen that is homologous to platelet-derived growth factor. Proc. Natl. Acad. Sci. USA 87, 2628–2634.

Conn, G., Soderman, D. D., Scaeffer, M. T., Wile, M., Hatcher, V., & Thomas, K. A. (1990). Purification of a glycoprotein vascular endothelial cell mitogen from a rat glioma-derived cell line. Proc. Natl. Acad. Sci. USA 87, 1323–1327.

Connolly, D. T., Heuvelman, D. M., Nelson, R., Olander, J., Eppley, B., Delfino, J. J., Siegel, R. N., Leingruber, R. S., & Feder, J. (1989). Tumor vascular permeability factor stimulates endothelial cell growth and angiogenesis. J. Clin. Invest. 84, 1470–1478.

Connolly, D. T., Olander, J. V., Heuvelman, D., Nelson, R., Monsell, R., Siegel, N., Haymore, B. L., Leimbruger, R., & Feder, J. (1989). Human vascular permeability factor. Isolation from U937 cells. J. Biol. Chem. 264, 20017–20024.

Connolly, D. T., Olander, J. V., Heuvelman, D., Nelson, R., Monsell, R., Siegel, N., Haymore, B. L., Leimgruber, R., & Feder, J. (1989). Vascular permeability factor, an endothelial cell mitogen related to PDGF. Science 246, 1309–1312.

Connolly, D. T. (1991). Vascular permeability factor: A unique regulator of blood vessel function. J. Cell Biochem. 47, 219–223.

Criscuolo, G. R., Merril, M. J., & Oldfield, E. H. (1988). Further characterization of malignant glioma-derived vascular permeability factor. J. Neurosurg. 69, 254–262.

Criscuolo, G. R., Lelkes, P. I., Rotrosen, D., & Oldfield, E. H. (1989). Cytosolic calcium changes in endothelial cells induced by a protein product of human gliomas containing vascular permeability factor activity. J. Neurosurg. 71, 884–891.

Cruess, R. L. (1981). Steroid-induced osteonecrosis: A review. Can. J. Surg. 24, 567–571.

Cullinan–Bove, K. & Koos, R. D. (1993). VEGF/VPF expression in the rat uterus: Rapid stimulation by estrogen correlates with estrogen-induced increases in uterine capillary permeability and growth. Endocrinology 133, 829–837.

Curran, T. & Franza, B. R., Jr. (1988). Fos and Jun: The AP-1 connection. Cell 55, 395–397.

D'Amore, P. A., Glaser, B. M., Bruson, S. K., & Fenselau, A. H. (1981). Angiogenic activity from bovine retina: Partial purification and characterization. Proc. Natl. Acad. Sci. USA 78, 3068–3072.

Detmar, M., Brown, L. F., Daffey, K. P., Yeo, K. T., Kocher, O., Jackman, R. W., Berae, B., & Pvorak, H. F. (1994). Overexpression of VEGF/VPF and its receptors in psoriasis. J. Exp. Med. 180, 1141–1146.

Devries, C., Escobedo, J. A., Ueno, H., Houck, K., Ferrera, N., & Williams, L. T. (1992). The fms-like tyrosine kinase, a receptor for vascular endothelial growth factor. Science 255, 989–991.

Dissen, G. A., Lara, H. E., Fahrenbach, W. H., Costa, M. E., & Ojeda, S. R. (1994). Immature rat ovaries become revascularized rapidly after autotransplantation and show a gonadotropin dependent increase in angiogenic factor gene expression. Endocrinology 134, 1146–1154.

Dvorak, H. F., Orenstein, N. S., Carvalho, A. C., Churchill, W. H., Dvorak, A. M., Galli, S. J., Feder, J., Bitzer, A. M., Rypysc, J., & Giovinco, P. (1979). Induction of a fibrin-gel investment: An early event in line 10 hepatocarcinoma growth mediated by tumor-secreted products. J. Immunol. 122, 166–174.

Dvorak, H. F., Sioussat, T. M., Brown, L. F., Berse, B., Nagy, J. A., Sotrel, A., Mannseau, E. J., Van de Water, L., & Senger, D. R. (1991). Distribution of vascular permeability factor vascular endothelial growth factor in tumors: Concentration in tumor blood vessels. J. Exp. Med. 174, 1275–1278.

Eichmann, A., Marcelle, C., Breant, C., & Le Douarin, N. M. (1993). Two molecules related to the VEGF receptor are expressed in early endothelial cells during avian embryonic development. Mech. Dev. 42, 33–48.

Ekblom, P., Sariola, H., Karkinen, M., & Saxen, L. (1982). The origin of the glomerular endothelium. Cell Differentiation 11, 35–39.

Elder, J. T., Fisher, G. T., Lindquist, P. B., Bennett, G. L., Pittelkow, M. R., Coffey, R. J., Ellingsworth, L., Derynck, R., & Voorhess, J. J. (1989). Overexpression of transforming growth factor α in psoriatic epidermis. Science 243, 811.

Engermann, R. I., Pfaffenbach, D., & Davis, M. D. (1967). Cell turnover of capillaries. Lab. Invest. 17, 738–743.

Enomoto, T., Okamoto, T., & Sato, J. D. (1994). VEGF induces the disorganization of actin stress fibers accompanied by protein tyrosine phosphorization and morphological changes in Balb/c 3T3 cells. Biochem. Biophys. Res. Comm. 202, 1716–1723.

Fava, R. A., Olsen, N. J., Spencer–Green, G., Yeo, K.–T., Yeo, T.–K., Berse, B., Jackman, R. W., Senger, D. R., Dvorak, H. F., & Brown, L. F. (1994). Vascular permeability factor/endothelial growth factor (VPF/VEGF): Accumulation and expression in human synovial fluids and rheumatoid synovial tissue. J. Exp. Med. 180, 341–346.

Ferrara, N. & Henzel, W. J. (1989). Pituitary follicular cells secrete a novel heparin-binding growth factor specific for vascular endothelial cells. Biochem. Biophys. Res. Commun. 161, 851–858.

Ferrara, N., Winer, J., & Burton, T. (1991). Aortic smooth muscle cells express and secrete vascular endothelial growth factor. Growth Factors 5, 1041–1048.

Ferrara, N., Houck, K., Jakeman, L., & Leung, D. W. (1992). Molecular and biological properties of the vascular endothelial growth factor family of proteins. Endocrin. Rev. 13, 18–32.

Findlay, J. K. (1986). Angiogenesis in reproductive tissues. J. Endocrinol. 111, 357–366.

Florkiewicz, R. Z., Shibata, F., Barankiewicz, T., Baird, A., Gonzalez, A. M., Florkiewicz, E., & Sha, N. (1991). Basic growth factor gene expression in the fibroblast growth factor family (Baird, A. & Klagsburn, M., eds.), Ann. Acad. Sci. 138, 109–126.

Folkman, J. (1986). How is blood vessel growth regulated in normal and neoplastic tissues? GHA Clowes Memorial Award Lecture. Cancer Res. 46, 467–473.

Folkman, J. & Klagsbrun, M. (1987). Angiogenic factors. Science 235, 442–447.

Fong, G.-H., Rossant, J., Gertsenstein, M., & Breitman, M. L. (1995). Role of the Flt-1 receptor tyrosine kinase in regulating the assembly of vascular endothelium. Nature 376, 66–70.

Form, D. M. & Auerbach, R. (1983). PGE_2 and angiogenesis. Proc. Soc. Exp. Biol. Med. 172, 214–218.

Friederici, H. H. R. (1967). The early response of uterine capillaries to estrogen stimulation: An electron microscopic study. Lab Invest. 17, 322–333.

Fry, L. (1988). Psoriasis. Br. J. Dermatol. 119, 445.

Garrido, C., Saule, S., & Gospodarowicz, D. (1993). Transcriptional regulation of vascular endothelial growth factor gene expression in ovarian granulosa cells. Growth Factors 8, 109–117.

Gitay–Goren, H., Soker, S., Vlodavsky, I., & Neufeld, G. (1992). The binding of vascular endothelial growth factor to its receptors is dependent on cell surface-associated heparin-like molecules. J. Biol. Chem. 267, 6093–6098.

Gitay–Goren, H., Halaban, R., & Neufeld, G. (1993). Human melanoma cells but not normal melanocytes express vascular endothelial growth factor receptors. Biochem. Biophys. Res. Commun. 190, 702–709.

Goldberg, M. A., Dunning, S. P., & Bunn, H. F. (1988). Regulation of the erythropoietin gene: Evidence that the oxygen sensor is a heme protein. Science 242, 1412–1416.

Goldberg, M. A. & Schneider, T. J. (1994). Similarities between the oxygen sensing mechanisms regulating the expression of VEGF and erythropoietin. J. Biol. Chem. 269, 4555–4359.

Gospodarowicz, D. & Thakral, K. K. (1978). Production of corpus luteum angiogenic factor responsible for proliferation of capillaries and neovascularization of the corpus luteum. Proc. Natl. Acad. Sci. USA 75, 847–851.

Gospodarowicz, D., Vlodavsky, I., Fielding, P., & Birdwell, C. R. (1978). The effects of the epidermal and fibroblast growth factors upon cell proliferation using vascular and corneal endothelial cells as a model. In: Birth Defects (Littlefield, J. W. and deGrouchy, J., eds.), pp. 233–271, Excerpta Medica, Amsterdam, The Netherlands.

Gospodarowicz, D. (1980). Growth and differentiation factors for cell populations in the normal ovaries. In: Ovarian Neoplasia—Workshops on the Biology of Human Cancer, Report No. 11 (Murphy, E. D. and Beamer, W. G., eds.), Vol. 50, pp. 20–21, Technical Report Series, Int. Union Against Cancer.

Gospodarowicz, D., Ferrara, N., Schweigerer, L., & Neufeld, G. (1987). Structural characterization and biological functions of fibroblast growth factor. Endocrine Reviews 8, 95–114.

Gospodarowicz, D., Abraham, J. A., & Schilling, J. (1989). Isolation and characterization of a vascular endothelial cell mitogen produced by pituitary-derived folliculo stellate cells. Proc. Natl. Acad. Sci. USA 86, 7311–7315.

Gospodarowicz, D. (1990). Growth factors for vascular endothelial cells. In: Cell Culture Techniques in Cardiovascular Research (Piper, H. M., ed.), pp. 230–244, Springer-Verlag, Berlin.

Goto, F., Goto, K., Weindel, K., & Folkman, T. (1993). Synergistic effects of VEGF and bFGF on the proliferation and cord formation of bovine capillary endothelial cells within collagen gels. Lab Investig. 69, 508–517.

Gottlieb, A. B., Chang, C. K., Posnett, D. N., Fanelli, B., & Tam, J. P. (1988). Detection of transforming growth factor α in normal, malignant, and hyperproliferative human keratinocytes. J. Exp. Med. 167, 670.

Guinn, T. P., Peters, K. C., DeVries, D., Ferrara, N., & Williams, L. T. (1993). Fetal liver kinase 1 is a receptor for VEGF and is selectively expressed in vascular endothelium. Proc. Natl. Acad. Sci. USA 90, 7533–7537.

Ham, K. N., Hurley, J. V., Lopata, A., & Ryan, G. B. (1970). A combined isotopic and electron microscopic study of the response of the rat uterus to exogenous oestradiol. J. Endocrinol. 46, 71–81.

Harada, S. I., Nagy, J. A., Sullivan, K. A., Thomas, K. A., Endo, N., Rodan, G. A., & Rodan, S. B. (1994). Induction of VEGF expression by prostaglandin E_2 and E_1 in osteoblasts. J. Clin. Invest. 93, 2490–2496.

Hart, C. E. & Bowen–Pope (1990). Platelet-derived growth factor receptor: Current views of the two-subunit model. J. Invest. Dermatology 94, 53S–57S.

Hechter, O., Krohn, L., & Harris, J. (1941). The effect of estrogen on the permeability of the uterine capillaries. Endocrinology 29, 386–392.

Heldin, C. H. (1995). Dimerization of cell surface receptors in signal transduction. Cell 80, 213–223.

Herschman, H. R. (1991). Primary response genes induced by growth factors and tumor promoters. Annu. Rev. Biochem. 60, 281–319.

Heuer, J. G., von Bartheld, C. S., Kinoshita, Y., Evers, P. C., & Bothwell, M. (1990). Alternating phases of FGF receptor and NGF receptor expression in the developing chicken nervous system. Neuron 5, 283–296.

Houck, K. A., Ferrara, N., Winer, J., Cachianes, G., Li, B., & Leung, D. W. (1991). The vascular endothelial growth factor family—Identification of a fourth molecular species and characterization of alternative splicing of RNA. Mol. Endocrinol. 5, 1806–1814.

Houck, K. A., Leung, D. W., Rowland, A. M., Winer, J., & Ferrara, N. (1992). Dual regulation of vascular endothelial growth bioavailability by genetic and proteolytic mechanisms. J. Biol. Chem. 267, 26031–26037.

Imagawa, M., Chiu, R., & Karin, M. (1987). Transcription factor AP-2 mediates induction by two different signal-transduction pathways: Protein kinase C and cAMP. Cell 51, 251–260.

Jakeman L. B., Armanini, M., Phillips, H. S., & Ferrara, N. (1993). Developmental expression of binding sites and messenger ribonucleic acid for vascular endothelial growth factor suggests a role for this protein in vasculogenesis and angiogenesis. Endocrinology 133, 848–859.

Jaye, M., Howk, R., Burgess, W., Ricca, G. A., Chiu, I. M., Ravera, M. W., O'Brien, S. J., Modi, W. E., Maciag, T., & Drohan, W. N. (1985). Human endothelial cell growth factors: Cloning, nucleotide sequence, and chromosome localization. Science 233, 541–545.

Jim Kim, K., Li, B., Winer, J., Armanini, M., Gillet, N., Phillips, H. S., & Ferrara, N. (1993). Inhibition of VEGF induced angiogenesis suppresses tumor growth *in vivo*. Nature 362, 841–844.

Joukov, V., Pajusola, K., Kaipainen, A., Chilov, D., Lahtinen, I., Kukk, E., Saksela, O., Kalkkinen, N., & Alitalo, K. (1996). A novel vascular endothelial growth factor, VEGF-C, is a ligand for the Flt-4 (VEGFR-3) and KDR (VEGFR-2) receptor tyrosine kinases. EMBO J. 15, 290–298.

Jungkunz, W., Eichhorn, M., Wörl, J., Marsch, W. C., & Holzmann, H. (1992). Carbonic anhydrase—A marker for fenestrated capillaries in psoriasis. Arch. Dermatol. Res. 284, 146.

Kasayama, S., Saito, H., Kouhara, H., Sumitani, S., & Sato, B. (1993). Suramin interrupts androgen-inducible autocrine loop involving heparin-binding growth factor in mouse mammary cancer (Shionogi Carcinoma-115) cells. J. Cell. Physiol. 154, 254–261.

Keck, P. J., Hauser, S. D., Krivi, G., Sanzo, K., Warren, T., Feder, J., & Connolly, D. T. (1989). Vascular permeability factor, an endothelial cell mitogen related to PDGF. Science 246, 1309–1312.

Kendall, R. L., Wang, G., Di Salvo, J., & Thomas, K. A. (1994). Specificity of VEGF receptor ligand binding domain. Biochem. Biophys. Res. Comm. 201, 326–330.

Kikkawa, U., Takai, Y., Minakuchi, R., Inohara, S., & Nishizuka, Y. (1982). Calcium-activated, phospholipid-dependent protein kinase from rat brain. J. Biol. Chem. 257, 13341–13348.

Klagsbrun, M. & Folkman, J. (1990). Angiogenesis. In: Peptide Growth Factors and their Receptors II (Sporn, M. B. & Roberts, A. B., eds.), pp. 459–586, Springer Publishing Company, New York.

Klagsbrun, M. & D'Amore, P. A. (1991). Regulators of angiogenesis. Annu. Rev. Physiol. 53, 217.

Klagsbrun, M. & Soker, S. (1993). VEGF/VPF: The angiogenesis factor found? Current Biology 3, 669–702.

Klein, P. S. & Melton, D. A. (1994). Hormonal regulation of embryogenesis: The formation of mesoderm in *Xenopus laevis*. Endocrine Rev. 15, 326–341.

Koch, A., Polverini, P. J., Kunkel, S. L., Harlow, L. A., DiPietro, L. A., Elner, V. M., Elner, S. G., & Strieter, R. M. (1992). Interleukin-8 as a macrophage-derived mediator of angiogenesis. Science 258, 1798–2000.

Koch, A. E., Harlow, L. A., Haines, G. K., Amento, E. P., Unemori, E. N., Wong, W. L., Pope, R. M., & Ferrara, N. (1994). Vascular endothelial growth factor: A cytokine modulation endothelial function in rheumatoid arthritis. J. Immunol. 152, 4149–4156.

Lange, R. A., Flores, E. D., & Hillis, L. D. (1991). Restenosis after coronary balloon angioplasty. Ann. Rev. Med. 42, 127–132.

Leibovich, S., Polverini, P. J., Shepard, H. M., Wiseman, D. M., Shivley, V., & Nuseir, N. (1987). Macrophage-induced angiogenesis is mediated by tumor necrosis α. Nature 329, 630–632.

Leung, D. W., Cachianes, G., Kuang, W. J., Goeddel, D. V., & Ferrara, N. (1989). Vascular endothelial growth factor is a secreted angiogenic mitogen. Science 246, 1306–1309.

Lindner, V., Majack, R. A., & Reidy, M.A. (1990). Basic fibroblast growth factor stimulates endothelial regrowth and proliferation in denuded arteries. J. Clin. Invest. 85, 2004–2008.

Lindner, V. & Reidy, M. A. (1991). Proliferation of smooth muscle cells after vascular injury is inhibited by an antibody against basic fibroblast growth factor. Proc. Natl. Acad. Sci. USA 88, 3739–3743.

Madan, A. & Curtin, P. T. (1993). A 24-base-pair sequence 3′ to the human erythropoietin gene contains a hypoxia-responsive transcriptional enhancer. Proc. Natl. Acad. Sci. USA 90, 3928–3932.

Maglione, D., Guerriero, V., Viglietto, G., Delli–Bovi, P., & Persico, M. G. (1991). Isolation of a human placenta cDNA coding for a protein related to the vascular permeability factor. Proc. Natl. Acad. Sci. USA 88, 9267–9271.

Maglione, D., Guerrero, V., Viglietto, G., Grazia–Ferraro, M., Aprelikova, O., Alitato, K., Del Vecchio, S., Lei, K. L., Yang Chou, T., & Graziella–Persico, M. (1993). Two alternative mRNAs coding for the angiogenic growth factor placenta growth factor (PIGF) are transcribed from a single gene of chromosome 14. Oncogene 8, 925–931.

Matthews, W., Jordan, C. T., Gavin, M., Jenkins, N. A., Copeland, N. G., & Lemischka, I. R. (1991). A receptor tyrosine kinase cDNA isolated from a population of enriched primitive hematopoietic cells and exhibiting close genetic linkage to c-kit. Proc. Natl. Acad. Sci. USA 88, 9026–9030.

Maxwell, I., Pugh, C. W., & Ratcliffe, P. J. (1993). Inducible operation of the erythropoietin 3′ enhancer in multiple cell lines: Evidence for a widespread oxygen-sensing mechanism. Proc. Natl. Acad. Sci. USA 90, 2423–2427.

McClure, N., Healy, D. L., Roger, P. A., Sullivan, J., Beaton, L., Haning, R. V., Connolly, D. T., & Robertson, D. M. (1994). VEGF as capillary permeability agent in ovarian hyperstimulation syndrome. Lancet 344, 235–236.

Michaelson, I. C. (1948). The mode of development of the vascular system of the retina, with some observations on its significance for certain retinal disease. Trans. Ophthalmol. Soc. UK 68, 137–180.

Middaugh, C. R., Mach, H., Burke, C. J., Volkin, D. B., Dabora, J. M., Tsai, P. K., Bruner, M. W., Ryan, J. A., & Marfia, K. E. (1992). Nature of the interaction of growth factors with suramin. Biochemistry 31, 9016–9024.

Midy, V. & Plouet, J. (1994). Vasculotropin/VEGF induces differentiation in cultured osteoblasts. Biochem. Biophys. Res. Comm. 380–386.

Millauer B., Wizigmann–Voss, S., Schnurch, H., Martinez, R., Moller, N. P. H., Risau, W., & Ullrich, A. (1993). High affinity VEGF binding and developmental expression suggest flk-1 as a major regulator of vasculogenesis and angiogenesis. Cell 73, 835–846.

Millauer, B., Shawver, L. K., Plate, K. H., Risau, W., & Ullrich, A. (1994). Glioblastoma growth inhibited *in vivo* by a dominant negative flk-1 mutant. Nature 367, 576–579.

Miller, J. W., Adamis, A. P., Shima, D. T., D'Amore, P. A., Moulton, R. S., O'Reilly, M. S., Folkman, J., Dvorak, H. F., Brown, L. F., Berse, B., Yeo, T.–K., & Yeo, K.–T. (1994). VEGF/VPF is temporally and spatially correlated with ocular angiogenesis in a primate model. Am. J. Pathol. 145, 574–584.

Minchenko, A., Bauer, T., Salceda, S., & Cairo, J. (1994). Hypoxic stimulation of VEGF expression *in vitro* and *in vivo*. Lab. Invest. 71, 374–379.

Mitchell, P. J. & Tijan, R. (1989). Transcriptional regulation in mammalian cells by sequence-specific DNA binding proteins. Science 245, 371–378.

Morris, B. & Sass, M. B. (1966). The formation of lymph in the ovary. Proc. R. Soc. Lond. (Biol) 164, 577–591.

Motro, B., Itin, A., Sachs, L., & Keshet, E. (1990). Pattern of interleukin 6 expression *in vivo* suggests a role in angiogenesis. Proc. Natl. Acad. Sci. USA 87, 3092–3096.

Mueller, G. C., Gorski, J., & Aizawa, Y. (1961). The role of protein synthesis in early estrogen action. Proc. Natl. Acad. Sci. USA 47, 164–169.

Mullins, D. E. & Rifkind, D. B. (1990). Induction of proteases and proteases inhibitor by growth factors. In: Peptide Growth Factors (Sporn, M. B. & Roberts, A. B., eds.), pp. 481–507, Springer-Verlag, New York.

Murray, D. W. & Rushton, N. (1990). The effect of strain on bone cell prostaglandin E_2 release: A new experimental method. Calcif. Tissue Int. 47, 35–39.

Myoken, Y., Kayada, Y., Okamoto, T., Kan, M., Sato, G. H., & Sato, J. D. (1991). Vascular endothelial cell growth factor (VEGF) produced by A-431 human epidermoid carcinoma cells and identification of VEGF membrane binding sites. Proc. Natl. Acad. Sci. USA 88, 5819–5823.

Neufeld, G., Tessler, S., Gitay Goren, H., Cohen, T., & Levi, B. Z. (1994). Vascular endothelial growth factor and its receptors. Prog. in Growth Factors Res. 5, 89–97.

Neulen, J., Yan, Z., Raczek, S., Weindel, K., Keck, C., Weich, H.A., Marmé, D., & Breckwoldt, M. (1995). Human chorionic gonadotropin-dependent expression of vascular endothelial growth factor/vascular permeability factor in human granulosa cells: Importance in ovarian hyperstimulation syndrome. J. Clin. Endocrin. 80, 1967–1971.

Noden, D. (1989). Embryonic origins and assembly of blood vessels. Am. Rev. Respir. Dis. 140, 1097–1103.

Öberg, C., Waltenberger, J., Claesson–Welsh, L., & Welsh, M. (1994). Expression of protein tyrosine kinase in islet cells: Possible role of the KDR-flk-1 receptor for β-cell maturation from duct cells. Growth Factors 10, 115–126.

Oelrichs, R. B., Reid, H. H., Bernard, O., Ziemiecki, A., & Wilks, A. F. (1993). NYK/FLK-1—A putative receptor protein tyrosine kinase isolated from E10 embryonic neuroepithelium is expressed in endothelial cells of the developing embryo. Oncogene 8, 11–18.

Olander, J. V., Connolly, D. T., & DeLarco, J. E. (1991). Specific binding of vascular permeability factor to endothelial cells. Biochem. Biophys. Res. Commun. 175, 68–76.

Pajusola, K., Aprelikova, O., Korhonen, J., Kipainen, A., Pertovaara, L., Alitalo, R., & Alitalo, K. (1992). FLT4 receptor tyrosine kinase contains seven immunoglobulin-like loops and is expressed in multiple human tissues and cell lines. Cancer Res. 52, 5738–5743.

Pardanaud, L., Yassine, F., & Dieterien–Lievre, F. (1989). Relationship between vasculogenesis, angiogenesis and haemopoiesis during avian ontogenesis. Development 105, 473–485.

Pe'er J., Shweiki, D., Itin, A., Hemo, I., Gneissen, H., & Keshet, E. (1995). Hypoxia induced expression of VEGF by retinal cells is a common factor in neovascularizing ocular disease. Lab Invest., in press.

Pekala, P., Marlow, M., Heuvilman, D., & Connolly, D. (1990). Regulation of hexose transport in aortic endothelial cells by vascular permeability factor and tumor necrosis factor α, but not by insulin. J. Biol. Chem. 265, 18051–18054.

Pepper, M. S., Ferrara, N., Orci, L., & Montesano, R. (1991). Vascular endothelial growth factor (VEGF) induces plasminogen activators and plasminogen activator inhibitor-1 in microvascular endothelial cells. Biochem. Biophys. Res. Commun. 181, 902–906.

Pepper, M. S., Ferrara, N., Orci, L., & Montesano, R. (1992). Potent synergism between VEGF and bFGF in the induction of angiogenesis *in vitro*. Biochem. Biophys. Res. Comm. 189, 824–831.

Pertovaara, L., Karpainen, A., Mustonen, T., Orpana, A., Ferrara, N., Saksela, O., & Alitalo, K. (1994). VEGF is induced in response to TGF-β in fibroblastic and epithelial cells. J. Biol. Chem. 269, 6271–6274.

Peters, K. G., De Vries, C., & Williams, L. T. (1993). VEGF receptor expression during embryogenesis and tissue repair suggests a role in endothelial differentiation and blood vessel growth. Proc. Natl. Acad. Sci. USA, in press.

Peters, K. G., Werner, S., Chen, G., & Williams, L. T. (1992). Two FGF receptor genes are differentially expressed in epithelial and mesenchymal tissues during limb formation and organogenesis in the mouse. Development 114, 233–243.

Phillips, H. S., Hains, J., Leung, D. W., & Ferrara, N. (1990). Vascular endothelial growth factor is expressed in rat corpus luteum. Endocrinology 127, 965–967.

Pictet, R. L. & Rutter, W. J. (1972). Development of the embryonic pancreas. In: Handbook of Physiology, Section 7: Endocrinology (Steiner, D. F., & Freinkel, N., eds.), pp 67–76. American Physiologic Society, Washington, DC.

Plate, K. H., Breier, G., Weich, H. A., & Risau, W. (1992). Vascular endothelial growth factor is a potential tumor angiogenesis factor in human gliomas *in vivo*. Nature 359, 845–847.

Plate, K. H., Breier, G., Millauer, B., Ullrich, A., & Risau, W. (1993). Upregulation of VEGF and its cognate receptors in a rat glioma model of tumor angiogenesis. Cancer Res. 53, 5822–5827.

Plouet, J., Schilling, J., & Gospodarowicz, D. (1989). Isolation and characterization of a newly identified endothelial cell mitogen produced by AtT-20 cells. EMBO J. 8, 3801–3806.

Plouet, J. & Moukadiri, H. (1990). Characterization of the receptor to vasculotropin on bovine adrenal cortex-derived capillary endothelial cells. J. Biol. Chem. 265, 22071–22074.

Quinn, T. P., Peters, K. G., DeVries, C., Ferrara, N., & Williams, L. T. (1993). Fetal liver kinase is a receptor for VEGF and is selectively expressed in vascular endothelium. Proc. Natl. Acad. Sci. USA 90, 7533–7537.

Raisz, L. G. & Martin, T. J. (1984). Prostaglandins in bone and mineral metabolism. In: Bone and Mineral Research. Vol. 3 (Peck, W. A., ed.) pp. 286–310, Elsevier, Amsterdam.

Rapraeger, A. C., Krufka, A., & Olwin, B. B. (1991). Requirement of heparan sulfate for bFGF-mediated fibroblast growth and myoblast differentiation. Science 252, 1705–1708.

Ravindranath, N., Little–Ihrig, L., Phillips, H. S., Ferrara, N., & Zeleznik, A. J. (1992). Vascular endothelial growth factor messenger ribonucleic acid expression in the primate ovary. Endocrinology 131, 254–260.

Risau, W. (1990). Angiogenic growth factors. Progress in growth factors research. Neuroendocrinol. Immunol. 2, 71–79.

Robertson, P. L., Dubois, M., Bowman, P. D., & Goldstein, G. W. (1985). Angiogenesis in developing rat brain: An *in vivo* and *in vitro* study. Dev. Brain Res. 23, 219–223.

Rondon, I. J., MacMillan, L., Beckman, B. S., Goldberg, M. A., Schneider, T., Bunn, H. F., & Malter, J. S. (1991). Hypoxia upregulates the activity of a novel erythropoietin mRNA binding protein. J. Biol. Chem. 266, 16594–16598.

Ross, R. (1986). The pathogenesis of atherosclerosis—An update. N. Engl. J. Med. 314, 488–500.

Ryan, T. J. (1980). Microcirculation in psoriasis. Pharmacol. & Ther. (B) 10, 27.

Schweigerer, L., Neufeld, G., Friedman, J., Abraham, J. A., Fiddes, J. C., & Gospodarowicz, D. (1987). Basic fibroblast growth factor is expressed in capillary endothelial cells. Nature 325, 257–259.

Semenza, G. L. & Wang, G. L. (1992). A nuclear factor induced by hypoxia via de novo protein synthesis binds to the human erythropoietin gene enhancer at a site required for transcriptional activation. Mol. Cell. Biol. 12, 5447–5454.

Senger, D. R., Galli, S. J., Dvorak, A. M., Peruzzi, C. A., Harvey, V. S., & Dvorak, H. F. (1983). Tumor cells secrete a vascular permeability factor that promotes accumulation of ascites fluid. Science 219, 983–985.

Senger, D. R., Perruzi, C. A., Feder, J., & Dvorak, H. F. (1986). A highly conserved vascular permeability factor secreted by a variety of human and rodent tumor cell lines. Cancer Res. 46, 5629–5632.

Senger, D. R., Van De Water, L., Brown, L. F., Nagy, J. A., Yeo, K.–T., Yeo, T. K., Berse, B., Jackman, R. W., Dvorak, A. M., & Dvorak, H. F. (1993). Vascular permeability factor (VPF, VEGF) in tumor biology. Cancer Met. Rev. 12, 303–324.

Shalaby, F., Rossant, J., Yamaguchi, T. P., Gertsenstein, M., Xiang-Fu, W., Breitman, M. L., & Schuh, A. C. (1995). Failure of blood-island formation and vasculogenesis in Flk-1 deficient mice. Nature 376, 62–66.

Shen, H., Clauss, M., Ryan, J., Schmidt, A.–M., Tyburg, P., Borden, L., Connolly, D., Stern, D., & Kao, J. (1993) Characterization of VPF/VEGF receptors on mononuclear phagocytes. Blood 81, 2767–2773.

Shibuya, M., Yamaguchi, S., Yamane, A., Ikdea, T., Tojo, A., Matsushime, H., & Sato, M. (1990). Nucleotide sequence and expression of a novel human receptor type tyrosine kinase gene (flt) closely related to the fms family. Oncogene 5, 519–524.

Shweiki, D., Itin, A., Soffer, D., & Keshet, E. (1992). Vascular endothelial growth factor induced by hypoxia may mediate hypoxia-initated angiogenesis. Nature 359, 843–845.

Shweiki, D., Itin, A., Neufeld, G., Gitay–Goren, H., & Keshet, E. (1993). Patterns of expression of vascular endothelial growth factor (VEGF) and VEGF receptors in mice suggest a role in hormonally regulated angiogenesis. J. Clin. Invest. 91, 2235–2243.

Spisni, E., Monica, F., & Tomasi, V. (1992). Involvement of prostanoids in the regulation of angiogenesis by polypeptide growth factors. Prostaglandins Leukotrienes Essenti. Fatty Acids 47, 111–115.

Swalingam, A., Kenney, J., Brown, G. C., Benson, W. E., & Donoso, L. (1990). Basic fibroblast growth factor levels in the vitreous of patients with proliferative diabetic retinopathy. Arch. Ophthalmol. 108, 869–872.

Takahashi, A., Sasaki, H., Kim, S. J., Tobisu, K., Kakizoe, T., Tsukamoto, T., Kumamoto, Y., Sugimura, T., & Terada, M. (1994). Markedly increased amounts of messenger RNAs for vascular endothelial growth factor and placenta growth factor in renal cell carcinoma associated with angiogenesis. Cancer Res. 54, 4233–4237.

Takeshita, S., Zheng, L. P., Brogi, E., Kearney, H., Pu, L. Q., Bunting, S., Ferrara, N., Symes, J. F., & Isner, J. M. (1994). Therapeutic angiogenesis. A single intra-arterial bolus of VEGF augments revascularization in a rabbit ischemic hind limb model. J. Clin. Invest. 93, 662–670.

Terman, B. I., Carrion, M. E., Kovacs, E., Rasmussen, B. A., Eddy, R. L., & Shows, T. B. (1991). Identification of a new endothelial cell growth factor receptor tyrosine kinase. Oncogene 6, 1677–1683.

Terman, B. I., Dougher–Vermazen, M., Carrion, M. E., Dimitrov, D., Armellino, D. C., Gospodarowicz, D., & Böhlen, P. (1992). Identification of the KDR tyrosine kinase as a receptor for vascular endothelial cell growth factor. Biochem. Biophys. Res. Commun. 187, 1579–1586.

Terman, B. I., Klandke, L., Dougher–Vermazan, M., Maglione, D., Lassam, N. J., Gospodarowicz, D., Grazella-Persico, M., Böhlen, P., & Eisinger, M. (1994). VEGF receptor subtypes KDR and flt-1 show different sensitivities to heparin and placenta growth factor. Growth Factors, in press.

Tessler, S., Rockwell, P., Hicklin, D., Cohen, T., Levi, B. Z., Witte, L., Lenischka, I. R., & Neufeld, G. (1994). Heparin modulates the interaction of VEGF with soluble and cell-associated KDR-flk-1 receptors. J. Biol. Chem. 269, 12456–12461.

Tischer, E., Gospodarowicz, D., Mitchell, R., Silva, M., Schilling, J., Lau, K., Crisp, T., Fiddes, J. C., & Abraham, J. A. (1989). Vascular endothelial growth factor: A new member of the platelet-derived growth factor gene family. Biochem. Biophys. Res. Commun. 165, 1198–1206.

Tischer, E., Mitchell, R., Hartman, T., Silva, M., Gospodarowicz, D., Fiddes, J. C., & Abraham, J. A. (1991). The human gene for vascular endothelial growth factor. Multiple protein forms are encoded through alternative exon splicing. J. Biol. Chem. 266, 11947–11954.

Ui, H. & Mueller, G. C. (1963). The role of RNA synthesis in early estrogen action. Proc. Natl. Acad. Sci. USA 50, 256–260.

Ullrich, A. & Schlessinger, J. (1990). Signal transduction by receptors with tyrosine kinase activity. Cell 61, 203–212.

Unemori, E. N., Ferrara, N., Bauer, E. A., & Amento, E. P. (1992). Vascular endothelial growth factor induces interstitial collagenase expression in human endothelial cells. J. Cell Physiol. 153, 557–562.

Vaisman, N., Gospodarowicz, D., & Neufeld, G. (1990). Characterization of the receptors for vascular endothelial growth factor. J. Biol. Chem. 265, 19461–19466.

Vlodavsky, I., Folkman, J., Sullivan, R., Fridman, R., Ishai–Michaeli, R., Sasse, J., & Klagsburn, M. (1987). Endothelial cell-derived basic fibroblast growth factor: Synthesis and deposition into subendothelial extracellular matrix. Proc. Natl. Acad. Sci. USA 84, 2292–2296.

Wahl, S. M., Hunt, D. A., Wakefield, L. M., McCartney, F. M., Wahl, L. M., Roberts, A. B., & Sporn, M. B. (1987). TGBβ induced monocyte chemotaxis and growth factor production. Proc. Natl. Acad. Sci. USA 84, 5788–5792.

Wanaka, A., Milbrandt, J., & Johnson, E. M. (1991). Expression of FGF receptor gene in rat development. Development 111, 455–468.

Wang, G. L. & Semenza, G. L. (1993). General involvement of hypoxia-inducible factor 1 in transcriptional response to hypoxia. Proc. Natl. Acad. Sci. USA 90, 4304–4308.

Weindel, K., Miringlane, J. R., Marmé, D., & Weich, H. A. (1994). Detection and quantification of VEGF/VPF in brain tumor tissue and cyst fluid: The key to angiogenesis. Neurosurgery 35, 439–448.

Weinstein, G. D., McCullough, J. L., & Ross, P. A. (1985). Cell kinetic basic for pathophysiology of psoriasis. J. Invest. Dermatol. 85, 579.

Yamaguchi, T. P., Dumont, D. J., Conion, R. A., Breitman, M. L., & Rossant, J. (1993). *Flk-1*, an *flt*-related receptor tyrosine kinase is an early marker for endothelial cell precursors. Development 118, 489–498.

Yarden, Y., Escobedo, J. A., Kuang, W. J., Yang Feng, T. L., Daniel, T. O., Tremble, P. M., Chen, E. Y., Ando, M. E., Harkins, R. N., & Francke, U. (1986). Structure of the receptor for platelet-derived growth factor helps define a family of closely related growth factor receptors. Nature 323, 226–232.

Yayon, A., Klagsbrun, M., Esko, J. D., Leder, P., & Ornitz, D. M. (1991). Cell surface, heparin-like molecules are required for binding of basic fibroblast growth factor to its high affinity receptor. Cell 64, 841–848.

Yen, S. S. (1991). Clinical endocrinology of reproduction. In: Hormones: From molecules to disease (Baulieu, E. E. & Kelly, P. A., eds.), pp. 445–481, Hermann Pub., New York.

Ziche, M., Jones, J., & Guillino, P. (1982). Role of prostaglandin E_1 and copper in angiogenesis. J. Natl. Cancer Inst. 69, 475–482.

THE HEPATOCYTE GROWTH FACTOR RECEPTOR (MET):

AN UNCONVENTIONAL TRANSDUCER OF MITOGENIC AND MOTOGENIC SIGNALS

Paolo M. Comoglio

I. HGF is a Mitogen, a Motogen and a Morphogen 466
A. HGF is a Large Heterodimeric Molecule 467
B. GF Is Activated by Extracellular Cleavage 467
II. There are Low- and High-Affinity Receptors for HGF 469
A. Structure, Biosynthesis, and Subcellular Distribution of p190^{MET}, the High-Affinity HGF Receptor 470
III. Posttranslational Processing Generates Dominant-Negative Truncated HGF Receptor Isoforms 472
IV. Positive and Negative Regulation of the HGF Receptor Kinase Activity by Tyrosine or Serine Phosphorylation 473
V. Signal Transduction by the HGF Receptor: The Docking "SuperSite" 474
VI. Tissue-Specific Regulation of HGF Receptor Expression 478
VII. The HGF Receptor Gene (*MET*) Can Transform Cells *In Vitro* 479

Growth Factors and Cytokines in Health and Disease
Volume 1B, pages 465–490.

ISBN: 0-7623-0091-4

VIII. The HGF Receptor Gene (*MET*) is a Harmful Oncogene *In Vivo* 481
References 482

I. HGF IS A MITOGEN, A MOTOGEN AND A MORPHOGEN

Hepatocyte growth factor (HGF) is a potent mitogen with a misleading name. It stimulates growth in parenchymal liver cells (Nakamura et al., 1986; Gohda et al., 1988; Zarnegar and Michalopoulos, 1989) and also in a variety of epithelial and nonepithelial cells including skin keratinocytes (Kan et al., 1991; Matsumoto et al., 1991), melanocytes (Rubin et al., 1991; Halaban et al., 1992), and kidney tubules (Santos et al., 1994). Moreover, HGF stimulates endothelial cells *in vitro*, and it is a powerful angiogenic factor "*in vivo*" (Bussolino et al., 1992). The meaning of the acronym HGF is consistent with the fact that this factor is, nonetheless, a major mediator of liver regeneration *in vivo* (Michalopoulos, 1990) and essential for liver development in the embryo (Schmidt et al., 1995; Yoshihiko et al., 1995).

HGF has chemotactic properties and dissociates epithelial cell layers, increasing their motility and invasiveness into extracellular matrices (Stoker et al., 1987; Gherardi et al., 1989; Weidner et al., 1990). The factor induces the spread of carcinoma cells *in vitro* and promotes the progression toward a malignant invasive phenotype (Weidner et al., 1990). Due to these unique properties it is also called "Scatter Factor." Paradoxically, the other name once used to identify HGF was "Sarcoma Necrosis Factor" for its growth-inhibitory effect on some tumor cell lines of mesenchymal origin (Higashio et al., 1990; Tajima et al., 1991).

Lastly, adding HGF to epithelial or endothelial cells grown in three-dimensional collagen networks induces tubule formation showing effective morphogenic activity (Montesano et al., 1991; Bussolino et al., 1992). When added to embryonal liver progenitor cells, it induces differentiation into primordial biliary-like tubular structures (Medico et al., 1995).

A possible role for HGF and its receptor in the developmental processes has been elucidated from the study of the expression of the genes that encode these two proteins during mouse embryogenesis (Sonnenberg et al., 1993a,b). *In situ* hybridization analysis showed that, during mouse development, *c-MET* mRNA is expressed in many epithelia, whereas the ligand is usually found in nearby mesenchymal cells, suggesting a paracrine relationship (Sonnenberg et al., 1993a). These data point toward a role of HGF and its receptor in mesenchymal-epithelial interactions. Recently we showed that HGF can activate and promote the phenotypic transition between monocytes and macrophages (Galimi et al., 1993) and stimulates the growth and differentiation of erythroid hematopoietic precursors (Galimi et al., 1994).

Due to these mitogenic, motogenic, and morphogenic properties, it is thought that HGF is critical in the development and physiological renewal of human epithelial organs and restricted cell populations of hemopoietic origin.

A. HGF is a Large Heterodimeric Molecule

HGF and Scatter Factor are now known to be identical. This became evident when the respective cDNAs were isolated from different sources (Miyazawa et al., 1989; Nakamura et al., 1989; Weidner et al., 1991; Naldini et al., 1991c) and a single locus was identified. The human gene encoding HGF is composed of 18 exons and 17 introns. It spans approximately 70 kb and is located on chromosome 7q11.1-21 (Fukuyama et al., 1991; Saccone et al., 1992). The overall organization of the human HGF gene is highly homologous to that of the serine proteases involved in blood coagulation and fibrinolysis, particularly plasminogen (38% amino acid sequence identity; Nakamura et al., 1989). The major transcript is 6 kb and encodes a polypeptide chain of 728 aminoacids.

The mature HGF protein is a disulfide-linked heterodimer with a 60-kDa (α) and a 32–36-kDa (β) chain (Nakamura et al., 1989; Gohda et al., 1988; Zarnegar and Michalopoulos, 1989; Gherardi et al., 1989; Weidner et al., 1990). The α chain consists of an N-terminal hydrophobic leader sequence, a putative hairpin loop and four "kringle" modules. The β chain is homologous to the catalytic domain of serine proteases but lacks enzymatic activity due to the substitution of two critical amino acids in the active site. A 1.2-kb cDNA was also isolated from a library of embryonic human fibroblasts. Its open reading frame encodes a molecule of 290 amino acids consisting of the signal sequence, the N-terminal hairpin loop, and the first two kringles of the α subunit (Chan et al., 1991; Miyazawa et al., 1991; Hartmann et al., 1992). The resulting two-kringle HGF variant is a monomer of 28 kDa, which (as the whole α subunit) stimulates tyrosine phosphorylation and induces scattering of epithelial cells but not mitogenesis (Hartmann et al., 1992). The minimal region necessary for receptor binding has been identified within the first kringle in the α subunit (Lokker and Godowsky, 1993). The β subunit, when separately expressed, is unable to bind to the receptor and is biologically inactive (Hartmann et al., 1992).

Interestingly, two independent reports have pointed to the existence of another HGF-related protein. This protein, known as HGF-like (Han et al., 1991) or MSP (Macrophage Stimulating Protein) (Yoshimura et al., 1993), shares the same overall two-chain, four-kringle/protease-like structure with HGF (Figure 1).

B. GF Is Activated by Extracellular Cleavage

The α and β chains originate from cleavage of the protein encoded in the primary transcript (Miyazawa et al., 1989; Nakamura et al., 1989). The N-terminal amino acid sequence of the isolated β chain corresponds to the residues that lie after Val_{495}. The precursor contains a putative proteolytic cleavage site following Arg_{494}. This site, including a basic residue followed by two hydrophobic amino acids (Arg-Val-Val), fits the consensus for cleavage by serine endoproteases.

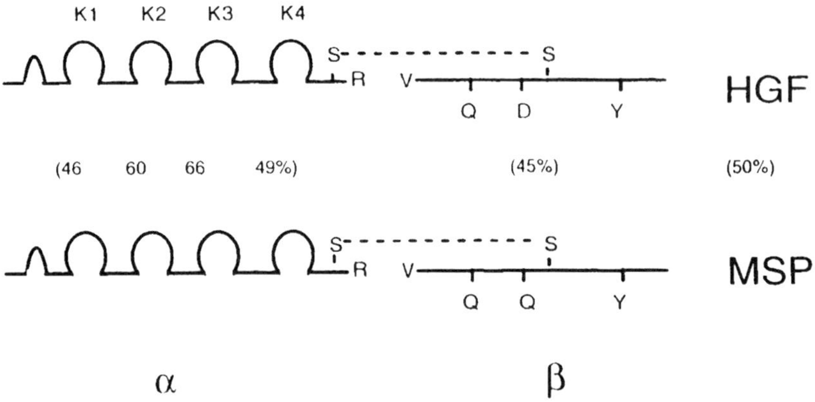

Figure 1. Schematic structure of hepatocyte growth factor (HGF) and of the so-called macrophage stimulating protein (MSP). The two molecules share the heterodimeric structure made of an α chain containing four "kringle" domains (K1-4) and a β chain highly homologue to serine proteases of the blood-clotting cascade, but devoid of catalytic activity. Numbers in parenthesis indicate the homology between the separate domains.

Site-directed mutagenesis of Arg_{494} generated an uncleaved, biologically inactive HGF mutant, suggesting that proteolytic cleavage is essential to obtain the active conformation (Hartmann et al., 1992; Lokker et al., 1992). We have shown that native HGF is secreted as a biologically inactive single-chain precursor having an apparent molecular weight of 92 kDa (pro-HGF), which is bound to the extracellular matrix. Activation takes place by proteolytic cleavage in the extracellular environment. Digestion of pro-HGF *in vitro* by pure enzyme and treatment with specific inhibitors *in vivo* indicate that urokinase is a pro-HGF convertase (Naldini et al., 1992). A serine protease isolated from serum, homologous to coagulation Factor XII, can also process the HGF precursor (Miyazawa et al., 1993). This enzyme must be proteolytically activated by thrombin (Shimomura et al., 1993).

One can, thus, postulate two different pathways for the biological activation of pro-HGF: one mediated by uPA, normally in plasma and tissues, the other activated only during the coagulation process. Interestingly, among the pleiotropic biological responses triggered by urokinase, some (e.g., cell growth, movement and matrix invasion) overlap those induced by HGF. Urokinase may, thus, operate by activating latent HGF bound to the extracellular matrix which in turn elicits growth, motogenic, and invasion signals.

We have shown that activation involves the formation of a stable complex between pro-HGF and uPA. This complex was isolated from the *in vitro* reaction of pure uPA with recombinant pro-HGF and from the membrane of target cells, after sequential addition of uPA and pro-HGF. The uPA-HGF complex was bound to the *Met* receptor on the cell membrane (see below). After adhesion, monocytes efficiently activated pro-HGF on their surface and in the culture medium. This activation occurred through a stoichiometric reaction and was inhibited by anticatalytic anti-uPA antibodies (Naldini et al., 1995). The stoichiometry of the activation reaction suggests that the biological effects of HGF can be titrated *in vivo* by the level of uPA activity. Adequate amounts of uPA can be locally provided by the macrophages, which would condition the tissue microenvironment by rendering HGF bioavailable to its target cells.

II. THERE ARE LOW- AND HIGH-AFFINITY RECEPTORS FOR HGF

The binding properties of HGF indicate that there are two distinct classes of sites with affinities one order of magnitude apart: the lower affinity, large capacity sites with a K_d in the range of 10^{-9} M, and the high affinity sites with a K_d around 2^{-10} nM. On epithelial or endothelial cells, the latter range between 1,000 and 10,000 per cell (Naldini et al., 1991c; Zarnegar et al., 1991; Lokker et al., 1992). The low-affinity sites can be eluted with excess heparin. They most likely correspond to matrix- or cell-associated heparansulfate proteoglycans (Naldini et al.,1991; Kobayashi et al., 1994), as suggested by the presence of a heparin-binding domain on the HGF α chain (Weidner et al., 1990). According to this model, the heparin-like receptors store and recruit inactive pro-HGF molecules to the cell membrane, in

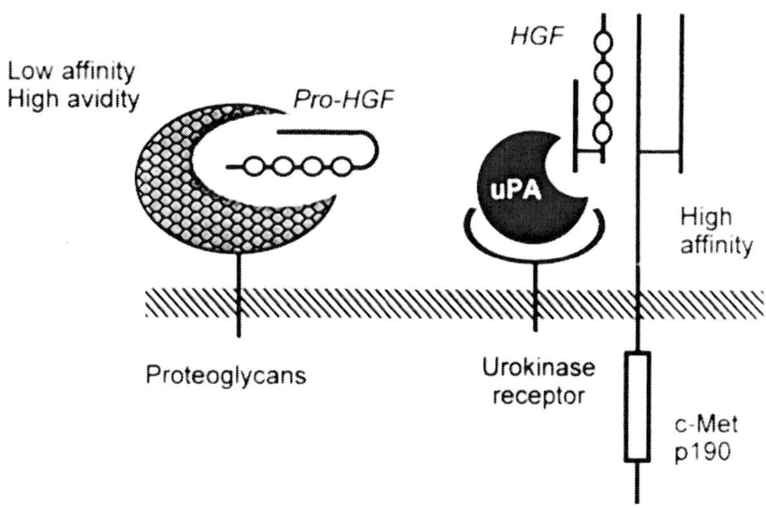

Figure 2. Low- and high-affinity sites for HGF. The biologically inactive single-chain proHGF binds to cell surface proteoglycans with low affinity, due to heparin-binding domain located in the α-chain. The tissue pro-HGF convertase (Urokinase) bound to its receptor generates the dimeric active form. Only the latter binds, with high affinity, to the transmembrane tyrosine kinase encoded by the proto-oncogene c-*MET*. The dashed line represents the cell plasma membrane.

close proximity to the high-affinity receptors. Urokinase, which is also bound to the cell surface through a specific receptor, is, thus, able to convert the inactive single-chain precursor to active HGF. Among the other possible functions of the low-affinity sites are ligand stabilization, induced fit for receptor binding, and dimerization (Ruoslahti and Yamaguchi, 1991). The high-affinity receptor for HGF is p190^{MET}, the protein encoded by the c-*MET* oncogene. This was proven by cross-linking experiments using purified HGF and by reconstituting the ligand binding site in insect cells infected by a recombinant baculovirus carrying the human *MET* cDNA (Naldini et al., 1992). In hepatocytes, the HGF high-affinity binding sites are located on a protein of 150 kDa, a size consistent with the molecular weight of the β subunit of p190^{MET} (Zarnegar et al., 1990). Moreover, the intrinsic tyrosine kinase activity of p190^{MET} is enhanced in cells stimulated by pure HGF (Naldini et al., 1991a) or in a molecule smaller than HGF, but with similar binding properties (Bottaro et al., 1991). Finally, NIH 3T3 fibroblasts transfected with the human *MET* cDNA express functional receptors and respond to HGF with increased motility and invasion of extracellular matrices (Giordano et al., 1993).

A. Structure, Biosynthesis, and Subcellular Distribution of p190^{MET}, the High-Affinity HGF Receptor

The tyrosine kinase encoded by the c-*MET* proto-oncogene has unique structural features: it is a two-chain oligomer, composed of a 50-kDa (α) chain disulfide-linked to a 145-kDa (β) chain in an αβ complex of 190 kDa (Giordano et al., 1989a). The α chain is exposed at the cell surface and the β chain spans the plasma membrane; the β chain contains a tyrosine kinase domain (Gonzatti–Haces et al., 1988) and sites for tyrosine autophosphorylation (Ferracini et al., 1991; Ponzetto et al., 1993). Based on the cDNA sequence derived from the major *MET* transcript (Ponzetto et al., 1991), the β chain would consist of 1085 amino acids, the extracellular domain of the β chain would include amino acid 306–932, the single transmembrane hydrophobic segment amino acid 933–955, and the intracellular domain amino acid 956–1390 (Fig. 3).

The unique p190^{MET} two-chain structure makes a prototype for a class of receptor-type tyrosine kinases that include the two receptors encoded by the oncogenes c-*Ron* and c-*Sea* (Ronsin et al., 1993; Huff et al., 1993; Gaudino et al., 1994). These receptors share significant sequential similarities: the extracellular cleavage site between the α and β chains and the location of the cysteine residues. In the cytoplasmic domain, these three family members are highly homologous in the kinase region and share an eleven amino acid motif containing two tyrosines acting as a "multifunctional docking site" for SH2-transducers (see below).

The HGF receptor is synthesized as a large precursor (pr170) that includes the α and β chains, undergoes cotranslational glycosylation, and is cleaved to form the mature subunits (Giordano et al., 1989a). Under normal conditions, the single-chain precursor is not exposed at the cell surface. Core glycosylation is crucial for further processing because the aglycoproreceptor, produced in the presence of tunicamy-

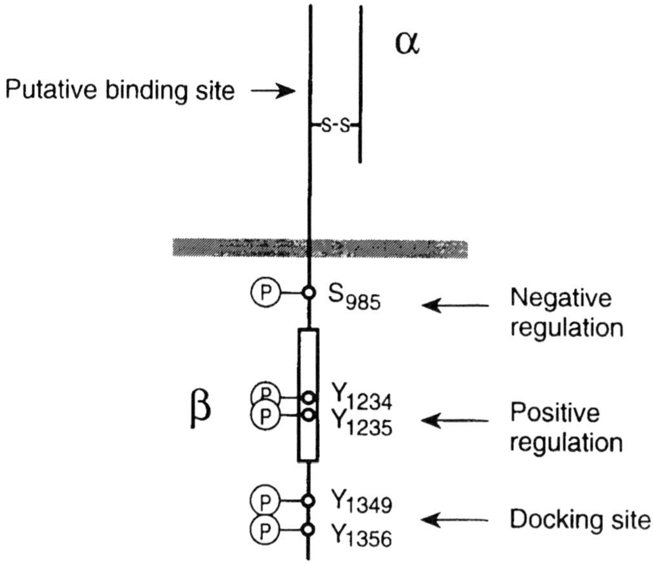

Figure 3. Structure of the HGF receptor. The heterodimer of 190 kDa is made of an extracellular α chain, disulfide-linked (S–S) to the transmembrane β chain, endowed with a tyrosine kinase domain in the cytoplasmic side (large open box). The juxtamembrane domain contains a serine residue (S) that is phosphorylated as a negative feedback control mechanism. Upon autophosphorylation, two tyrosines (Y) in the kinase domain upregulate, the catalytic activity. The two phosphotyrosines of the receptor tail generate the docking site for intracellular signal transducers.

cin, is neither cleaved into the mature chains nor inserted into the plasma membrane. Disulfide bond formation between α and β chains takes place before the cleavage (Giordano et al., 1989b). The cleavage site[1] (K^{303} RKKR-S^{308}) is a canonical consensus for the endoplasmic reticulum protease furin (Barr, 1991). Site-directed mutagenetic experiments confirmed that the critical residues for cleavage are the two arginines in positions 304 and 307 (Mark et al., 1993). The HGF receptor is selectively exposed at the baso-lateral plasma membrane domain of polarized epithelial cells and is targeted after synthesis to that surface by direct delivery from the trans-golgi network. Fluorescence-tagged or radiolabeled monoclonal antibodies against the β chain stained the baso-lateral, but not the apical, cell surface of epithelia lining the lumens of human organs. They also selectively bind the baso-lateral surface of MDCK cells forming polarized monolayers *in vitro*. The receptor is concentrated around the cell–cell contacts, showing a distribution pattern overlapping that of the cell-adhesion molecule E-cadherin and, in polarized cells, interacts with detergent-insoluble cytoskeleton components (Crepaldi et al., 1994b).

III. POSTTRANSLATIONAL PROCESSING GENERATES DOMINANT-NEGATIVE TRUNCATED HGF RECEPTOR ISOFORMS

Neoplastic cells overexpressing the c-*MET* oncogene, and to a lesser extent, normal epithelial cells, express two different HGF receptor isoforms at their cell surface both devoid of kinase activity (Prat et al., 1991a). The molecular weights of these truncated forms are 140 and 130 kDa, respectively. The first, $p140^{MET}$, is composed of the canonical α chain linked to an 85-kDa β chain, lacking most of the cytoplasmic domain. The second, $p130^{MET}$, is a soluble protein released from the cell, differing from the former in that it lacks the transmembrane domain. NIH3T3 fibroblasts transfected with the full-length human *MET* cDNA express the full-size $p190^{MET}$ and both $p140^{MET}$ and $p130^{MET}$. This indicates that posttranslational events, and not alternative mRNA splicing, are responsible for generating the receptor isoforms (Giordano et al., 1993). The truncated forms maintain the functional HGF binding site and, upon ligand binding, can make both homodimers and heterodimers using an intact, fully functional, transmembrane receptor molecule as a partner (Prat et al., unpublished results). Heterodimerization between the intact and truncated receptors lacking the kinase domain results in inactive com-

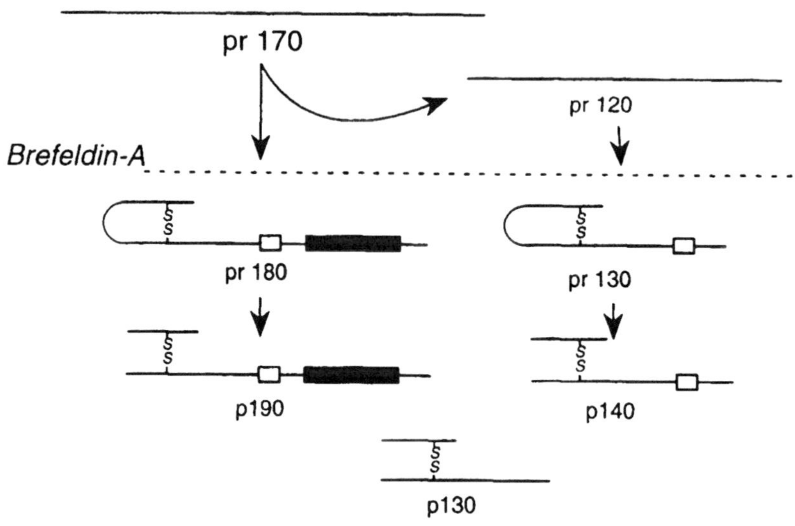

Figure 4. The HGF receptor is synthesized as a single-chain precursor (pr 170) which is glycosylated and cleaved into the mature α-β-heterodimer (p 190). If inappropriately phosphorylated in the cytoplasm (see text), the precursor is cleaved by cytosolic proteases to a shorter form of 120 kDa. The latter is further processed and generates truncated receptor isoforms exposed at the cell surface or secreted.

plexes, probably, due to their inability to activate the kinase domain by transphosphorylation (see below). Thus, the generation of truncated receptor isoforms may represent a dominant-negative control mechanism to regulate signal transduction by p190^{MET}.

The generation of truncated receptors was investigated in detail (Crepaldi et al., 1994a). The soluble p130^{MET} is generated by proteolytic cleavage at the cell surface. In the endoplasmic reticulum a fraction of the single-chain p170^{MET} precursor is cleaved at the cytosolic side, generating a second precursor of 120 kDa. A further proteolytic event, occurring in the trans-Golgi network, converts both precursors into the mature heterodimers, p190^{MET} and p140^{MET}, respectively (Fig. 4). Under physiological conditions, the p170^{MET} precursor is inactive. Conditions that cause its accumulation in the endoplasmic reticulum (e.g., overexpression) lead to activation of the tyrosine kinase. Consequently, an accumulation of p120^{MET} is observed. Overproduction of p120^{MET} does not occur in cells overexpressing a transfected kinase-dead receptor mutant. Proteolytic cleavage of the cytoplasmic domain of p170^{MET} and generation of dominant-negative truncated isoforms may, thus, represent a safety mechanism aimed at preventing the effect of ligand-independent intracellular activation of the receptor kinase (Crepaldi et al., 1994a).

IV. POSITIVE AND NEGATIVE REGULATION OF THE HGF RECEPTOR KINASE ACTIVITY BY TYROSINE OR SERINE PHOSPHORYLATION

The earliest detectable consequence of HGF stimulation of target cells is the tyrosine autophosphorylation of its receptor (Naldini et al., 1991b). As with other growth factor receptors, this indicates activation of the receptor kinase. Autophosphorylation of the receptor β subunit was observed both in cells showing growth as response to HGF and in cells showing increased motility. The response is time- and dose-dependent and could be observed as early as two minutes after stimulation.

Unlike other growth factor receptors, such as those for EGF or PDGF, the catalytic activity of the p190^{MET} kinase is strongly activated by tyrosine autophosphorylation (Naldini et al., 1991a). When exposed to ATP and Mg salts, immunoprecipitated p190^{MET} incorporates phosphate into the β subunit with first order kinetics and displays a time-dependent increase in the phosphorylation activity of a peptide substrate. Kinase activation is maximal when phosphorylation reaches the plateau. The increase in phosphorylation rate is due to a several fold increase in the V_{max} of the enzyme-catalyzed phosphotransfer reaction. In the absence of ligand, the K_m for ATP is not affected by autophosphorylation, and a slight increase is observed in the K_m for the peptide substrate. Other enzymatic parameters calculated for p190^{MET} were similar to those reported for other tyrosine kinases. Analogy with the EGF receptor showed a 10-fold difference between the K_m [ATP] for autophosphorylation and substrate phosphorylation (Bertics et al., 1985; Weber et al., 1984).

The major tyrosine autophosphorylation site of the p190^{MET} receptor was identified with the tyrosine residue 1235 (Ferracini et al., 1991). Tyr^{1235} is located within the tyrosine kinase domain (Figure 3) and is part of a "three-tyrosine" motif, including Y^{1230}, Y^{1234} and Y^{1235}, conserved in other tyrosine kinase receptors (Hanks et al., 1988). The HGF receptor's role in positive regulation of these tyrosines has been studied with site-directed mutagenesis by replacing the tyrosine with phenylalanine (Longati et al., 1994). We observed that both Y^{1234} and Y^{1235} are essential for fully activating the enzyme. Phosphorylation of both residues may be required for maximally activating the HGF receptor kinase or, given that Y^{1235} has been identified as the major phosphorylation site, Y^{1234} may facilitate phosphorylation of the nearby residue or may stabilize the activated form of the enzyme via its phenolic side group. This tyrosine doublet is present at homologous locations in the proteins encoded by *RON* and *SEA*, the other two members of the *MET* family (Ronsin et al., 1993; Huff et al., 1993). The doublet is also conserved and phosphorylated in other tyrosine kinases, including the insulin receptor (Tornqvist et al., 1987, 1988).

Several membrane receptors are modulated by protein kinase-C, an enzyme involved in signal transduction, via phosphorylation on serine and/or threonine residues (for a review see Nishizuka et al., 1988). Stimulation of protein kinase-C by phorbol esters is followed by serine phosphorylation and consequent decrease in tyrosine kinase activity of the HGF receptor (Figure 2) (Gandino et al., 1990). An independent, negative, regulatory pathway is triggered by increasing the intracellular Ca^{2+} concentration, either experimentally or by physiological agonists. The enzyme involved is a serine kinase with the biochemical properties of Ca^{2+}-calmodulin kinase III (Gandino et al., 1991). Interestingly, the target residue phosphorylated in both instances is Ser^{985}, located within the juxtamembrane domain of the HGF receptor, in a canonical consensus sequence for phosphorylation by protein kinase-C and Ca^{2+}-calmodulin kinase III (Kennelly and Krebs, 1991) (Figure 3). The inhibitory effects of either TPA or calcium ionophores on the HGF receptor activity were lost when the juxtamembrane Ser^{985} was substituted with an alanine residue by site-directed mutagenesis (Gandino et al., 1994).

V. SIGNAL TRANSDUCTION BY THE HGF RECEPTOR: THE DOCKING "SUPERSITE"

Elucidation of the signal-transduction pathways leading to the diverse cellular responses triggered by HGF is a biological enigma. Cell division and motility must ultimately result from modification of distinct cellular targets. These effects may be elicited by coupling the HGF receptor to specific transducers or to a different combination of the same set of transducers.

The currently accepted mechanism for coupling tyrosine kinase receptors with cytoplasmic signalling molecules involves ligand binding, receptor dimerization, and reciprocal transphosphorylation of each monomer at multiple sites. Some of the phosphorylated residues are docking sites for cytoplasmic transducers, which bind specific phosphotyrosines via conserved structural modules known as Src homology 2 (SH2) domain (Cantley et al., 1991; Koch et al., 1991). SH2 domains are found in one or two copies in molecules involved in signal transduction, such as PI 3-kinase, *Ras* GAP, PLCγ, *Src*-related tyrosine kinases, the tyrosine phosphatase SHPTP2, and the *Shc* and *Grb*-2 adaptors (for a review see Pawson and Schlessinger, 1993). In a number of tyrosine kinase receptors, distinct phosphotyrosine residues have been identified, responsible for binding PI 3-kinase, PLCγ, *Ras* GAP, or $pp60^{c\text{-}Src}$. Comparison of the binding sites in different receptors (Cantley et al., 1991; Pawson, 1995) and selecting the optimal phosphopeptide for all of the available SH2 domains from a synthetic library (Songyang et al., 1993), have led to the identification of consensus sequences for individual SH2 domains. Studies based on peptide competitions and site-directed mutagenesis have shown that the amino acids in the +1, +2 and +3 positions are critical in determining the selectivity of the different SH2 domains (Fantl et al., 1992; Cantley et al., 1991; Rotin et al., 1992; Songyang et al., 1993; Waksman et al., 1992, 1993; Eck et al., 1993).

HGF is a pleiotropic factor, capable of evoking complex biological responses, such as mitogenesis, motogenesis and morphogenesis. Accordingly, its receptor activates multiple signal-transduction pathways. Stimulation of responsive cells induces activation of PI 3-kinase (Graziani et al., 1991; Ponzetto et al., 1993), of a *Ras* nucleotide exchanger (Graziani et al., 1993) and of a tyrosine phosphatase (Villa–Moruzzi et al., 1993). Moreover, in epithelial cells, upon phosphorylation of the receptor, PLCγ and MAP kinase become phosphorylated on tyrosine (Ponzetto et al., 1994). For PLCγ this modification has been found to be essential for activating the enzyme and hydrolysis of inositol phospholipids leading to generations of second intracellular messengers (Kim et al., 1991). MAP kinase phosphorylation has been correlated to activation of the Ras pathway (Pelech and Sanghera, 1992). HGF also induces tyrosine phosphorylation and stimulation of the cytoplasmic tyrosine kinase $pp60^{Src}$ (Ponzetto et al., 1994). Finally, after ligand-induced tyrosine phosphorylation, the HGF receptor associates with the *Shc* adaptor, via the SH2 domain. *Shc* is phosphorylated on Y^{317}VNV, generating a high-affinity binding site for *Grb2* (K_d = 15 nM). This duplicates the high-affinity binding site for *Grb2* on the HGF receptor (Y^{1356}VNV; see below). Thus, HGF stimulation can trigger the *Ras* pathway by recruiting *Grb2* directly through the receptor and indirectly through *Shc*. Overexpression of wild-type *Shc*, but not of the $Y^{317} \rightarrow F$ mutant, enhances cell migration and growth in response to HGF, showing that *Shc* is a relevant substrate of the HGF receptor and works as an "amplifier" of the motogenic as well as the mitogenic response.

We have shown that two phosphotyrosines located in the HGF receptor C-terminal tail are responsible for mediating its interactions with SH2-containing cytoplasmic effectors (Fig. 5) (Ponzetto et al., 1994). Unlike the docking sites identified in other tyrosine kinase receptors (for a review, see Pawson, 1995), the HGF receptor sequence Y^{1349}VHVNATY^{1356}VNV can interact with high affinity and activate PI 3-kinase, PLCγ, pp60$^{c\text{-}Src}$, the SHC adaptor and the *Grb-2*/SOS complex. All of these molecules, with the exception of *Grb-2*, interact directly with either phosphotyrosine Y^{1349}VHV or Y^{1356}VNV in the "bidentate" site. Comparing the sequence *YVH/NV* (the docking "super-site") with the optimal binding motifs listed by Songyang et al. (1993) indicates that it represents a degenerate consensus potentially permissive for a number of SH2 domains. Using synthetic phosphopeptides and a BIAcore biosensor to measure intermolecular binding, we showed that the SH2 domains of p85, PLCγ, and pp60$^{c\text{-}Src}$ interact directly with either version

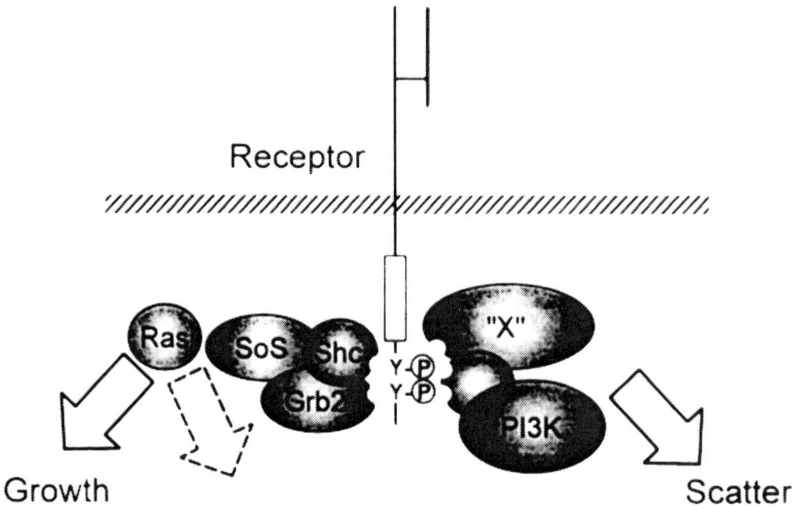

Figure 5. Transduction of mitogenic and motogenic signals by the HGF receptor. Upon binding to the specific ligand, the receptor tail is autophosphorylated on two tyrosines that generate a "docking" site for cytoplasmic transducers containing SH2 domains. The tandemly repeated sequence YVH/NV (the "*supersite*") allows high-affinity binding with PI3 kinase (PI3K) and the adaptor molecules Shc and Grb2. The latter binds and activates the GTP-exchanger SoS, triggering the Ras pathway. Although incompletely understood, the signaling pathway leading to the motogenic response ("scattering") is driven by PI3 kinase and by yet unidentified transducers ("X"). The cell growth is driven by Ras, whose activation is, however, also required for "scattering."

(YVHV or YVNV) of the "supersite". *Grb-2*, which has a strong requirement for asparagine in the +2 position (Songyang et al., 1993), specifically interacts with the sequence YVNV. All bindings are characterized by fast association and dissociation rates. When the kinetic parameters measured for the HGF receptor sites were compared to those for the respective optimal sequences, the affinities were similar for all the SH2's studied, with the exception of p85 which showed a K_d one order of magnitude apart from that for the optimal sequence YXXM (Ponzetto et al., 1993).

The kinetic parameters of these interactions (see also Panayotou et al., 1993; Felder et al., 1993) are consistent with the notion of a multifunctional docking site capable of binding a host of rapidly exchanging, signaling molecules. The close spacing of the two phosphotyrosines suggests that reiteration of the degenerate consensus may increase binding affinities. An SH2 dissociating from one site may have a high likelihood of rebinding to either one or the other in the activated receptor. Moreover, molecules containing two SH2 domains (like p85 and PLCγ) may be able to engage both SH2 at the same time.

The multifunctional "supersite" represents a variation from the common theme of sequential specificity in the recognition process between SH2 domains and phosphotyrosine residues in tyrosine kinase receptors. To explain how such a site may work, we propose that a number of factors will concur in determining which transducer binds at any moment. Listed among these factors are differences in affinity, variations in local concentration of effectors, and levels of receptor phosphorylation. At very low levels of receptor phosphorylation (early times after stimulus), only the highest affinity interactions will occur and, thus, only a few transductional pathways will be activated. As the number of phosphorylated receptors increases, the formation of additional complexes involving molecules capable of binding with lower affinities may become possible. Through this mechanism, additional pathways can be superimposed onto those already operating. Different cell types contain significantly different amounts of receptor and transducers. This may explain why HGF activates some pathways (e.g., mitogenesis or motogenesis) in some cells and not in others.

To demonstrate just how crucial the docking "supersite" is in signal transduction, the two tyrosines in the oncogenic counterpart of the receptor were mutated. This is a constitutively active hybrid molecule (TPR-MET), which includes a dimerization motif fused to the intracellular domain of the HGFR (Cooper et al., 1984; Rodrigues and Park, 1993). Substitution of Y^{1349} and Y^{1356} with phenylalanine in TPR-MET, abolished its transforming ability without interfering with its kinase activity (Ponzetto et al., 1994). Individual mutations inhibited transformation to a different extent. Substitution of Y^{1356}, which has the unique ability over Y^{1349} to bind *Grb-2*, determined a drastic reduction in the number of foci. This result highlights the importance for transformation by tyrosine kinases of the link with the *Ras* pathway provided by the *Grb-2* adaptor (see also Pendergast et al., 1993). However, Y^{1356} alone was not sufficient to bring about the full transforming potential of the oncogenic Met protein. Mutation of Y^{1349} significantly reduced the number of foci. This indicates that the maximal signaling efficiency results from the cooperation between the two phosphotyrosines.

The bidentate motif Y-hydrophobic-X-hydrophobic-$(X)_3$-Y-hydrophobic-N-hydrophobic is conserved in the otherwise divergent C-terminal regions of the transmembrane kinases *Sea* and *Ron*, the other two members of the HGF receptor family. Synthetic phosphopeptides derived from *Sea* and *Ron* inhibit interaction of the SH2 domains, previously considered, with the HGF receptor "supersite" (Ponzetto el al., 1994). The newly described multifunctional docking site (in the subtle variations found in the three members of the family) may, thus, represent the main transductional switch for all of these related receptors.

VI. TISSUE-SPECIFIC REGULATION OF HGF RECEPTOR EXPRESSION

The HGF receptor is expressed in adult epithelial tissues, including liver, intestine, thyroid, and kidney (Prat et al., 1991a; Di Renzo et al., 1991; Di Renzo et al., 1992). It has been reported to be expressed at early stages in development of epithelial organs (Sonnenberg et al., 1993), and it is often overexpressed in epithelial cancer (Prat et al., 1991a; Di Renzo et al., 1991). We have shown that the receptor is also expressed in endothelial cells and that HGF is a potent angiogenic factor both *in vitro* and *in vivo* (Bussolino et al., 1992; Grant et al., 1993). HGF is a paracrine growth factor that governs mesenchymal–epithelial interactions. It is produced by the nonparenchymal cells of liver, kidney, and lung, and parenchymal (epithelial) cells in these organs express high levels of HGF receptor (Noji et al., 1990). A similar expression pattern is found during development of epithelial organs, where the factor is secreted by mesenchymal cells and the receptor is expressed by adjacent epithelial cells (Sonnenberg et al., 1993a).

Similarly, HGF seems to be a paracrine mediator between stromal and hemopoietic cells during embryogenesis and adult life (Galimi et al., 1995). Recombinant HGF stimulates hemopoietic progenitors to form colonies *in vitro*. The growth stimulatory activity is prominent for the erythroid lineage and includes multipotent CFU-GEMM precursor cells.

From flow cytometric analysis, the receptor for HGF was found expressed on the cell surface in a fraction of CD34+ progenitors. Moreover, *in situ* hybridization experiments showed that HGF receptor mRNA is highly expressed in embryonic erythroid cells (megaloblasts). Another cell population of hemopoietic origin expressing the receptor is the macrophage-derived brain microglia (Di Renzo et al., 1993).

We have shown that expression of the HGF receptor is inducible at transcriptional level (Boccaccio et al., 1994). The amount of the HGF receptor expressed by epithelial cells varies in different growth conditions *in vitro*, being lower in growth-arrested confluent monolayers and higher in growing sparse cells. Recruitment in the growing state by stimulation with fresh serum results in a significant increase of both mRNA and protein. The insulin receptor behaves contrarily. The receptor mRNA level rises in growth arrested cells (Levy and Hug, 1992). Protein kinase-C activation upregulates receptor synthesis, inducing both mRNA accumulation and protein translation. A similar late effect of protein kinase-C on the EGF receptor has been described by Bjorge and Kudlow (1987), who showed that the receptor synthesis rate increases fivefold in cells stimulated by phorbol esters.

Interestingly, in cells grown *in vitro*, HGF receptor expression is also induced by HGF itself. Similar behavior was again shown in the EGF receptor. EGF stimulation increases receptor mRNA levels by three to fivefold and stimulates receptor synthesis within a few hours (Clark et al.,1985; Earp et al., 1986; Bjorge and Kudlow, 1987). It would appear that, in the case of HGF receptor, induction by the cognate growth factor also operates *in vivo*. In this respect, it has been found that partial hepatectomy, or CCl_4-induced hepatocellular necrosis both cause a dramatic rise in the level of circulating HGF (Lindroos et al., 1991). Within few hours, this rise in HGF is followed by increased expression of HGF receptor mRNA in the regenerating hepatic tissue, in the cases of hepatectomy (Gaudino et al., unpublished observation) and of CCl_4 intoxication (Ito et al., 1993).

The HGF receptor is encoded by the c-*MET* proto-oncogene. The cloning of the c-*MET* promoter (Gambarotta et al., 1994) (Figure 6) gave some insight into the mechanism(s) likely to operate to increase expression of the receptor after serum, phorbol esters, or HGF stimulation. The promoter, encompassing at least 297 base pairs upstream to the start site of transcription, contains *cis* acting elements including a consensus sequence for the transcriptional factor PEA3 at position –82, and 2 AP2 sites, at positions –84 and –166. The PEA3 motif responds to TPA and to growth factors (Xin et al., 1992). AP2 regulatory elements mediate the induction of transcriptional activity after stimulation by TPA (Imagawa et al., 1987). The transcriptional activity of the isolated *MET* promoter, transfected in epithelial lines, was low in resting cells and was stimulated by serum, TPA, or HGF. The data strongly suggest that these factors may also operate at transcriptional level on the endogenous promoter.

Interestingly, a GATA-1 consensus sequence was found within 300 bp upstream from the transcription start site (Gambarotta et al., 1994). The GATA motif is known to identify genes preferentially expressed during erythroid differentiation (Martin et al., 1989). This finding confirms the reported expression of HGF receptor in hemopoietic cell lineages (Galimi et al., 1995).

VII. THE HGF RECEPTOR GENE (*MET*) CAN TRANSFORM CELLS *IN VITRO*

MET was originally described as a human oncogene that becomes activated *in vitro* after treatment of a cell line with a chemical carcinogen (Cooper et al., 1984). The cloned oncogene is a hybrid between 5′ sequences of a gene named *TPR*, derived from chromosome 1, and 3′ sequences of *MET*, located on chromosome 7 (Park et al., 1986; Tempest et al., 1986; Dean et al., 1987; Mitchell et al., 1992a,b). Moreover, it has been reported that transfection of a full size *MET* cDNA can induce transformation upon establishment of an autocrine circuit (Rong et al., 1993).

To elucidate the molecular mechanisms that unleash the oncogenic potential of the *MET* proto-oncogene, the biochemical and biological features of a number of deletion or point mutants of the HGF receptor were studied (Zhen et al., 1994). The HGF receptor lacking the extracellular and the transmembrane domains has deregulated tyrosine kinase activity and is both transforming and tumorigenic. Truncation of the extracellular domain activates other transmembrane tyrosine kinases, such as the EGF receptor (Downword et al., 1984; Khazaie et al., 1988) and the SCF receptor (Qiu et al., 1988). Similarly, proteolytic truncation of the extracellular domain of the insulin receptor releases the kinase activity of the β subunit from inhibitory control (Shoelson et al., 1988). Catalytic activity of the HGF receptor is normally latent unless the extracellular domain binds the cognate ligand (Naldini et al., 1991a,b). Although deletion of the α chain is not sufficient to induce deregulation of the tyrosine kinase activity (Zhen et al., 1994), lack of cleavage between the α and the β chain can cause constitutive activation (Mondino et al., 1991).

Constitutive kinase activation and transforming ability are lost when portions of the juxtamembrane domain are removed. Full oncogenic potential is restored by replacing the juxtamembrane domain with 5′ sequences derived from the *TPR* gene (Mitchell et al., 1992a; Mitchell et al., 1992b). Comparing the first 39 amino acids of the HGF receptor juxtamembrane domain to the *TPR* moiety does not reveal obvious structural similarities. It is generally agreed that receptor tyrosine kinases are activated by conformational changes induced by ligand binding, resulting in receptor oligomerization which stabilizes the interactions between contiguous cytoplasmic domains (Yarden et al., 1987a; Yarden et al., 1987b; Cochet et al., 1988; Greenfield et al., 1989; Hammacher et al., 1989; Heldin et al., 1989; Seifert et al., 1989). According to this activation mechanism, it is conceivable that a critical segment of the juxtamembrane domain may be necessary to stabilize the formation of kinase dimers leading to *trans* activation. A role of *TPR* sequences in the dimerization of the *Tpr-Met* hybrid kinase has been shown (Rodriguez et al., 1993).

The transforming potential of the intracellular domain of the HGF receptor relies on its tyrosine kinase activity. This has been established in all receptor and nonreceptor tyrosine kinases so far studied. Substitution of the critical lysine residue responsible for ATP binding invariably results in a loss of catalytic function and transforming activity (Snyder et al., 1985). Moreover, to unleash the oncogenic potential of the HGF receptor kinase, additional amino acid residues located in the catalytic domain are required. These are the two tyrosines Y^{1234} and Y^{1235}, known to positively regulate the receptor enzymatic activity upon autophosphorylation (Ferracini et al., 1991; Longati et al., 1994). The HGF receptor is "autocatalytic" because autophosphorylation dramatically increases the V_{MAX} of the phosphate transfer reaction (Naldini et al., 1991c). This mechanism of up-regulation is critical for unleashing the oncogenic potential of the MET oncogene.

VIII. THE HGF RECEPTOR GENE (*MET*) IS A HARMFUL ONCOGENE *IN VIVO*

The initial observation of amplification of the *MET* proto-oncogene was made in a cell line derived from a metastasis of a human gastric carcinoma (Giordano et al., 1989a). The gene was found amplified and overexpressed, and the tyrosine kinase was constitutively activated. No modification in the sequence has been found in the cloned cDNA (Ponzetto et al., 1991), showing that constitutive activation of the *Met* kinase may be brought about by simple overexpression. Since then, *MET* has been found amplified and/or overexpressed in a significant number of human tumors of epithelial origin (Di Renzo et al., 1991; Prat et al., 1991a; Di Renzo et al., 1992a; Liu et al., 1992). In thyroid tumors, overexpression of *MET* is observed in a significant proportion (~70%) of carcinomas derived from the follicular epithelium and correlates with the aggressive phenotype (Di Renzo et al., 1992b). Although the c-*MET*-encoded receptor is detectable only in the acinar cells of the normal human exocrine pancreas, it is upregulated in the majority of pancreatic ductal adenocarcinomas (Di Renzo et al., 1995a). In most cases the HGF receptor found in the malignant cells has features of the normal receptor. In overexpressing cell lines, the *Met*/HGF receptor is phosphorylated in the absence of ligand, suggesting that activation of the c-*MET* gene may be involved in the growth of pancreatic cancer and may contribute to the ductal phenotype of these tumors.

The changes of c-*MET* gene expression that occur during the progression of colorectal tumors have been studied in detail in a large number of adenomas, in primitive carcinomas and in liver metastases. In several examples, it was possible to compare same patient samples of normal colon mucosa to primary tumor and primary carcinoma to synchronous metastasis (Di Renzo et al., 1995b). The expression of the c-*MET* gene was increased from five to fiftyfold in about 50% of tumors, at any stage of progression, and in 70% of liver metastases. Overexpression was associated with amplification of the c-*MET* gene in only few primary carcinomas, but in a significant proportion of the metastases examined. These data suggest that overexpression of the c-*MET* oncogene contributes a selective growth advantage to neoplastic colorectal cells at any stage of tumor progression. Moreover, amplification appears to give a further selective advantage for the acquisition of metastatic potential.

DNA level and expression of the Met/HGF receptor gene were also studied in human ovary, benign ovarian tumors, and epithelial ovarian carcinomas (Di Renzo et al., 1994). The Met/HGF receptor was detectable in the surface epithelium in normal ovary. The level of expression was unchanged in benign ovarian tumors of various origin. About 30% of malignant carcinomas showed a 3 to 10-fold increase in expression. In a few cases, the Met/HGF protein was overexpressed more than

50-fold without *MET* gene amplification. Clinically, overexpression was associated with disease at any pathological stage, but it was significantly correlated to particularly aggressive behavior by the malignant cells.

Although c-*MET* is considered, in particular, to be an oncogene affecting epithelial tumors, other reports have suggested that the *Met*/HGF receptor might also have a role in the growth of sarcomas. c-*MET* was originally identified as an oncogene activated by artificial rearrangement in a human osteosarcoma cell line (Cooper et al., 1984; Park et al., 1986). The murine *met* homologue was found to be amplified and overexpressed in spontaneously transformed fibroblasts grown *in vitro* (Cooper et al., 1986). In addition, it was reported that NIH/3T3 fibroblasts that coexpress the *Met* receptor and its ligand HGF are tumorigenic (Rong et al., 1992, 1993). It has been recently shown that the creation of an HGF autocrine loop either in fibroblasts or in epithelial cells induces invasive properties *in vitro* and metastatic ability *in vivo* (Bellusci et al., 1994; Rong et al., 1994).

This prompted us to study the expression of the *Met*/HGF receptor and its ligand in human bone tumors. We found that the expression of the *Met*/HGF receptor is turned on in a significant percentage of human osteosarcomas and in all the osteosarcoma cell lines examined. HGF expression was also detectable in some cases. HGF stimulated cell motility and penetration in extracellular matrices ("scattering"), suggesting that inappropriate *Met*/HGF receptor activation might provide osteosarcoma cells with the capacity to invade neighboring tissues (Ferracini et al., 1995).

REFERENCES

Appella, E., Robinson, E. A., Ullrich, S. J., Stoppelli, M. P., Corti, A., Cassani, G., & Blasi, F. (1987). The receptor-binding sequence of urokinase. J. Biol. Chem. 262, 4437–4440.

Barr, J. (1991). Mammalian subtilisins: the long-sought dibasic processing endoproteases. Cell 66, 1–3.

Bellusci, S., Moens, G., Gaudino, G., Comoglio, P. M., Nakamura, T., Thiery J. P., & Jouanneau, J. (1994). Creation of an hepatocyte growth factor/scatter factor autocrine loop in carcinoma cells induces invasive properties associated with increased tumorigenicity. Oncogene 9, 1091–1099.

Bertics, P. J. & Gill, G. N. (1985). Self-phosphorylation enhances the protein-tyrosine kinase activity of the epidermal growth factor receptor. J. Biol. Chem. 260, 14642–14647.

Bjorge, J. D. & Kudlow, J. E. (1987). Epidermal growth factor receptor synthesis is stimulated by phorbol ester and epidermal growth factor. J. Biol. Chem. 262, 6615–6622.

Boccaccio, C., Gaudino, G., Gambarotta, G., Galimi, F., & Comoglio, P. M. (1994). Hepatocyte growth factor (HGF) receptor expression is inducible and is part of the delayed early response to HGF. J. Biol. Chem. 269, 12846–12851.

Bottaro, D. P., Rubin, J. S., Faletto, D. L., Chan, A. M. L., Kmiecick, T. E., Van de Woude, G. F., & Aaronson, S. A. (1991). Identification of the hepatocyte growth factor receptor as the c-met proto-oncogene. Science 251, 802–804.

Bussolino, F., Di Renzo, M. F., Ziche, M., Bocchietto, E., Olivero, M., Naldini, L., Gaudino, G., Tamagnone, L., Coffer, A., Marchisio, P. C., & Comoglio, P. M. (1992). Hepatocyte growth factor is a potent angiogenic factor which stimulates endothelial cell motility and growth. J. Cell Biol. 119, 629–641.

Cantley, L. C., Auger, K., Carpenter, C., Duckworth, B., Graziani, A., Kapeller, R., & Soltoff, S. (1991). Oncogenes and signal transduction. Cell 64, 281–302.

Chan, A. M.–L., Rubin, J. S., Bottaro, D. P., Hirschfield, D. W., Chedid, M., & Aaronson, S. A. (1991). Identification of a competitive HGF antagonist encoded by an alternative transcript. Science 254, 1382–1385.

Clark, A. J. L., Ishii, S., Richert, N., Merlino G. T., & Pastan, I. (1985). The human hematopoietic colony-stimulating factors. Proc. Natl. Acad. Sci. USA 82, 8374–8378.

Cochet, C., Kashles, O., Chambaz, E. M., Borello, I., King, C. R., & Schlessinger, J. (1988). Demonstration of epidermal growth factor-induced receptor dimerization in living cells using a chemical covalent cross-linking agent. J.Biol.Chem. 263, 3290–3295.

Cooper, C. S., Park, M., Blair, D. G., Tainsky, M. A., Huebner, K., Croce, C. M., & Van de Woude, G. (1984). Molecular cloning of a new transforming gene from a chemically transformed human cell line. Nature 311, 29–33.

Cooper, C. S.,Tempest, P. R., Beckman, M. P., Heldin, C.–H., & Brooks, P. (1986). Amplification and overexpression of the *met* gene in spontaneously transformed NIH3T3 mouse fibroblasts. EMBO 5, 2623–2628.

Crepaldi, T., Prat, M., Giordano, S., Medico, E., & Comoglio, P. M. (1994a). Generation of a truncated hepatocyte growth factor receptor in the endoplasmic reticulum. J. Biol. Chem. 269, 1750–1755.

Crepaldi, T., Pollack, A. L., Prat, M., Zborek, A., Mostov, K., & Comoglio, P. M. (1994b). Targeting of the SF/HGF receptor to the basolateral domain of polarized epithelial cells. J. Cell Biol. 125, 313–320.

Dean, M., Park, M., & Van de Woude, G. F. (1987). Characterization of the rearranged TPR-MET oncogene breakpoint. Mol. Cell. Biol. 7, 921–924.

Di Renzo, M. F., Narsimhan, R. P., Olivero, M., Bretti, S., Giordano, S., Medico, E., Gaglia, P., Zara, P., & Comoglio, P. M. (1991). Expression of the MET/HGF receptor in normal and neoplastic human tissues. Oncogene 6, 1997–2003.

Di Renzo, M. F., Olivero, M., Ferro, S., Prat, M., Bongarzone, I., Pilotti, S., Pierotti, M. A., & Comoglio, P. M. (1992). Overexpression of the c-MET/HGF receptor gene in human thyroid carcinomas. Oncogene 7, 2549–2553.

Di Renzo, M. F., Olivero, M., Bertolotto, A., Crepaldi, T., Schiffer, D., Pagni, C. A., & Comoglio, P. M. (1993). Selective expression of the MET/HGF receptor in human central nervous system microglia. Oncogene, 8, 219–222.

Di Renzo, M. F., Olivero, M., Katsaros, D., Crepaldi, T., Gaglia, P., Zola, P., Sismondi P., & Comoglio, P. M. (1994). Overexpression of the MET/HGF receptor in ovarian cancer. Int. J. Cancer 58, 658–662.

Di Renzo, M. F., Poulsom, R., Olivero, M., Comoglio, P. M., and Lemoine, N. R. (1995a). Expression of the MET/hepatocyte growth factor receptor in human pancreatic cancer. Cancer Res. 55, 1129–1138.

Di Renzo, M. F., Olivero, M., Giacomini, A., Porte, H., Chastre, E., Mirossay, L., Nordlinger, B., Bretti, S., Bottardi, S., Giordano, S., Plebani, M., Gespach, C., & Comoglio, P. M. (1995b). Overexpression and amplification of the MET/HGF receptor gene during the progression of colorectal cancer. Clin. Cancer Res. 1, 147–154.

Downword, J., Yarden, Y., Mayes, E., Scrace, G., Totty, N., Stockwell, P., Ullrich, A., Schlessinger, J., & Waterfield, M. D. (1984). Close similarity of epidermal growth factor receptor and *v-erb-B* oncogene protein sequences. Nature 307, 521–527.

Earp, H. S., Austin, K. S., Blaisdell, J., Rubin, R. A., Nelson, K. G., Lee, L. W., & Grisham, J. W. (1986). Epidermal growth factor (EGF) stimulates EGF receptor synthesis. J. Biol. Chem. 261, 4777–4780.

Eck, M. J., Shoelson, S. E., & Harrison, S. C. (1993). Recognition of a high affinity phosphotyrosyl peptide by the Src homology 2 domain of $p56^{lck}$. Nature 362, 87–91.

Fantl, W. J., Escobedo, J. A., Martin, G. A., Turck, C. W., Rosario, M., McCormick, F., & Williams, L. T. (1992). Distinct phosphotyrosines on a growth factor receptor bind to specific molecules that mediate different signaling pathways. Cell 69, 413–423.

Felder, S., Zhou, M., Hu, P., Ureña, J., Ullrich, A., Chaudhuri, M., White, M., Shoelson, S. E., & Schlessinger, J. (1993). SH2 domains exhibit high-affinity binding to tyrosine-phosphorylated peptides yet also exhibit rapid dissociation and exchange. Mol. Cell. Biol. 13, 1449–1455.

Ferracini, R., Longati, P., Naldini, L., Vigna, E., & Comoglio, P. M. (1991). Identification of the major phosphorylation site of the MET/hepatocyte growth factor receptor tyrosine kinase. J. Biol. Chem. 266, 19558–19564.

Ferracini, R., Di Renzo, M. F., Scotlandi, K., Baldini, N., Olivero, M., Lollini, P., Cremona, O., Campanacci, M., & Comoglio, P. M. (1995). The MET/HGF receptor is overexpressed in human osteosarcomas and is activated by either a paracrine or an autocrine circuit. Oncogene 10, 739–749.

Fukuyama, R., Ichijoh, Y., Minoshima, S., Kitamura, N., & Shimuzu, N. (1991). Regional localization of the hepatocyte growth factor (HGF) gene to human chromosome 7 band q21.1. Genomics 11, 410–415.

Galimi, F., Brizzi, M. F., & Comoglio, P. M. (1993). The hepatocyte growth factor and its receptor. Stem Cells 11, 22–30.

Galimi, F., Bagnara, G. P., Bonsi, L., Cottone, E., Follenzi, A., & Simeone, A. (1994). Hepatocyte growth factor induces proliferation and differentation of multipotent and erythroid hemopoietic progenitors. J. Cell Biol. 127 (6), 1743–1754.

Gambarotta, G., Pistoi, S., Giordano, S., Comoglio, P. M., & Santoro, C. (1994). Structure and inducible regulation of the human met promoter stem cells, 11, 22–30. 1993 J. Biol. Chem. 269, 12852–12857.

Gandino, L., Di Renzo, M. F., Giordano, S., Bussolino, F., & Comoglio, P. M. (1990). Protein kinase-C activation inhibits tyrosine phosphorylation of the c-*MET* protein. Oncogene 5, 721–725.

Gandino, L., Munaron, L., Naldini, L., Ferracini, R., Magni, M., & Comoglio, P. M. (1991). Intracellular calcium regulates the tyrosine kinase receptor encoded by the MET oncogene. J. Biol. Chem. 266, 16098–16104.

Gandino, L., Longati, P., Medico, E., Prat, M., & Comoglio, P. M. (1994). Phosphorylation of ser^{985} negatively regulates the hepatocyte growth factor receptor kinase. J. Biol. Chem. 269, 1815–1820.

Gaudino, G., Follenzi, A., Naldini, L., Collesi, C., Santoro, M., Gallo, K. A., Godowski, P. J., & Comoglio, P. M. (1994). Ron is a heterodimeric tyrosine kinase receptor activated by the HGF homolog MSP. EMBO J. 13, 3524–3532.

Gherardi, G., Gray, J., Stoker, M., Perryman, M., & Furlong, R. (1989). Purification of scatter factor, a fibroblast-derived basic protein that modulates epithelial interaction and movement. Proc. Natl. Acad. Sci. USA 86, 5844–5848.

Giordano, S., Ponzetto, C., Di Renzo, M. F., Cooper, C. S., & Comoglio, P. M. (1989a). Tyrosine kinase receptor indistinguishable from the c-*MET* protein. Nature 339, 155–156.

Giordano, S., Di Renzo, M. F., Narsimhan, R. P., Cooper, C. S., Rosa, C., & Comoglio, P. M. (1989b). Biosynthesis of the protein encoded by the c-*MET* proto-oncogene. Oncogene 4, 1383–1388.

Giordano, S., Zhen, Z., Medico, E., Gaudino, G., Galimi, F., & Comoglio, P. M. (1993). Transfer of motogenic and invasive response to scatter factor/hepatocyte growth factor by transfection of human MET protooncogene. Proc. Natl. Acad. Sci. USA 90, 649–653.

Gohda, E., Tsubouchi, H., Nakayama, H., Hirono, S., Sakiyama, O., Takahashi, K., Miyazaki, H., Hashimoto, S., & Daikuhara, Y. (1988). Purification and partial characterization of hepatocyte growth factor from plasma of a patient with hepatic failure. J. Clin. Invest. 81, 414–419.

Gonzatti–Haces, M., Seth, A., Park, M., Copeland, T., Oroszlan, S., & Van de Woude, G. F. (1988). Characterization of the TPR-MET oncogene p65 and the MET proto-oncogene p140 protein tyrosine kinases. Proc. Natl. Acad. Sci. USA 85, 21–25.

Grant, D. S., Kleinmann, H. K., Goldberg, I. D., Bhargava, M., Nickoloff, B. J., Kinsella, J. K., Polverini, P. J., & Rosen, E. M. (1993). Scatter factor induces blood vessel formation *in vivo*. Proc. Natl. Acad. Sci. USA 90, 1937–1941.

Graziani, A., Gramaglia, D., Cantley, L. C., & Comoglio, P. M. (1991). The tyrosine phosphorylated hepatocyte growth factor/scatter factor receptor associates with phosphatidylinositol 3-kinase. J. Biol. Chem. 266, 22087–22090.

Graziani, A., Gramaglia, D., dalla Zonca, P., & Comoglio, P. M. (1993). Hepatocyte growth factor/scatter factor stimulates the ras-guanine nucleotide exchanger. J. Biol. Chem. 268, 9165–9168.

Greenfield, C., Hils, I., Waterfield, M. D., Federwisch, W., Wollmer, A., Blundell, T. L., & McDonald, N. (1989). Epidermal growth factor binding induces a conformational change in the external domain of its receptor. EMBO J. 8, 4115–4124.

Halaban, R., Rubin, J., Funasaka, Y., Cobb, C., Boulton, T., Faletto, D., Rosen, E., Chan, A., Yoko, K., White, W., Cook, C., & Moellmann, G. (1992). Met and HGF/SF signal transduction in normal melanocytes and melanoma cells. Oncogene 7, 2195–2206.

Hammacher, A., Mellstrom, K., Heldin, C.-H., & Westermark, B. (1989). Isoform-specific induction of actin reorganization by platelet-derived growth factor suggests that the functionally active receptor is a dimer. EMBO J. 8, 2489–2495.

Heldin, C. H., Ernlund, A., Rorsman, C., & Ronnstrand, L. (1989). Dimerization of B-type platelet-derived growth factor receptors is closely associated with receptor kinase activation. J. Biol. Chem. 264, 8905–8912.

Han, S., Stuart, L. A., & Friezner–Degen, S. J. (1991). Characterization of the DNF15S2 locus on human chromosome 3: Identification of a gene coding for four kringle domains with homology to hepatocyte growth factor. Biochemistry 30, 9768–9780.

Hanks, S. K., Quinn, A. M., & Hunter, T. (1988). The protein kinase family: Conserved features and deduced phylogeny of the catalytic domains. Science 241, 42–52.

Hartmann, G., Naldini, L., Weidner, M., Sachs, M., Vigna, E., Comoglio, P., & Birchmeier, W. (1992). A functional domain in the heavy chain of scatter factor/hepatocyte growth factor binds the c-*MET* receptor and induces cell dissociation but not mitogenesis. Proc. Natl. Acad. Sci. USA 89, 11574–11578.

Higashio, K., Shima, N., Goto, M., Itagaki, Y., Nagao, M., Yasuda, H., & Morinaga, T. (1990). Identity of a tumor cytotoxic factor from human fibroblasts and hepatocyte growth factor. Biochem. Biophys. Res. Comm. 170 (1), 397–404.

Huff, J. I., Jelinek, A. M., Borgman, C. A., Lansing, T. J., & Parsons, J. T. (1993). The proto-oncogene c-sea encodes a transmembrane protein-tyrosine kinase related to the MET/hepatocyte growth factor/scatter factor receptor. Proc. Natl. Acad. Sci. USA 90, 6140–6144.

Imagawa, M., Chiu, R., & Karin, M. (1987). Transcription factor AP-2 mediates induction by two different signal transduction pathways: Protein kinase C and cAMP. Cell 51, 251–260.

Ito, T., Hayashi, N., Horimoto, M., Sasaki, Y., Tanaka, Y., Kaneko, A., Fusamoto, H., & Kamada, T. (1993). Expression of the c-met/hepatocyte growth factor receptor gene during rat liver regeneration induced by carbon tetrachloride. Biochem. Biophys. Res. Comm. 190, 870–874.

Kan, M., Zhang, G. H., Zarnegar, R., Michalopoulos, G., Myoken, Y., Mckeehan, W. L., & Stevens, J. L. (1991). Hepatocyte growth factor/hepatopoietin A stimulates the growth of rat kidney proximal tubule epithelial cells (RPTE), rat nonparenchymal liver cells, human melanoma cells, mouse keratinocytes and stimulates anchorage-independent growth of SV40-transformed RPTE. Biochem. Biophys. Res. Commun. 174, 331–337.

Kennelly, P. J. & Krebs, E. G. (1991). Consensus sequences as substrate specificity determinants for protein kinases and protein phosphatases. J. Biol. Chem. 266, 15555–15558.

Khazaie, K., Dull, T. J., Graf, T., Schlessinger, J., Ullrich, A., Beug, H., & Vennstrom, B. (1988). Truncation of the human EGF receptor leads to differential transforming potentials in primary avian fibroblasts and erythroblasts. EMBO J. 7, 3061–3071.

Kim, H. K., Kim, J. W., Zilberstein, A., Margolis, B., Kim, J. G., Schlessinger, J., & Rhee, S. G. (1991). PDGF stimulation of inositol phospholipid hydrolysis requires PLCγ1 phosphorylation on tyrosine residues 783 and 1254. Cell 65, 435–441.

Kmiecik, T. E., Keller, I. R., Rosen, E., & Van de Woude, G. F. (1992). Hepatocyte growth factor is a synergic factor for the growth of hematopoietic progenitor cells. Blood 16, 2454–2457.

Kobayashi, T., Honke, K., Miyazaki, T., Matsumoto, K., Nakamura, K., Ishizuka, I., & Matika, A. (1994). Hepatocyte growth factor specifically binds to sulfoglycololipids. J. Biol. Chem. 269, 9817–9821.

Koch, C. A., Anderson, D., Moran, M. F., Ellis, C., & Pawson, T. (1991). SH2 and SH3 domains: Elements that control interactions of cytoplasmic signaling proteins. Science 252, 668–674.

Levy, J. R. & Hug, V. (1992). Regulation of insulin receptor gene expression. J. Biol. Chem. 267, 25289–25295.

Lindroos, P. M., Zarnegar, R., & Michalopoulos, G. K. (1991). Hepatocyte growth factor (hepatopoietin A) rapidly increases in plasma prior to DNA synthesis and liver regeneration stimulated by partial hepatectomy and CCl_4 administration. Hepatology 13, 743–750.

Liu, C., Park, M., & Tsao, M.-S. (1992). Overexpression of c-*MET* proto-oncogene but not epidermal growth factor receptor or c-*erb*B-2 in primary colorectal carcinomas. Oncogene 7, 181–185.

Lokker, N. A., Mark, M. R., Luis, E. A., Bannet, G. L., Robbins, K. A., Baker, J. B., & Godowski, P. J. (1992). Structure-function analysis of hepatocyte growth factor: Identification of variants that lack mitogenic activity yet retain high affinity receptor binding. EMBO J. 11, 2503–2510.

Lokker, N. & Godowski, P. J. (1993). Generation and characterization of a competitive antagonist of human hepatocyte growth factor, NGF/NK1. J. Biol. Chem. 268, 17145–17150.

Longati, P., Bardelli, A., Ponzetto, C., Naldini, L., & Comoglio, P. M. (1994). Tyrosines 1234-1235 are critical for activation of the tyrosine kinase encoded by the MET proto-oncogene (HGF receptor). Oncogene 9, 49–57.

Mark, M. R., Lokker, N. A., Zioncheck, T. F., Luis, E. A., & Godowski, P. J. (1992). Expression and characterization of hepatocyte growth factor receptor-IgG fusion proteins. J. Biol. Chem. 267, 26166–26171.

Martin, D. I., Tsai, S. F., & Orkin, S. H. (1989). Increased γ-globin expression in a nondeletion HPFH mediated by an erythroid-specific DNA-binding factor. Nature 338, 435–438.

Matsumoto, K., Hashimoto, K., Yoshikawa, K., & Nakamura, T. (1991). Marked stimulation of growth and motility of human keratinocytes by hepatocyte growth factor. Exp. Cell Res. 196, 114–120.

Medico, E., Mongiovi', A. M., Follenzi, A., Huff, J., Jelinex, M. A., Gaudino, G., Parsons, J. T., & Comoglio, P. M. (1996). The *met*-related tyrosine kinase receptors *ron* and *sea* control motility, growth and morphogenesis in liver progenitor cells *in vitro*. Mol. Cell. Biol., in press.

Michalopoulos, G. K. (1990). Liver regeneration: Molecular mechanisms of growth control. FASEB J. 4, 176–187.

Mitchell, P. J. & Cooper, C. (1992a). Nucleotide sequence analysis of human *tpr* cDNA clones. Oncogene 7, 383–388.

Mitchell, P. J. & Cooper, C. (1992b). The human *tpr* gene encodes a protein of 2094 amino acids that has extensive coiled-coil regions and an acidic C-terminal domain. Oncogene 7, 2329–2333.

Miyazawa, K., Tsubouchi, H., Naka, D., Takahashi, K., Okigaki, M., Arakaki, N., Nakayama, H., Hirono, S., Sakiyama, O., Takahashi, K., Gohda, E., Daikuhara, Y., & Kitamura, N. (1989). Molecular cloning and sequence analysis of cDNA for human hepatocyte growth factor. Biochem. Biophys. Res. Commun. 163, 967–973.

Miyazawa, K., Kitamura, A., Naka, D., & Kitamura, N. (1991). An alternatively processed mRNA generated from human hepatocyte growth factor gene. Eur. J. Biochem. 197, 15–22.

Miyazawa, K., Shimomura, T., Kitamura, A., Kondo, J., Morimoto, Y., & Kitamura, N. (1993). Molecular cloning and sequence analysis of the cDNA for a human serine protease responsible for activation of hepatocyte growth factor. J. Biol. Chem. 268, 10024–10028.

Mondino, A., Giordano, S., & Comoglio, P. M. (1991). Defective posttranslational processing activates the tyrosine kinase encoded by the *MET* proto-oncogene (hepatocyte growth factor receptor). Mol. Cell. Biol. 11, 6084–6092.

Montesano, R., Matsumoto, K., Nakamura, T., & Orci, L. (1991). Identification of a fibroblast-derived epithelial morphogen as hepatocyte growth factor. Cell 67, 901–908.

Nakamura, T., Teramoto, H., & Ichihara, A. (1986). Purification and characterization of a growth factor from rat platelets for mature parenchymal hepatocytes in primary cultures. Proc. Natl. Acad. Sci. USA 86, 6489–6493.

Nakamura, T., Nishizawa, T., Hagiya, M., Seki, T., Shimonishi, M., Sugimura, A., Tashiro, K., & Shimizu, S. (1989). Molecular cloning and expression of human hepatocyte growth factor. Nature 342, 440–443.

Naldini, L., Vigna, E., Ferracini, R., Longati, P., Gandino, L., Prat, M., & Comoglio, P. M. (1991a). The tyrosine kinase encoded by the MET proto-oncogene is activated by autophosphorylation. Mol. Cell. Biol. 11, 1793–1803.

Naldini, L., Vigna, E., Narshiman, R. P., Gaudino, G., Zarnegar, R., Michalopoulos, G., & Comoglio, P. M. (1991b). Hepatocyte growth factor (HGF) stimulates the tyrosine kinase activity of the receptor encoded by the proto-oncogene c-MET. Oncogene 6, 501–504.

Naldini, L., Weidner, K. M., Vigna, E., Gaudino, G., Bardelli, A., Ponzetto, C., Narsimhan, R. P., Hartmann, G., Zarnegar, R., Michalopoulos, G. K., Birchmeier, W., & Comoglio, P. M. (1991c). Scatter factor and hepatocyte growth factor are indistinguishable ligands for the *MET* receptor. EMBO J. 10, 2867–2878.

Naldini, L., Tamagnone, L., Vigna, E., Sachs, M., Hartmann, G., Birchmeier, W., Daikuhara, Y., Tsubouchi, H., Blasi, F., & Comoglio, P. M. (1992). Extracellular proteolytic cleavage by urokinase is required for activation of hepatocyte growth factor/scatter factor. EMBO J. 11, 4825–4833.

Naldini, L., Vigna, E., Bardelli, A., Follenzi, A., Galimi, F., & Comoglio, P. M. (1995). Biological activation of pro-HGF (hepatocyte growth factor) by urokinase is controlled by a stoichiometric reaction. J. Biol. Chem. 217 (2), 603–611.

Nishizuka, Y. (1988). The molecular heterogeneity of protein kinase C and its implications for cellular regulation. Nature 334, 661–665.

Noji, S., Tashiro, K., Koyama, E., Nohno, T., Ohyama, K., Taniguchi, S., & Nakamura, T. (1990). Expression of hepatocyte growth factor gene in endothelial and Kupffer cells of damaged rat livers, as revealed by *in situ* hybridization. Biochem. Biophys. Res. Comm. 173, 42–47.

Panayotou, G., Gish, G., End, P., Truong, O., Gout, I., Dhand, R., Fry, M. J., Hiles, I., Pawson, T., & Waterfield, M. D. (1993). Interactions between SH2 domains and tyrosine-phosphorylated platelet-derived growth factor β receptor sequences: analysis of kinetic parameters by a novel biosensor-based approach. Mol. Cell. Biol. 13, 3567–3576.

Park, M., Dean, M., Cooper, C. S., Schmidt, M., O'Brien, S. J., Blair, D. G., & Van de Woude, G. F. (1986). Mechanism of *MET* oncogene activation. Cell 45, 895–904.

Pawson, T. & Schlessinger, J. (1993). SH2 and SH3 domains. Curr. Biol. 3, 434–442.

Pawson, T. (1995). Protein modules and signaling networks. Nature 373, 573–576.

Pelech, S. L. & Sanghera, J. S. (1992). MAP kinases: charting the regulatory pathways. Science 257, 1355–1356.

Pendergast, A. M., Quilliam, L. A., Cripe, L. D., Bassing, C. H., Dai, Z., Li, N., Batzer, A., Rabun, K. M., Der, C., Schlessinger, J., & Gishizky, M. L. (1993). BCR-ABL-induced oncogenesis is mediated by direct interaction with the SH2 domain of the *Grb-2* adaptor protein. Cell 75, 175–185.

Ponzetto, C., Giordano, S., Peverali, F., Della Valle, G., Abate, M., Vaula, G., & Comoglio, P. M. (1991). c-*MET* is amplified but not mutated in a cell line with an activated *MET* tyrosine kinase. Oncogene 6, 553–559.

Ponzetto, C., Bardelli, A., Maina, F., Longati, P., Panayotou, G., Dhand, R., Waterfield, M. D., & Comoglio, P. M. (1993). A novel recognition motif for phosphatidylinositol 3-kinase binding mediates its association with the hepatocyte growth factor/scatter factor receptor. Mol. Cell. Biol. 13, 4600–4608.

Ponzetto, C., Bardelli, A., Zhen, Z., Maina, F., dalla Zonca, P., Giordano, S., Graziani, A., Panayotou, G., & Comoglio, P. M. (1994). A multifunctional docking site mediates signaling and transformation by the hepatocyte growth factor/scatter factor receptor family. Cell 77, 261–271.

Prat, M. P., Narsimhan, R. P., Crepaldi, T., Nicotra, M. R., Natali, P. G., & Comoglio, P. M. (1991a). The receptor encoded by the human c-*MET* oncogene is expressed in hepatocytes, in epithelial cells and in solid tumors. Int. J. Cancer 49, 323–328.

Qiu, F., Ray, P., Brown, K., Barker, E., Jhanwar, S., Ruddle, F. H., & Besmer, P. (1988). Primary structure of *c-kit*: Relationship with the CSF-1/PDGF receptor kinases family oncogenic activation of *v-kit* involves deletion of extracellular domain and C terminus. EMBO. J. 7, 1003–1011.

Rodriguez, G. A. & Park, M. (1993). Mol. Cell. Biol. 13, 6711–6722.

Rong, S., Bodescot, M., Blair, D., Dunn, J., Nakamura, T., Mizuno, K., Park, M., Chan, A., Aaronson, S., & Van de Woude, G. F. (1992). Tumorigenicity of the MET proto-oncogene and the gene for hepatocyte growth factor. Mol. Cell. Biol. 12(11), 5152–5158.

Rong, S., Bodescot, M., Blair, D., Dunn, J., Nakamura, T., Mizuno, K., Park, M., Chan, A., Aaronson, S., & Van de Woude, G.F. (1993). Mol. Cell. Biol., 12, 5152–5158.

Rong, S., Segal, S., Anver, M., Resau, J. H., & Van de Woude, G. F. (1994). Invasiveness and metastasis of NIH 3T3 cells induced by MET hepatocyte growth factor/scatter factor autocrine stimulation. Proc. Natl. Acad. Sci. USA 91, 4731–4745.

Ronsin, C., Muscatelli, F., Mattei, M. G., & Breathnach, R. (1993). A novel putative receptor protein tyrosine kinase of the MET family. Oncogene 8, 1195–1202.

Rubin, J. S., Chan, A. M. L., Bottaro, D. P., Burgess, W. H., Taylor, W. G., Cech, A. C., Hirschfield, D. W., Wong, J., Miki, T., Finch, P. W., & Aaronson, S. A. (1991). Proc. Natl. Acad. Sci. USA 88, 415–419.

Rotin, D., Margolis, B., Mohammadi, M., Daly, R. J., Daum, G., Li, N., Fischer, E. H., Burgess, W. H., Ullrich, A., & Schlessinger, J. (1992). SH2 domains prevent tyrosine dephosphorylation of the EGF receptor: Identification of tyr992 as the high-affinity binding site for SH2 domains of phospholipase Cγ. EMBO J. 11, 559–567.

Ruoslahti, E. & Yamaguchi, Y. (1991). Proteoglycans as modulators of growth factor activities. Cell 64, 867–869.

Saccone, S., Narsimhan, R. P., Gaudino, G., Dalpra', L., Comoglio, P. M., & Della Valle, G. (1992). Regional mapping of the human hepatocyte growth factor-scatter factor gene to chromosome 7q21.1. Genomics 13, 912–914.

Santos, O. F. P., Barros, E. J. G., Yang, X.–M., Matsumoto, K., Nakamura, T., Park, M., & Nigam, S. K. (1994). Dev. Biol. 163, 525–529.

Schmidt, C., Bladt, F., Goedecke, W., Brinkmann, V., Zschiesche, W., Sharpe, M., Gherardi, E., & Birchmeier, C. (1995). Scatter factor/hepatocyte growth factor is essential for liver development. Nature 373, 699–702.

Seifert, R. A., Hart, C. E., & Philips, P. E. (1989). Two different subunits associate to create isoform-specific platelet-derived growth factor. J. Biol. Chem. 264, 8771–8778.

Shimomura, T., Kondo, J., Ochiai, M., Naka, D., Miyazawa, K., Morimoto, Y., & Kitamura, N. (1993). Activation of the zymogen of hepatocyte growth factor activator by thrombin. J. Biol. Chem. 268, 22927–22932.

Shoelson, S. E., White, M. F., & Kahn, C. R. (1988). Tryptic activation of the insulin receptor. J. Biol. Chem. 10, 4852–4860.

Snyder, M. A., Bishop, J. M., McGrath, J. P., & Levinson. (1985). A mutation at the ATP-binding site of pp60$^{v\text{-}src}$ abolishes kinase activity, transformation and tumorigenicity. Mol. Cell. Biol. 5, 1772–1779.

Sonnenberg, E., Meyer, D., Weidner, K. M., & Birchmeier, C. (1993a). Scatter factor/hepatocyte growth factor and its receptor, the c-*MET* tyrosine kinase, can mediate a signal exchange between mesenchyme and epithelia during mouse development. J. Cell Biol. 123, 223–235.

Sonnenberg, E., Weidner, K. M., & Birchmeier, C. (1993b). In: Hepatocyte Growth Factor-Scatter Factor and the c-Met Receptor (Goldberg, I. D. & Rosen, E. M., eds.) pp. 382–394. Birkhaüser Verlag, Basel.

Songyang, Z., Shoelson, S. E., Chaudhuri, M., Gish, G., Pawson, T., Haser, W. G., King, F., Roberts, T., Ratnofsky, S., Lechleider, R. J., Neel, B. G., Birge, R. B., Fajardo, J. E., Chou, M. M., Hanafusa, H., Schaffhausen, B., & Cantley, L. C. (1993). SH2 domains recognize specific phosphopeptide sequences. Cell 72, 1–20.

Stoker, M., Gherardi, E., Perryman, M., & Gray, J. (1987). Scatter factor is a fibroblast-derived modulator of epithelial cell mobility. Nature 327, 239–242.

Tajima, H., Matsumoto, K., & Nakamura, T. (1991). Hepatocyte growth factor has potent antiproliferative activity in various tumour cells. FEBS 291, 229–232.

Tempest, P. R., Reeves, B. R., Spurr, N. K., Rance, A. J., Chan, M.–L., & Brooks, P. (1986). Activation of the *MET* oncogene in the human mnng-hos cell line involves a chromosomal rearrangement. Carcinogenesis 7, 2051–2057.

Tornqvist, H. E., Pierce, M. W., Frackleton, A. R., Nemenoff, R. A., & Avruch, J. (1987). Identification of insulin receptor tyrosine residues autophosphorylated *in vitro*. J. Biol. Chem. 262, 10212–10219.

Tornqvist, H. E. & Avruch, J. (1988). Relationship of site-specific β subunit tyrosine autophosphorylation to insulin activation of the insulin receptor (tyrosine) protein kinase activity. J. Biol. Chem. 263, 4593–4601.

Villa–Moruzzi, E., Lapi, S., Prat, M., Gaudino, G., & Comoglio, P. M. (1993). A protein tyrosine phosphatase activity associated with the hepatocyte growth factor/scatter factor receptor. J. Biol. Chem. 268, 18176–18180.

Waksman, G., Kominos, D., Robertson, S. R., Pant, N., Baltimore, D., Birge, R. B., Cowburn, D., Hanafusa, H., Mayer, B. J., Overduin, M., Resh, M. D., Rios, C. B., Silverman, L., & Kuriyan, J. (1992). Crystal structure of the phosphotyrosine recognition domain SH2 of v-src complexed with tyrosine-phosphorylated peptides. Nature 358, 646–653.

Waksman, G., Shoelson, S., Pant, N., Cowburn, D., & Kuriyan, J. (1993). Binding of a high-affinity phosphotyrosyl peptide in the src SH2 domain: Crystal structures of the complexed and peptide-free forms. Cell 72, 779–790.

Weber, W., Bertics, P. J., & Gill, G. N. (1984). Immunoaffinity purification of the epidermal growth factor receptor. Stoichiometry of binding and kinetics of self-phosphorylation. J. Biol. Chem. 259, 14631–14636.

Weidner, K. M., Behrens, J., Vandekerckove, J., & Birchmeier, W. (1990). Scatter factor: Molecular characteristics and effect on the invasiveness of epithelial cells. J. Cell Biol. 111, 2097–2108.

Weidner, K. M., Arakaki, N., Vandekerckove, J., Weingart, S., Hartmann, G., Rieder, H., Fonatsch, C., Tsubouchi, H., Hishida, T., Daikuhara, Y., & Birchmeier, W. (1991). Evidence for the identity of human scatter factor and hepatocyte growth factor. Proc. Natl. Acad. Sci. USA 88, 7001–7005.

Xin, J. H., Cowie, A., Lachance, P., & Hassel, J. A. (1992). Molecular cloning and characterization of PEA3, a new member of the ETS oncogene family that is differentially expressed in mouse embryonic cells. Genes Dev. 6, 481–496.

Yarden, Y., & Schlessinger, J. (1987a). Self-phosphorylation of epidermal growth factor receptor: Evidence for a model of intermolecular allosteric activation. Biochemistry 26, 1434–1442.

Yarden, Y., & Schlessinger, J. (1987b). EGF induces rapid, reversible aggregation of the purified epidermal growth factor receptor. Biochemistry 26, 1443–1451.

Yoshihiko, U., Minowa, O., Mori, C., Shiota, K., Kuno, J., Noda, T., & Kitamura, N. (1995). Placental defect and embryonic lethality in mice lacking hepatocyte growth factor/scatter factor. Nature 373, 702–705.

Yoshimura, T., Yuhki, N., Wang, M.–H., Skeel, A., & Leonard, E. J. (1993). Cloning, sequencing, and expression of human macrophage stimulating protein (MSP, MST1) confirms MSP as a family of kringle proteins and locates the MSP gene on chromosome 3. J. Biol. Chem. 268, 15461–15468.

Zarnegar, R. & Michalopoulos, G. (1989). Purification and biological characterization of human hepatopoietin A, a polypeptide growth factor for hepatocytes. Cancer Res. 49, 3314–3320.

Zarnegar, R., Muga, S., Rahija, R., & Michalopoulos, G. (1990). Tissue distribution of hepatoprotein-A: A heparin-binding polypeptide growth factor for hepatocytes. Proc. Natl. Acad. Sci. USA 87, 1252–1256.

Zarnegar, R., DeFrances, M. C., Kost, D., Lindroos, P., & Michalopoulos, G. K. (1991). Expression of hepatocyte growth factor mRNA in regenerating rat liver after partial hepatectomy. Biochem. Biophys. Res. Commun. 177, 559–565.

Zhen, Z., Giordano, S., Longati, P., Medico, E., Campiglio, M., & Comoglio, P. M. (1994). Structural and functional domains critical for constitutive activation of the HGF receptor (MET). Oncogene 9, 1691–1697.

BOMBESIN/GRP AND VASOACTIVE INTESTINAL PEPTIDE/PACAP AS GROWTH FACTORS

Terry W. Moody and Robert T. Jensen

I. Introduction . . . 492
II. Bombesin/GRP . . . 492
  A. GRP Structure . . . 492
  B. GRP Biosynthesis and Processing . . . 496
  C. GRP Secretion . . . 498
  D. GRP Receptor Ligands . . . 499
  E. GRP Receptors . . . 499
  F. GRP Second Messengers . . . 504
  G. GRP Gene Expression . . . 508
  H. GRP Proliferation . . . 510
  I. GRP Cancer Growth . . . 512
III. VIP/PACAP . . . 513
  A. VIP Structure . . . 513
  B. VIP Biosynthesis and Processing . . . 514
  C. VIP Receptor Ligands . . . 515

Growth Factors and Cytokines in Health and Disease
Volume 1B, pages 491–535.

ISBN: 0-7623-0091-4

D. VIP Receptors 516
E. VIP Second Messengers and Gene Expression 519
F. VIP Proliferation and Cancer Growth 521
IV. Summary 521
Acknowledgments 524
References 524

I. INTRODUCTION

Bombesin (BB) was the first peptide identified as a cancer cell autocrine growth factor. High levels of immunoreactive BB were detected in small-cell lung cancer (SCLC) cells and tumors (Moody et al., 1981; Wood et al., 1981; Erisman et al., 1982). SCLC is a neuroendocrine cancer and the BB-like peptide in SCLC was identified as gastrin releasing peptide (GRP). GRP was secreted from SCLC cells into the extracellular fluids and blood (Sorenson et al., 1982; Pert and Schumacher, 1982; Wood et al., 1982; Korman et al., 1986). High-affinity GRP binding sites are present on the surface of SCLC cells (Moody et al., 1985) and BB stimulates the growth of SCLC cells (Carney et al., 1987) and tumors (Alexander et al., 1988). The growth of SCLC is inhibited by neutralizing antibodies against GRP or GRP receptor antagonists (Cuttitta et al., 1985; Mahmoud et al., 1991; Trepel et al., 1988). Recent studies show that BB-related peptides are growth factors for a number of normal tissues and tumors.

Vasoactive intestinal peptide (VIP) is an autocrine growth factor for epithelial cancers such as non-SCLC (NSCLC). VIP mRNA and VIP immunoreactivity is present in NSCLC cells (Moody et al., 1993c). VIP is secreted from certain NSCLC cells and ^{125}I-VIP binds with high affinity to NSCLC cells (Shaffer et al., 1987) Also, pituitary adenylate cyclase activating peptide (PACAP), which is structurally related to VIP (Miyata et al., 1989, Arimura et al., 1992), binds with high affinity to lung cancer cells (Moody et al., 1993b). The increase in cAMP caused by VIP in NSCLC cells is inhibited by the receptor antagonist VIPhybrid (VIPhyb). VIP stimulates the growth of NSCLC *in vitro* and NSCLC proliferation is inhibited by VIPhyb. Also, VIPhyb slows NSCLC xenograft growth in nude mice *in vivo* (Moody et al., 1993c). These data suggest that VIP is an autocrine growth factor for NSCLC. This review will focus on GRP and VIP as peptide growth factors.

II. BOMBESIN/GRP

A. GRP Structure

BB, a 14 amino acid peptide initially isolated from frog skin, represents one family of peptides biologically active in the CNS and periphery (Anastasi et al., 1973). Table 1 shows that the main mammalian members of the BB family are GRP and neuromedin B (NMB) isolated from porcine stomach and spinal cord, respectively (McDonald et al., 1979; Minamino et al., 1983). GRP and BB have 9 of the

Table 1. Structure of BB-Like Peptides

GRP	Ala-Pro-Val-Ser-Val-Gly-Gly-Gly-Thr-Val-Leu-Ala-Lys-Met-Tyr-Pro-Arg-Gly-Asn-His-Trp-Ala-Val-Gly-His-Leu-Met-NH2
BB	pGlu-Gln-Arg-Leu-Gly-Asn-Gln-Trp-Ala-Val-Gly-His-Leu-Met-NH2
NMB	Gly-Asn-Leu-Trp-Ala-Thr-Gly-His-Phe-Met-NH2
(Psi-13,14)BB	Pyr-Gln-Arg-Leu-Gly-Asn-Gln-Trp-Ala-Val-Gly-His-Leu=Leu-NH2
(FBB^{6-13})ME	D-Phe-Gln-Trp-Ala-Val-Gly-His-Leu-methylester

Note: Sequential homologies relative to BB are underlined. Standard amino acid abbreviations are used; =, reduced peptide bond (CH_2-NH) instead of (C(O)-NH).

10 same terminal amino acid residues whereas NMB and BB have 7 of the same 10 amino acids. Nonetheless the genes for NMB and GRP are distinct (Krane et al., 1988; Sausville et al., 1986; Spindel et al., 1984). Similarly distinct receptors for NMB and GRP exist (Wade et al., 1991; Battey et al., 1991; Spindel et al., 1990, Corjay et al., 1991). The NMB or BB_1 receptor subtype is present on C6 rat glioma cells which binds NMB with high affinity and binds BB and GRP with moderate affinity (Wang et al., 1992). The GRP or BB_2 receptor subtype is present on U138 human glioblastoma cells which bind BB and GRP with high affinity and NMB with moderate affinity (Moody et al., 1989). SCLC cells have both BB_1 and BB_2 receptors (Moody et al., 1985; Moody et al., 1992). Each of the peptides listed above are agonists, however, high-affinity peptide antagonists for the GRP or BB_2 receptor exist. In particular, reduced peptide bond antagonists, such as ($Psi^{13,14}$, Leu^{14})BB (Psi-13,14)BB or (Des-Met^{14})BB antagonists, such as (D-Phe^6)BB^{6-13}methylester (FBN^{6-13}ME), bind to and block GRP receptors with high affinity (Coy et al., 1988; Wang et al., 1990b).

The biological potency of numerous GRP receptor agonists has been evaluated. In particular, the C-, but not the N-terminal, is essential for biological activity. Figure 1 shows that GRP, but not NMB, inhibits ^{125}I-GRP binding with high affinity to SCLC cells. The effects of GRP were dose-dependent with 1 uM GRP inhibiting all specific binding of ^{125}I-GRP to SCLC cell line NCI-H345. Two nM GRP inhibited 50% (IC_{50}) of the specific ^{125}I-GRP binding. Similarly, GRP^{14-27} and BB strongly inhibited ^{125}I-GRP binding whereas GRP^{1-16} does not. NMB was less potent with an IC_{50} value of 200 nM. The C-terminal of GRP could be further shortened to Ac-GRP^{20-27}, which contains the C-terminal octapeptide of GRP, with retention of high-affinity binding activity (Jensen et al., 1978; Moody et al., 1985). Further shortening of the peptide to GRP^{21-27} resulted in two orders of magnitude lower affinity binding (Heimbrook et al., 1988). Substitutions of D- for L-amino acids at the 8,10 or 12 positions resulted in loss of high-affinity binding activity (Moody et al., 1978; Rivier and Brown, 1978). Because BB-G (BB with a C-terminal glycine extension) bound with three orders of magnitude lower affinity than did BB, an amidated C-terminal is essential for high-affinity binding of BB to GRP

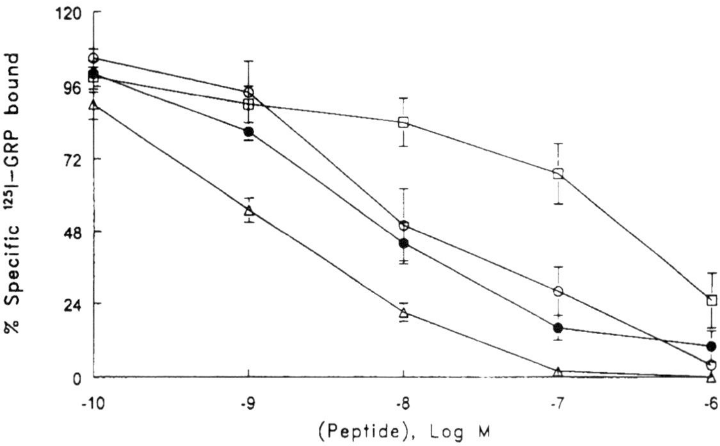

Figure 1. Specific binding of ^{125}I-GRP. The % specific ^{125}I-GRP bound was determined as a function of BW22258 (o), GRP (Δ), (F)BB$^{6-13}$ME (•) and NMB (□) concentration to SCLC cell line NCI-H345. The mean value ± S.E. of four determinations is indicated.

receptors (Mervic et al., 1991). Because all agonists are positively charged, they may interact electrostatically with the acidic GRP receptor (Zachary and Rozengurt, 1987a). It has been proposed (Coy et al, 1988; Wang et al., 1990a) that the active configuration of BB or GRP for interacting with the GRP receptor is a β-sheet structure in the COOH terminal position, with a turn at positions 10 to 13 of BB, and that hydrogen bonds exist between the terminal Leu^{14} amide NH_2 and the Trp^{8} C=O, between Leu^{13} C=O and Val^{10} N-H, and between Leu^{13} N-H and the Val^{10} C=O. Support for this model by structure-function studies and computer modeling has recently been obtained (Kull et al., 1992).

The NMB receptor prefers NMB relative to BB or GRP (Table 2). NMB differs from BB at the 7 (His for Gln), 10 (Thr for Val), and 13 (Phe for Leu) positions of BB. Thus, the BB_1 receptor may prefer NMB due to increased H-bonding between the receptor and the His as well as Thr of NMB. The NMB receptor binds NMB with approximately two orders of magnitude higher affinity than GRP or BB (Von Schrenck et al., 1989; Von Schrenck et al., 1990; Benya et al., 1995).

Numerous GRP receptor antagonists have been synthesized. (D-Arg^{1}, D-Pro^{2}, D-Trp^{7,9}, Leu^{11}) substance P ((RPWWL)SP), a SP antagonist, was found to weakly inhibit (IC_{50} = 1 uM) (^{125}I-Tyr^{4})BB binding to GRP receptors on pancreatic acinar cells (Jensen et al., 1984). Subsequently it was found to inhibit binding to both GRP and NMB receptors on SCLC cells (Moody et al., 1985; Moody et al., 1992). Although (RPWWL)SP has little sequence homology to GRP or NMB, it is a

Table 2. Binding of BB-Like Peptide Agonists

Peptide	*IC$_{50}$(nM)*	
	^{125}I-GRP binding	*(^{125}I-Tyr^{0})NMB*
BB	1	100
GRP	3	50
GRP$^{14-27}$	2	60
GRP$^{1-16}$	>10,000	>10,000
NMB	100	1
BBG	1000	>10,000

Note: The IC$_{50}$ required to inhibit ^{125}I-GRP or (^{125}I-Tyr^{0})NMB binding to NCI-H345 cells is indicated.

competitive antagonist and slows the rate of (^{125}I-Tyr^{4})BB association to GRP receptors, having no effect on the rate of dissociation. Numerous other analogues of SP and SP(4-11) have been shown to inhibit the action of GRP at GRP receptors. However, their affinity is generally low and in most cases they also inhibit interaction with NK$_1$ (substance P) receptors (Jensen et al., 1984; Jensen et al., 1988). (D-Phe^{12})BB was subsequently found to inhibit (^{125}I-Tyr^{4})BB binding to guinea pig pancreatic acini and rat brain homogenates (Heinz–Erian et al., 1987). Although (D-Phe^{12})BB resembles BB in that an essential His at the 12 position was replaced by D-Phe, it had an IC$_{50}$ value of 1 μM and, thus, bound to GRP receptors with low affinity. (RPWWL)SP and (D-Phe^{12})BB represent first-generation, BB-receptor antagonists.

Subsequently a second generation of high-affinity GRP-receptor antagonists was developed having either a reduced peptide bond at the penultimate position or lacking a C-terminal methionine (Jensen et al., 1993). Table 3 shows that (Psi-13,14)BB, which was the initial reduced peptide bond antagonist, had an IC$_{50}$ of 30 nM. Subsequently (Psi-13,14)BB fragments were synthesized including (3-phenylpropanoyl-His^{20}, D-Ala^{24}, Pro^{26}, Psi^{26,27}, Phe^{27})GRP$^{20-27}$ (BW2258U89) and (Tpi^{6}, Psi^{13,14}, Tpi^{14})BN$^{6-14}$. Tpi is 2.3.4.9-tetrahydro-1H-pyrido(3,4-b)indol-

Table 3. Binding of BB Antagonists

Peptide	*IC$_{50}$(nM)*	
	^{125}I-GRP binding	*(^{125}I-Tyr^{0})NMB binding*
(Psi-13,14)BB	30	>1000
BW2258U89	10	>1000
(FBB$^{6-13}$)ME	5	>1000
(RPWWL)SP	1000	1000

Note: The ability to inhibit ^{125}I-GRP and (^{125}I-Tyr^{0})NMB binding to SCLC cell line NCI-H345 is indicated.

3-carboxylic acid, a conformationally constrained secondary amine derivative of tryptophan (Coy et al., 1988; Singh et al., 1992; Cai et al., 1992). BW2258U89 was reported to inhibit ^{125}I-GRP binding to Swiss 3T3 cells with extremely high affinity (IC_{50} = 1 pM (Leban et al., 1993)) but it binds with lower affinity to the SCLC GRP receptor (IC_{50} = 10 nM). Nonetheless BW2258U89 represents one of the most potent GRP-receptor peptide antagonists for SCLC cells. (Des-Met^{14})BB antagonist fragments synthesized include (D-PentaF,Phe^{6}, D-Ala^{11})BN$^{6-13}$methylester (BIM26226), Ac-GRP$^{20-26}$ ethylester, and (isobutyryl-His^{20}, D-Ala^{24}) GRP$^{20-26}$methylamide, and these inhibit ^{125}I-GRP binding to Swiss 3T3 cells with high affinity (IC_{50} = 1, 1, and 1 nM respectively (Wang et al., 1990a; Camble et al., 1989; Heimbrook et al., 1989). They are less potent at inhibiting (^{125}I-Tyr^{4})BB binding to human SCLC cells and have IC_{50} values of 5,10, and 100 nM, respectively. These data suggest that the human GRP receptor may have a slightly smaller binding site than the mouse receptor in that if bulky aromatic groups, such as 3-phenylpropanoyl or isobutyryl, are at the N-terminal of the antagonist, the binding affinity is greatly reduced.

The NMB receptor (RPWWL)SP was a weak blocker of (^{125}I-DTyr^{0})NMB binding to rat esophagus or SCLC cells (Von Schrenck et al., 1990). The various (des-Met^{27})GRP or (des-Met^{14})BB analogues and pseudopeptides, that had a high affinity for GRP receptors, have low affinity for NMB receptors (von Schrenck et al., 1990; Benya et al., 1995). Also, certain somatostatin (SS) analogues moderately inhibit binding to NMB receptor (Orbuc et al., 1993). (Nal^{6},Thr^{8})cyclo-SS-octapeptide inhibited (^{125}I-DTyr^{0})NMB binding to C6 cells with an IC_{50} value of 0.2 uM, but had little effect on binding to GRP receptors (IC_{50} >10 uM). High-affinity antagonists for NMB are currently not available. Peptide mimetic or nonpeptide antagonists remain to be developed for the GRP and NMB receptors.

B. GRP Biosynthesis and Processing

GRP is synthesized as a high molecular weight precursor protein (preproGRP) and the gene for GRP contains two introns (Sauaville et al., 1986; Spindel et al., 1984). Surprisingly, one intron occurs at the Val amino acid residue in the C-terminal of GRP. Hence, if there is not proper splicing of the GRP mRNA at the first splice junction, biologically active GRP will not be produced. The second intron occurs near the C-terminal of preproGRP in the GRP gene-associated peptide (GGAP). There is alternative splicing at the second splice junction resulting in three different human preproGRP proteins. Each of these isoforms, however, has a single biologically active GRP. GRP mRNA (0.9 kb) is present in high densities in numerous SCLC cell lines (Giaccone et al., 1991). By radioimmunoassay, high levels of GRP immunoreactivity are present in extracts derived from SCLC cell NCI-H209 and H345 (12- and 4-pmol/mg protein, respectively (Moody et al., 1981).

The three forms of human preproGRP have been identified and termed type I, type II, and type III. Type I preproGRP is the most abundant form and has 148

PreproGRP-type I (148 aa)

↓ Signal Peptidase

ProGRP-type I (124 aa)

↓ Trypsin-like enzyme

GRP-GKK (30 aa) + GGAP (94 aa)

↓ Carboxypeptidase B-like enzymes

GRP-G (28 aa)

↓ PAM enzymes

GRP (27 aa)

↓ Trypsin-like enzyme

GRP^{14-27} (14 aa)

Figure 2. Processing of preproGRP. The peptide products of preproGRP are indicated on the left and the processing enzymes on the right. The number of amino acids (aa) in the peptide product is indicated in parentheses. GRP and GRP^{14-27} have an amidated C-terminal.

amino acid residues (Figure 2). It is metabolized by a signal protease to 124 amino acid proGRP. The proGRP is metabolized by trypsin-like enzymes to the 30 amino acid GRP-glycyl-lysyl-lysine (GRP-GKK). GRP-GKK is metabolized by carboxypeptidase B-like enzymes to GRP-GK and GRP-G which contain 29 and 28 amino acids, respectively. GRP-G is metabolized by PHM and PAL enzymes to the 27 amino acid GRP which has an amidated C-terminal. Because an amidated C-terminal is essential for biological activity of GRP at the GRP receptor, the precursor protein for GRP may have little biological activity (Mervic et al., 1991). A recent study (Seva et al., 1994) reports that with gastrin, which is also a COOH-terminal active peptide which requires amidation for interaction with its classical receptor, the CCK_B receptor, the glycine extended form of gastrin has growth effects at a distinct receptor from the CCK_B receptor. At present, there is no data to suggest that either GRP-glycine-extended or NMB-glycine-extended peptides have growth effects at distinct receptors.

The gene for NMB was cloned and low levels of NMB mRNA (0.8 kb) are present in lung cancer cell lines (Giaccone et al., 1991; Cardona et al., 1991). NMB is made from a different precursor protein than GRP (Krane et al., 1988). PreproNMB contains 116 amino acids and is metabolized by signal proteases, trypsin-like enzymes, carboxypeptidase B-like enzymes, and peptidyl-α-monooxygenase (PAM) enzymes to NMB-32. In SCLC cells, GRP and NMB-32 are further metabolized by trypsin-like enzymes to GRP^{14-27} and NMB which are biologically active.

C. GRP Secretion

GRP immunoreactivity is widely distributed in the CNS, nerve fibers, and cell bodies throughout the alimentary tract. The release of neuropeptides, such as GRP, has been difficult to measure because of their rapid degradation. GRP is rapidly metabolized in the blood with a half-life of 1.4 minutes in the pig (Knutsen et al., 1984). It may be metabolized by exopeptidases, such as aminopeptidase (Bishop et al., 1986), or neutral endopeptidases, such as enkephalinase (Bunnet et al., 1985). The degradation of BB was reduced by the protease inhibitor phosphoramidon. GRP is probably stored in the dense-core neurosecretory granules associated with SCLC. Elevation of the intracellular cAMP may cause phosphorylation of synapsin-I causing granule exocytosis. VIP (10 nM) elevates the cAMP 10-fold in the SCLC cell line NCI-H345 or H209 within 5 minutes, and the GRP secretion rate into the tissue culture medium is increased twofold (Korman et al., 1986). The VIP receptor interacts with Gs, stimulating adenylate cyclase activity. The increase in cAMP caused by VIP was reversed by somatostatin (SS) (Kee et al., 1988; Taylor et al., 1991). SS-2 receptors are present on SCLC cells which couple to Gi, inhibiting adenylate cyclase activity (Eden and Taylor, 1993). The increased secretion of GRP and cAMP caused by VIP is reversed by SS which acts as a physiological antagonist.

Also, corticotropin releasing factor (CRF) receptors are present on SCLC cells. CRF elevates the cAMP fivefold and increases the secretion rate of GRP from SCLC cells by 50% (Moody et al., 1994b). The actions of CRF are reversed by CRF receptor antagonists, such as α-helical CRF(9–41). Recently ^{125}I-CRF was found to bind with high affinity to SCLC cells (Dieterich et al., 1994).

GRP-like peptides are secreted into the blood by SCLC cells. The basal level of GRP in the plasma of SCLC patients with limited disease is 30 pM (Pert and Schumacher, 1982; Wood et al., 1982). In patients with extensive SCLC, however, the levels are as high as 2 nM (Korman et al., 1986). BB-like peptides have been purified from human plasma and used to detect SCLC (Maruno et al., 1989). Infusion of secretin, which binds with low affinity to VIP receptors, increased the plasma GRP levels fivefold after 1 hour in a patient with extensive disease (Korman et al., 1986). The GRP levels returned to baseline after 8 hours. These data indicate that biologically active VIP receptors may be present on SCLC cells *in vivo*. Further, using *in vitro* autoradiography techniques, high-affinity ^{125}I-VIP binding sites are present on SCLC biopsy specimens (Moody et al., 1993a).

GRP is neutralized by monoclonal antibody (mAb) 2A11. This mAb, similar to the GRP receptor, recognizes the C-terminal octapeptide of GRP. Hence, when GRP is neutralized by the mAb it cannot bind to the receptor. The GRP receptor binds GRP with slightly higher affinity than does the mAb and, in low concentrations of mAb 2A11, (^{125}I-Tyr^4)BB will bind to the receptor (Cuttitta et al., 1985). When high concentrations of 2A11 are added, however, (^{125}I-Tyr^4)BB will not bind to the receptor. MAb 2A11 was not toxic in phase I clinical trials (Avis et al., 1991) and is now being tested in phase II clinical trials (Kelley et al., 1993).

D. GRP Receptor Ligands

(^{125}I-Tyr^4)BB was initially used as a receptor probe. It bound with high affinity to rat brain membranes (Kd = 4 nM; Moody et al., 1978), guinea pig pancreatic acinar cells (Kd = 2 nM; Jensen et al., 1978), Swiss 3T3 cells (Kd = 0.5 nM; Zachary and Rozengurt, 1985) and human small cell lung cancer (Kd = 1 nM; Moody et al., 1985) cells. Unfortunately (^{125}I-Tyr^4)BB can also bind with moderate affinity to NMB receptors. Subsequently ^{125}I-GRP became commercially available and bound with high affinity to GRP receptors and low affinity to NMB receptors (Zachary and Rozengurt, 1985). ^{125}I-GRP bound to a single class of sites using Swiss 3T3 cells (Bmax = 100,000/cell) and was cross-linked to a 75-kDa protein using 1 mM disuccinimidylsuberate (DSS) (Zachary and Rozengurt, 1987a; Kris et al., 1987). (^{125}I-DTyr^0)NMB was found to bind with high affinity to NMB and moderate affinity to GRP receptors (Lee et al., 1990a; Ladenheim et al., 1990). (^{125}I-DTyr^0)NMB bound to a single class of sites using C6 cells (Bmax = 2200/cell; Wang et al., 1992) and SCLC cells (Bmax = 800/cell, Moody et al., 1992). Recently a radiolabeled antagonist, (^{125}I-D-Tyr^6)BN$^{6-13}$ME bound with high affinity to GRP but not NMB receptors (Mantey et al., 1993). The radiolabeled GRP antagonist only bound to the cell surface. In contrast, radiolabeled agonists, such as ^{125}I-GRP, bound to the cell surface at 4 °C but internalized at 37 °C. Studies in numerous cell types show that both the GRP and NMB receptors rapidly internalize radiolabeled agonists (Brown et al., 1988; Zachary and Rozengurt, 1987; Zhu et al., 1991; Benya et al., 1992; Wang et al., 1993). A recent study (Slice et al., 1994), using GRP and GRP receptor antibodies labeled with fluorescent probes, demonstrates that both the GRP receptor and agonist ligands are internalized rapidly. After 4 hours the internalized GRP receptors recycled back to the cell surface, whereas the agonist ligand remained in lysosomes and other intracellular compartments.

E. GRP Receptors

Physiological and pharmacological studies provided evidence for at least two classes of receptors mediating the action of bombesin-related peptides (Falconieri et al., 1988; Ladenheim et al., 1992; Ladenheim et al., 1993; Jensen et al., 1993; von Schrenck et al., 1989; von Shrenck et al., 1990; Moody et al., 1992). The GRP receptor had high affinity for GRP and the nonmammalian peptides litorin, ranatensin, alytesin and a low affinity for NMB (Table 2). In contrast, the NMB receptor had a high affinity for NMB and the nonmammalian peptides litorin and ranatensin, an intermediate affinity for bombesin, and low affinity for GRP.

Cloning studies have now demonstrated that three structurally-related receptors in mammals mediate the action of this family of peptides. The three receptors are the NMB receptor or BB_1 receptor, the GRP receptor or BB_2 receptor, and the

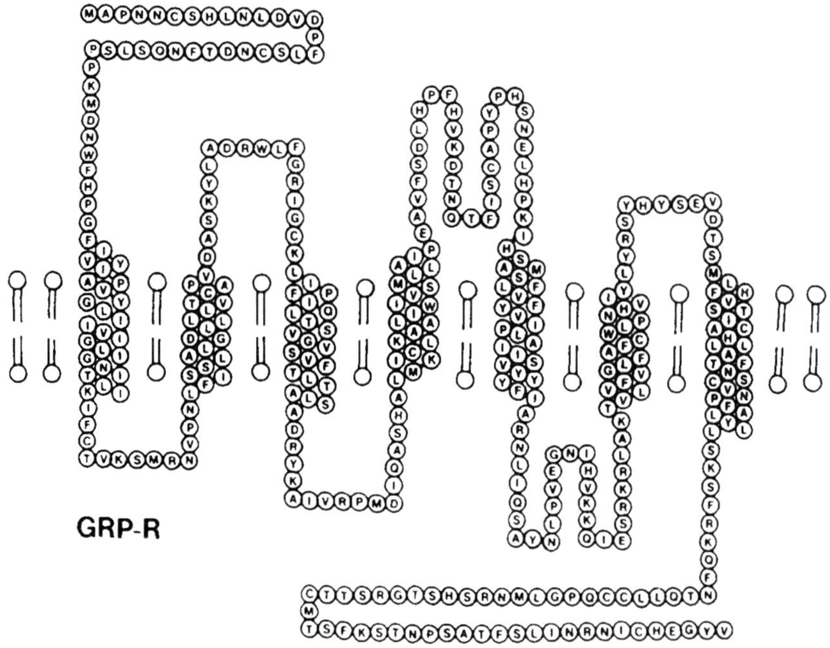

Figure 3. Schematic illustration of the GRP receptor. The GRP receptor amino acid sequence (384 amino acids) is shown using standard single letter abbreviations (Battey et al., 1991; Spindel et al., 1990).

BRS-3 or bombesin-related subtype-3 receptor or BB_3 receptor. The GRP receptor was initially cloned from Swiss 3T3 cells. In the mouse and human, the GRP receptor has 384 amino acids with a molecular weight of 43 kDa and is a member of the G protein-coupled superfamily with 7 transmembrane-scanning regions (Battey et al., 1991; Spindel et al., 1990; Corjay et al., 1991) (Figure 3). The NMB receptor was cloned from a cDNA library made from rat esophageal muscle which possesses high-affinity NMB receptors (Wada et al., 1991; Corjay et al., 1991; von Schrenck et al., 1989). The rat and human NMB receptor have 390 amino acids and a molecular weight of 43 kDa. The human NMB receptor shares 89% amino acid identity with the rat NMB receptor and 55% identity with the human GRP receptor Corjay et al., 1991). Recently, using low stringency Southern blot analysis of Eco RI-digested human genomic DNA with GRP and NMB receptor probes, an additional receptor with 47–51% amino acid identity with the human GRP and NMB receptors was described (Fathi et al., 1993b; Jensen et al., 1994). This receptor was called BRS-3 for bombesin receptor subtype 3 because the natural ligand of this receptor remains unknown. This receptor probably belongs to the bombesin receptor family because, in general, when G protein-coupled receptor subtypes are within the same family, the amino acid identities are about 50% or higher, whereas, with unrelated receptors, it is less than 30%. Although bombesin-related peptides cause activation of the BRS-3 when it is expressed in *Xenopus* oocytes, the affinity of the receptor for all members of the BB family examined is low, suggesting that none of the described naturally occurring BB-related peptides was the natural ligand for this receptor. This receptor was also subsequently cloned from guinea pig uterus and was found expressed mainly in the uteri of pregnant animals (Gorbulev et al., 1992). The gene for the BRS-3 receptor subtype maps to the human X chromosome (Fathi et al., 1993b), whereas the genes for the human NMB receptor and human GRP receptors mapped to chromosomes 6 and X, respectively (Spindel et al., 1993). Studies using *in situ* hybridization and Northern analysis demonstrate that the GRP and NMB receptors are widely present in the CNS and various gastrointestinal tissues, whereas BRS-3 has a much more localized distribution (Corjay et al, 1991; Wada et al., 1991; Battey et al., 1991; Spindel et al., 1990). The BRS-3 in rat tissues was expressed only in secondary spermatocytes in the testis (Fathi et al., 1993). In human small-cell lung cancer cells, BRS-3 as well as GRP and NMB receptors have been identified using RT-PCR or RNAase protection assays (Corjay et al., 1991; Fathi et al., 1993b).

Numerous studies demonstrate that the human GRP and NMB receptors, as well as those on other species, are glycosylated (Benya et al., 1995a; Kris et al., 1987; Brown et al., 1988; Feldman et al., 1990; Kusui et al., 1994). Figure 4 shows that the human GRP receptor has a molecular weight of 60 ± 1 kDa by cross-linking studies and 35% of the molecular weight represents oligosaccharides (Kris et al., 1987; Benya et al., 1995). In contrast, the human NMB receptor has a molecular weight of 72 ± 1 kDa of which 38% represents carbohydrate moieties (Benya et al., 1995). The native mouse GRP receptor has a higher molecular weight of 82,000 than the human GRP receptor, likely reflecting the fact that the mouse GRP receptor has four potential glycosylation sites, all of which are utilized. The human GRP receptor has only two potential glycosylation sites (Benya et al., 1995a; Kusui et al., 1994). In contrast, the rat NMB receptor has a lower molecular weight of 63,000 than the human NMB receptor, likely reflecting the fact that the rat NMB receptor has only two glycosylation sites which are both utilized. The human receptor has three glycosylation sites. A recent detailed study demonstrated that both the NMB and GRP receptors have only N-linked glycosylation with tri- and/or tetra-antennary complex oligosaccharide chains (Kusui et al., 1994). The rat NMB receptor contains terminal sialic acids, whereas the mouse GRP receptor does not. Numerous studies have demonstrated that receptor glycosylation is important in protein folding, maintaining receptor stability, trafficking, and targeting of receptors to cell membranes. In classical hormone receptors, such as LH/HCG or TSH, and receptors for neurotransmitters, such as adrenergic agents or muscarinic agents, where receptor glycosylation has no effect on high affinity receptor binding. A recent study, however, demonstrates that, with the GRP receptor but not the NMB receptor, glycosylation is required for high-affinity binding or G-protein coupling. At pre-

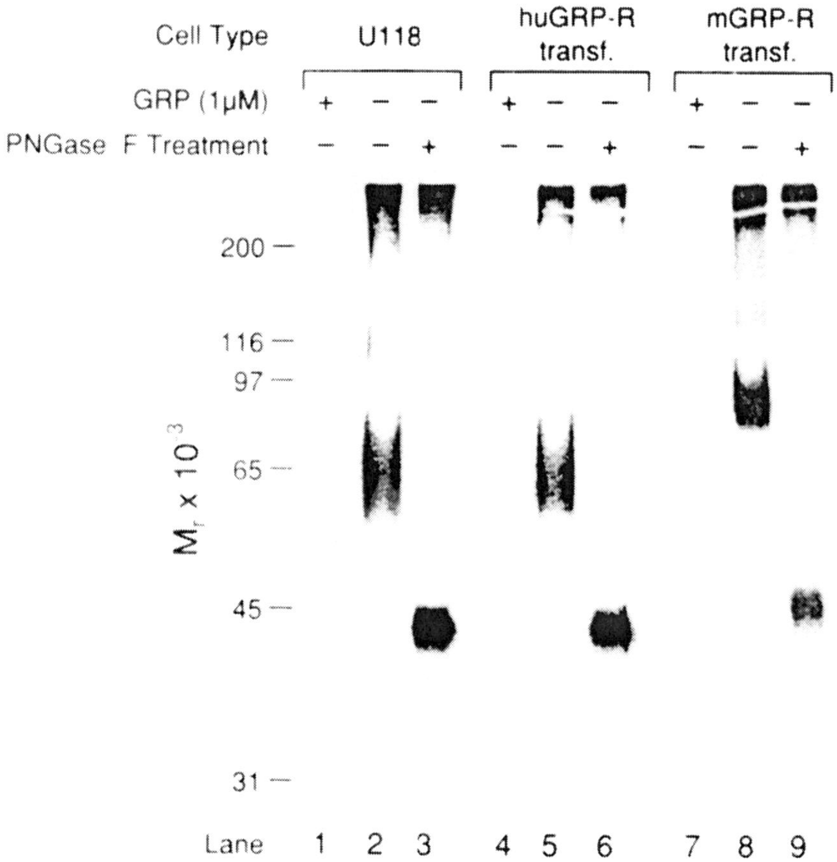

Figure 4. Affinity labeling and deglycosylation of the human and mouse GRP receptor (left panels) and human and rat NMB receptor (right panel). In the left panel, ^{125}I-GRP was cross-linked with DSS to GRP receptor on human glioblastoma U-118 cells, to BALB 3T3 cells stably transfected with the human GRP receptor (huGRP-R), or to BALB 3T3 cells transfected with murine GRP receptors (mGRP-R). Cross-linking was in the presence or absence of 1-μM GRP. After cross-linking, an aliquot was incubated with PNGase to deglycosylate the receptor. In the right panels, the human NMB receptor (huNMB-R) or rat NMB receptor (rNMB-R) were affinity labeled using ^{125}I-[DTyr^0]NMB and MBS in the presence of absence of 1-μM NMB. A portion was treated with PNGase to deglycosylate the receptor. The huGRP-R on U-118 cells and in huGRP-R transfected cells had a molecular weight of 60 ± 1 kDa and with the murine GRP-R it was 82 ± 2kDa. In both cases treatment with PNBGase reduced the molecular mass to 43 ± 1 kDa. The human and ratNMB-R had molecular masses of 72 ± 1 and 63 ± 1 kDa, respectively, and deglycosylation reduced the molecular mass to 43 ± 1 kDa.

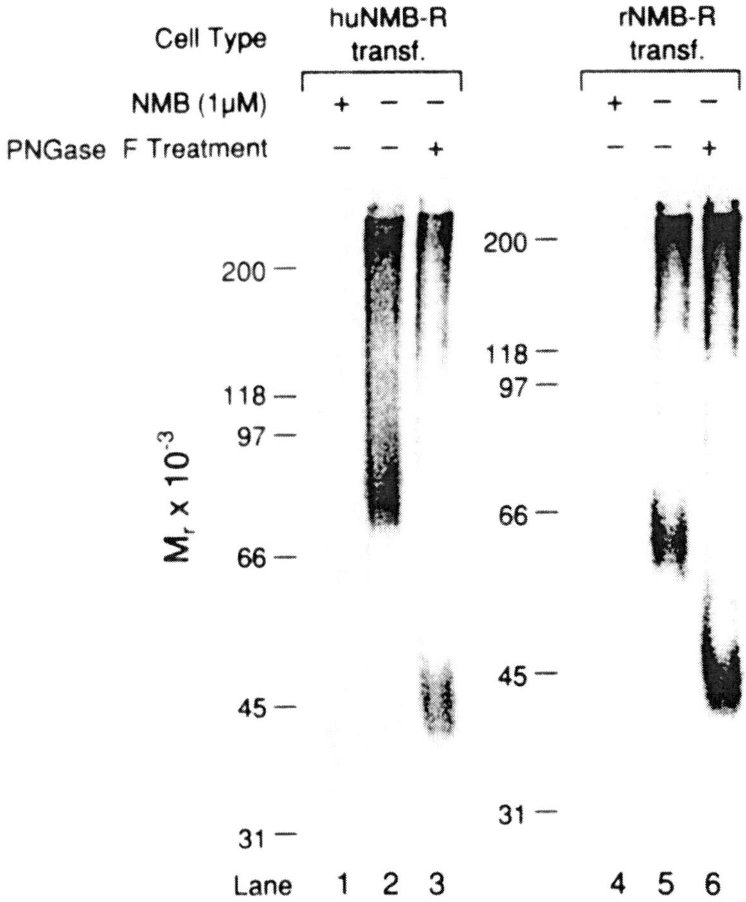

Figure 4. (Continued)

sent, it is unknown which or what combination of the four potential glycosylation sites on the mouse GRP receptor is responsible for high-affinity coupling to occur.

In addition to binding ligand with high affinity, activation of both the NMB and GRP receptors by agonists results in activation of phospholipase C, ligand internalization, and receptor modulation including desensitization and downregulation (Jensen et al., 1994; Wang et al., 1993; Wang et al., 1992; Rozengurt and Sinnett–Smith, 1983; Millar and Rozengurt, 1990a; Walsh et al., 1993; Frankel et al., 1993; Swope and Schonbrunn, 1990; Benya et al., 1994c). Recent receptor structure-function studies have provided some insights into the important receptor components and processes mediating these effects. With the NMB receptor, it was recently demonstrated that high-affinity NMB-receptor binding is critically dependent on the amino acid sequences in transmembrane (TM) domain V of the NMB receptor (Fathi et al., 1993a). Of the seven divergent amino acid residues between the NMB and GRP receptor in TM domain V, the substitution of a Ser^{216} for Ile^{216} in the NMB receptor abolished high-affinity NMB-receptor binding and abolished the ability of the mutant receptor to active PLC (Fathi et al., 1993a). At present the critical residues in determining GRP receptor selectivity are unknown. Structure-function studies of the GRP receptor demonstrate that residues on the COOH terminus, but not the 7th TM, are essential for receptor internalization (Benya et al., 1993; Slice et al., 1994). Both deletion of a PKC consensus sequence as well as of other serines and threonines in the GRP receptor's COOH terminus decreases internalization, suggesting that both PKC activation and phosphorylation of serines and threonine residues in this area may be important for mediating internalization (Benya et al., 1993; Benya et al., 1994a). A recent study demonstrates that acute desensitization of the GRP receptor does not involve activation of PKC (Walsh et al., 1993). However, chronic desensitization as well as downregulation of the GRP receptor is critically dependent on the presence of a PKC consensus sequence in the COOH tail of the GRP receptor (Benya et al., 1994a), suggesting that PKC activation is essential for this process (Benya et al., 1995).

F. GRP Second Messengers

Activation of both NMB and GRP receptors causes increases in PLC activity. Previously it was shown that guanine nucleotides, such as GPP(NH)P inhibit binding of (^{125}I-Tyr^4)BB or (^{125}I-DTyr^0)NMB to membranes possessing either GRP receptors or NMB receptors (Westendorf, and Schonbrun, 1983; Wang et al., 1993; Mantey et al., 1993). The guanine nucleotides increased the dissociation rate of bound (^{125}I-Tyr^4)BB or (^{125}I-DTyr^0)NMB from the GRP or NMB receptor. Binding of (^{125}I-Tyr^4)BB to GRP receptors GH_3 membranes was not pertussis-toxin or cholera-toxin sensitive suggesting that GRP receptors are not coupled to Gs or Gi. Two recent studies using different methodologies demonstrate differing results for the ability of GRP receptor activation to couple and activate different G proteins. In one study, activation of GRP receptors in Swiss 3T3 cells resulted in activating both Gp and G_{11} (Offermanns et al., 1994). In contrast, using antisense probes against αq and α_{11}, when the GRP and NMB receptors were expressed in *Xenopus* oocytes, the αq antisense inhibited the responses mediated by the NMB receptor by 74% but not by the GRP receptor (Shapira et al., 1994). The α_{11} antisense had little effect on activating either receptor (Shapira et al., 1994). In another study (Profrock et al., 1992) using a radiolabeled GTP photo-affinity analogue, demonstrated activation of the GRP receptor in rat pancreatic acini resulted in coupling to both G_{i1}, G_{i2}, and G_{i3}, as well as to monomeric low molecular weight GTP binding proteins.

Phosphatidylinositol-4,5-diphosphate (PIP_2)is metabolized by phospholipase C (PLC) to inositol-1,4,5-trisphosphate (IP_3) and diacylglycerol (DAG). The IP_3 released elevates cytosolic Ca^{2+}. BB elevates cytosolic Ca^{2+} in a dose-dependent manner and the ED_{50} is 1 nM. In Swiss 3T3 cells which have 100,000 receptors/cell, the cytosolic Ca^{2+} increased from 150 to 400 nM, 1 minute after the addition of 10 nM BB (Mendoza et al., 1986). Figure 5 shows that the cytosolic Ca^{2+} rapidly increased within 15 seconds after the addition of BB, was maximal after 1 minute, and, then, declined, returning to basal values after 4 minutes, possibly due to receptor desensitization of the NCI-H345 cells. Because the SCLC cells have only 1500 receptors/cell the cytosolic Ca^{2+} increased from 150 to 190 nM (Moody et al., 1987; Sausville et al., 1988). These data indicate that the increase in cytosolic Ca^{2+} caused by BB is a function of the number of receptors activated. NMB (10 nM) similarly increased the cytosolic Ca^{+2} in NCI-H345 cells. However, the increase in cytosolic Ca^{2+} caused by NMB was minimally affected by (F)BN^{6-13}ME (Figure 5). In contrast, 1 uM (F)BN^{6-13}ME totally antagonized the increase in cytosolic Ca^{2+} caused by 10 nM BB. These data indicate that (F)BN^{6-13}ME antagonizes GRP but not NMB receptors.

Activation of the GRP receptor in Swiss 3T3 cells results in activation of a complex cascade including, besides PLC activation, mobilization of cellular calcium, changes in Na^+ and K^+ fluxes, transmodulation of the EGF receptor, increased c-fos and c-myc gene expression, cAMP production, arachidonic acid release and the phosphorylation of numerous intracellular proteins including tyrosine phosphorylation (Von Lint, 1993) (Figure 6).

When PIP_2 is metabolized, the released DAG activates PKC resulting in phosphorylation of protein substrates. In Swiss 3T3 cells an 80-kDa cytosolic protein is serine phosphorylated (Zachary et al., 1986) when GRP receptors are activated. Also, PKC is translocated in Swiss 3T3 from the cytosol to the membrane. Using Western blot techniques and SCLC cells an 80-kDa PKC isoform is translocated from the cytosol to the membrane within 5 minutes after addition of 1 μM PMA (Draoui et al., 1993). Also, addition of 10 nM BB to SCLC cells causes translocation of some of the PKC to the membrane, and the effect caused by 10 nM BB is reversed by GRP receptor antagonists.

BB causes phosphorylation of MAP kinase. Ten-nM BB phosphorylated MAP kinase on Thr^{183} and Tyr^{185}, 3 minutes after addition to Swiss 3T3 cells (Pang et al., 1993; von Lint et al., 1993). It is possible that PKC may phosphorylate substrate proteins, such as MAP kinase kinase, which, in turn, phosphorylates MAP kinase (MAP-K). MAP-K may phosphorylate various proteins in the cytosol and nucleus, altering gene expression.

Figure 6 shows that BB causes tyrosine phosphorylation of a 125-kDa protein. Previously it was hypothesized that p125 was the GRP receptor and that the receptor was a tyrosine kinase (Cirillo et al., 1986). Subsequently p125 was identified as FAK kinase (Sinnett-Smith et al., 1993). FAK kinase may interact with cytoskeletal proteins altering the morphology of cells which have GRP receptors. It is not clear how addition of GRP to Swiss 3T3 cells increases tyrosine phosphorylation of p115.

Addition of 10-nM BB to Swiss 3T3 cells increases cAMP levels fourfold (Rozengurt, 1991). This effect is not observed in SCLC cells which lack cyclooxy-

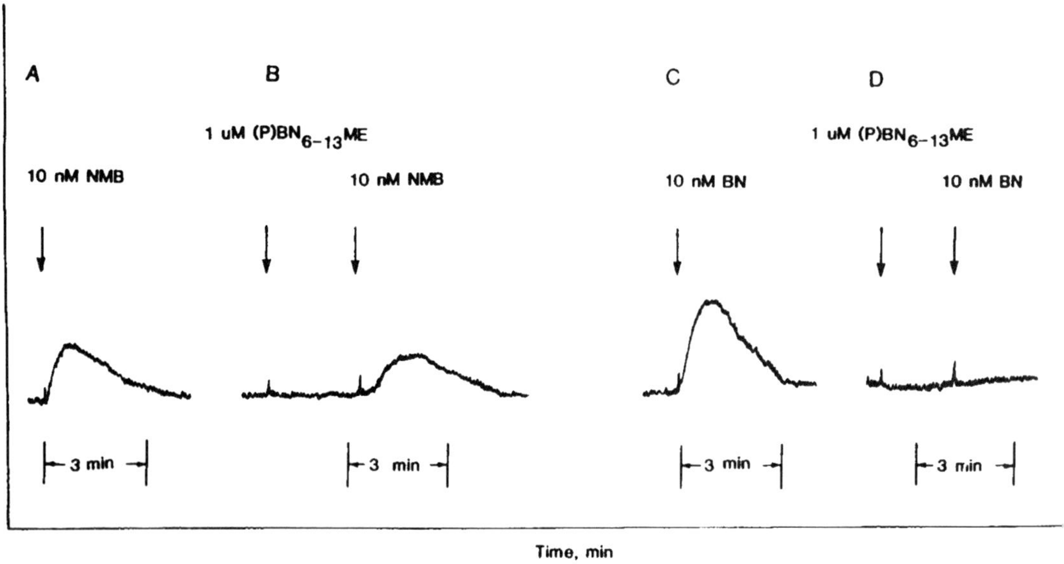

5. Effect of BB and NMB on cytosolic Ca^{2+}. SCLC cell line NCI-H345 was loaded with Fura 2-AM and ained after the addition of (A) 10-nM NMB, (B) 1-uM (F)BB^{6-13}ME plus 10-nM NMB, (C) 10-nM BB and (D) ed by 10-nM BB.

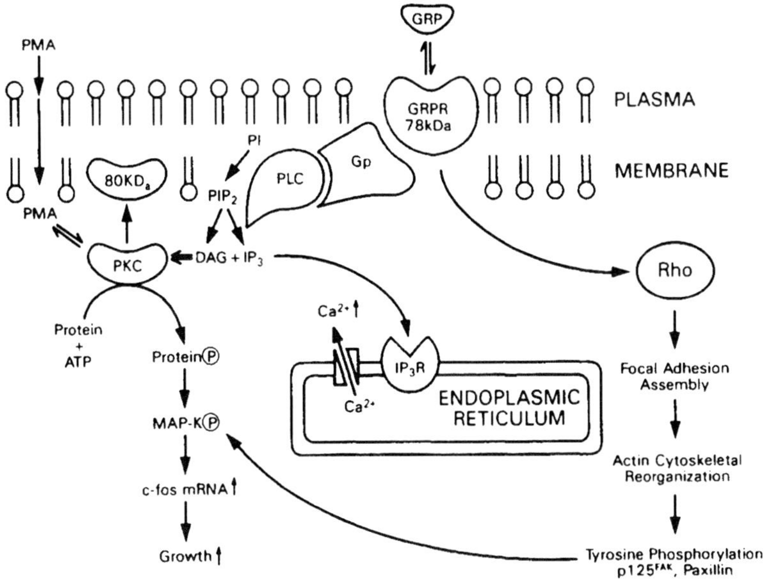

Figure 6. Schematic illustration of SCLC cells. When GRP binds to the receptor, Gp is activated, stimulating PLC. PIP_2 is metabolized, and the IP_3 released binds to receptors in the endoplasmic reticulum, releasing Ca^{2+} into the cytosol. The DAG released activated PKC which, in turn, phosphorylates proteins, ultimately resulting in activation of MAP-K, which stimulates c-fos gene expression and growth. Also, the GRP receptor can activate Rho resulting in focal adhesion assembly, actin cytoskeletal reorganization, tyrosine phosphorylation of $p125^{FAK}$ and other proteins, such as MAP-K, ultimately simulating growth.

genase activity. It is possible that activation of GRP receptors stimulates phospholipase A_2 (PLA_2) activity, resulting in arachidonic acid (AA) release (Millar and Rozengurt, 1990b). The AA may, then, be metabolized to prostaglandin E2 by cyclooxygenase. The PGE_2 released may, then, bind to prostaglandin receptors stimulating adenylate cyclase. The mechanism by which BB elevates cAMP may be very indirect. In SCLC cells, which have lipoxygenase activity, however, GRP transiently increases 5-HETE levels (Avis et al., 1994). The GRP receptor may stimulate PLA_2 activity, resulting in AA release. The AA released may, then, be metabolized by lipoxygenase, resulting in 5-HETE production.

The exact cellular mechanisms by which activation of GRP receptors or NMB receptors stimulate growth remains unclear. One early study (Zachary et al., 1987) demonstrated that pertussis toxin partially blocked bombesin-stimulated mitogenesis in Swiss 3T3 cells. However it had no effect on PLC activation, $[Ca^{2+}]_i$ mobilization, or PKC activation. These data suggest that the growth effect is at least partially mediated by pertussis-toxin-sensitive G-proteins and partially by either pertussis-toxin-insensitive G-proteins or not by G-proteins. The dissociation of growth inhibition from effects on PLC activation demonstrate that growth is at least partially mediated by non-PLC pathways. Several recent results suggest that the stimulation of tyrosine phosphorylation, particularly of p125^{FAK}, may be a particularly important pathway in causing GRP-receptor-mediated mitogenic effects. Tryphostin has been shown to inhibit BB-stimulated tyrosine phosphorylation in Swiss 3T3 cells and cause a 48% decrease in tyrosine phosphorylation of p125^{FAK} (Seckl and Rozengurt, 1993). Tryphostin treatment prevented both BB-stimulated increases in c-fos and DNA synthesis in Swiss 3T3 cells (Seckl and Rozengurt, 1993). Tryphostin had no effect on GRP activation of PLC or on PKA activation, demonstrating that its effect was specifically on tyrosine phosphorylation. In another study, the ability of GRP-receptor activation to stimulate tyrosine phosphorylation of p125^{FAK} was shown not to be mediated by activation of either PKC or the Ca^{2+} pathway (Sinnett-Smith et al., 1993). These results demonstrate that GRP receptor activates the tyrosine phosphorylation cascade parallel to the PLC pathway and that its activation is a mitogenic signal. Botulinium C3 exoenzyme has been shown to block GRP receptor-induced tyrosine phosphorylation of p125^{FAK} which suggests that this is mediated by the small GTP-binding protein, p21 (Rankin et al., 1994).

The relationship of BB-stimulated increases in cAMP to growth remains unclear. BB activation of the GRP receptor in Swiss 3T3 cells in the presence of cholera toxin or forskolin results in increases in cAMP (Millar and Rozengurt, 1988). This increase was partially inhibited by either PKC downregulation or indomethacin, and the combination resulted in complete inhibition. The increase in cAMP was inhibited 75% by pertussis toxin. The dose-inhibition curves for PTX-induced cAMP accumulation, thymidine incorporation, and c-myc induction were similar, suggesting a possible association. Recently (Benya et al., 1994c), GRP receptor-induced changes in cells transfected with the murine GRP receptor were examined. Although the transfected GRP receptor behaved identically to the native receptor in Swiss 3T3 cells for PLC activation, receptor modulation, and G-protein coupling, no activation of cAMP, c-fos, or thymidine incorporation was seen in the transfected cells. Conversely, inhibition of the cAMP increase in Swiss 3T3 cells by the PKA inhibitor H89 inhibited the ability of BB to increase c-fos or thymidine incorporation in these Swiss 3T3 cells. These results are consistent with cAMP playing an important role in BB-induced growth. This conclusion was further supported by a study in which the cAMP-mediated response was diminished in Swiss 3T3 cells by either overexpressing a phosphodiesterase inhibitor or a defective PKA regulatory subunit (Huang et al., 1994). The ability of BB in these transfected cells to induce growth was markedly reduced, further supporting the importance of activation of PKA in mediated BB-induced growth.

G. GRP Gene Expression

When MAP kinase is activated, it may cause phosphorylation of nuclear proteins, such as elk-1, increasing c-fos gene expression. In Swiss 3T3 cells, BB caused a transient increase in mRNA of early oncogenes, such as c-fos and c-myc (Palumbo

Figure 7. SCLC c-fos mRNA. NCI-H345 cells were treated with the indicated stimuli for 1 hour and the c-fos mRNA determined by Northern analysis.

et al., 1986; Mahmet et al., 1990). In SCLC cells which do not express the c-myc oncogene, BB maximally increased c-fos gene expression after 1 hour. The effects of BB were dose-dependent and the ED_{50} for BB was 1 nM (Draoui et al., 1995). The increase in c-fos gene expression caused by 10-nM BB was reversed by 1-μM BW2258U89. Also, Figure 7 shows that 1-nM GRP increases c-fos mRNA in SCLC cell line NCI-H345. The increase in c-fos gene expression caused by GRP was reversed by 1-uM (FA)BB6-13ME. These data suggest that GRP receptors regulate nuclear oncogene expression in SCLC cells.

Using SCLC cell line NCI-H345, 10-nM BB increased c-fos gene expression fivefold. Also, BB increased c-jun gene expression twofold, and the increase in

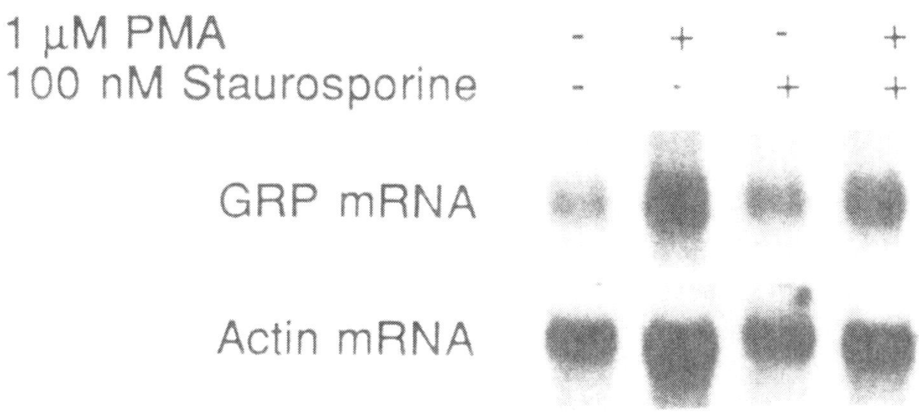

Figure 8. SCLC GRP mRNA. NCI-H209 cells were treated with the indicated stimuli for 8 hours, and the GRP and β-actin mRNA were determined by Northern analysis.

c-jun gene expression caused by BB was reversed by 1-μM BW2258U89. The c-fos and c-jun mRNAs may be translated into proteins which form heterodimers activating AP-I sites. Some growth factor genes, such as GRP, have AP-I sites (TGACTCA) in the 5′ upstream regulatory region. Previously, we found that 1-μM PMA increased GRP gene expression threefold after 8 hours (Draoui et al., 1993). Figure 8 shows that 1-uM PMA increases GRP but not β-actin mRNA in NCI-H209 cells threefold after 8 hours. Staurosporine had no effect on GRP mRNA in the presence or absence of 1-uM PMA. Previously we found that the increase in GRP gene expression caused by PMA was reversed by H7. These data suggest that SCLC cells have a PKC isoform inhibited by H7 but not by staurosporine.

H. GRP Proliferation

By interacting at GRP receptors, bombesin stimulates the growth of a number of normal tissues such as Swiss 3T3 cells, normal bronchial epithelial cells and human endometrial stomal cells (Endo et al., 1991; Willey et al., 1984; Rozengurt and Sinett-Smith, 1983). Activation of NMB receptors is reported to cause growth of the zone glomerula of the adrenal cortex (Markowska et al., 1993).

In quiescent Swiss 3T3 cells containing GRP receptors, BB stimulates ^{3}H-thymidine uptake after 24 hours. Also, insulin (1-μg/ml) or IGF-I (10-ng/ml) slightly stimulated ^{3}H-thymidine uptake, whereas addition of BB + insulin or BB + IGF-I strongly stimulates ^{3}H-thymidine uptake (Table 4). These data demonstrate that BB and IGF-I or insulin are synergistic at stimulating proliferation of Swiss 3T3 cells. This synergism may occur at the level of MAP kinase where BB receptors indirectly phosphorylate Thr^{183} as a result of PKC activation. In contrast, IGF-I receptors may phosphorylate Tyr^{185} indirectly as a result of tyrosine kinase activation. BB in the presence of insulin stimulated ^{3}H-thymidine uptake 10-fold causing

Table 4. Effect of Peptide Growth Factors on Swiss 3T3 Cells

Addition	*^{3}H-Thymidine Incorporation*
None	8,094 ± 1,040
Insulin, 1 ug/ml	9,178 ± 1,357
IGF-I, 10 ng/ml	10,003 ± 947
BB, 1 nM	12,854 ± 1,232*
BB + insulin	63,053 ± 3,276**
BB + IGF-I	72,284 ± 2,654**
(Psi-13,14)BB, 10 uM	7,525 ± 698
BB + insulin + (Psi-13,14)BB	8,519 ± 724
BB + IGF-I + (Psi-13,14)BB	9,042 ± 1,436
10% FBS	127,141 ± 18,110**

Notes: The Swiss 3T3 cells were incubated with peptide growth factors in the presence of ^{3}H-thymidine for 16 hr at 37 °C. The TCA precipitable DNA was then isolated and counted in a β-counter.

$^*p < 0.05$; $^{**}p < 0.01$

Table 5. Ability of BW2258U89 to Alter SCLC Growth *In Vitro*

Addition	*NCI-H345*
none	211 ± 14
BB, 10 nM	$440 \pm 78^{*}$
BW2258U89, 1 μM	$90 \pm 3^{**}$
BB + BW2258U89	$180 \pm 6^{*}$

Note: The mean value ± S.E. of 3 determinations is indicated. $^{*}p < 0.05$. $^{**}p < 0.01$.

the cells to go from the Go to the S-growth phase. The increase in ^{3}H-thymidine uptake caused by BB was reversed by GRP-receptor antagonists such as (Psi-13,14)BB.

In contrast to Swiss 3T3 cells, which are mouse epithelial cells, human SCLC cells do not rest in the Go-growth phase. Hence clonogenic growth assays are conducted to investigate proliferation of lung cancer cells. Single cells are plated into soft agarose using a serum-free medium containing 5-ug/ml insulin, 10-ug/ml transferrin and 30-nM M Na_2SeO_3. After 2 weeks, colonies greater than 50 cells were counted and BB-stimulated colony number as well as size. The effects of BB were dose-dependent and the ED_{50} was 1 nM for BB (Carney et al. 1987). Maximal colony number was obtained using 10-nM BB and colony number decreased at 100-nM BB possibly due to receptor desensitization. The increase in colony formation caused by BB was reversed by GRP antagonists, such as BW2258U89 (Table 5). Also 1-μM BW2258U89 decreased basal colony formation by 50%. These data suggest that BW2258U89 inhibits SCLC growth caused by endogenous or exogenous GRP-like peptides.

Besides altering proliferation *in vitro*, BB-like peptides alter proliferation *in vivo*. Using nude mice bearing SCLC xenografts, BB increased the growth rate of SCLC tumors (Alexander et al., 1988). In contrast, GRP-receptor antagonists, such as (Psi-13,14)BB, slowed proliferation of SCLC tumors by 50% (Mahmoud et al., 1991). More recently, BW2258U89 [3PhPr[DAla^{24},Pro^{26},psi 26–27]P(20–27)] has proven even more potent, slowing SCLC tumor growth by 75% (Figure 9). Reduced peptide bond analogues of BB significantly inhibited SCLC xenograft growth when 10 μg/day of BB2258U89, but not as little as 0.1 μg/day, was injected. Thus, the effects of BW2258U89 were dose-dependent and 0.4 mg/kg day was an effective concentration (Draoui et al., 1995). Recently, we found that, when BW2258U89 was put into microspheres, it strongly inhibited SCLC xenograft growth by 90%. Here the microspheres were injected i.m. monthly and BW2258U89 was slowly released into the body as opposed to subcutaneous injection where the BW2258U89 concentrations were initially very high (100 μM), but, then, the concentrations decreased during the day as BW2258U89 was degraded by endogenous proteases.

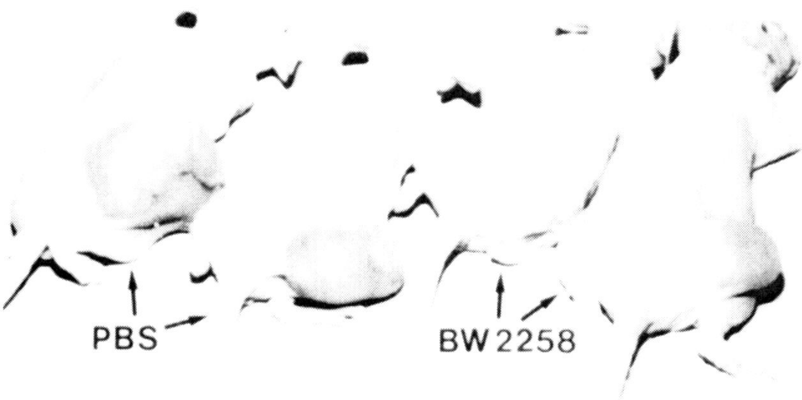

Figure 9. SCLC xenografts and nude mice. SCLC cells were injected subcutaneously, and, after 1 week, NCI-H510 tumors formed. Animals were then injected daily with 100 μl of PBS s.c. or 100 μl of PBS containing BW2258U89 (10 μg). The tumors grew rapidly in the PBS injected animals and grew slowly in the animals receiving the GRP receptor antagonist.

Because SCLC uses multiple growth factors, it may be necessary to use multiple blockers to stop SCLC growth.

I. GRP Cancer Growth

Recently in phase II clinical trials, mAb 2A11 was found to cause tumor regression in a SCLC female patient (Kelley et al., 1993). MAb 2A11 neutralizes secreted GRP so that it is unable to bind to the GRP receptor. These data suggest that blockade of the GRP autocrine growth cycle may inhibit cancer proliferation.

In preclinical trials, GRP-receptor antagonists, such as (Psi-13,14)BB, BW2258U89, and (Tpi6,Psi13,14,Tpi14)BB6-14 (RC-3095) (Radulovic et al., 1991) slowed xenograft formation in nude mice and were effective at a dose of 0.4 mg/kg daily (Mahmoud et al., 1991). In contrast, much higher doses of a SP antagonist (40 mg/kg) were required to slow SCLC xenograft formation in nude mice (Sethi et al., 1992). The pharmacokinetics and biodistributions of these antagonists are being investigated. Previously thiorphan, an inhibitor of neutral endopeptidase E.C. 3.4.24.11, was found to inhibit degradation of (Psi-13,14)BB (Davis et al., 1992). A goal is to obtain a high-affinity GRP receptor antagonist that is not readily degraded by endogenous proteases.

Activation of GRP receptors on numerous tumor tissues stimulates growth. These include such varied tumors as SCLC, prostatic adenocarcinomas, breast adenocarcinomas, human gliablastomas, various gastric carcinomas, and human pancreatic carcinomas (Qin et al., 1994a and 1994b; Farre et al., 1993; Pinski et al., 1994a and 1994b; Radulovic et al., 1994; Yano et al., 1994; Yano et al., 1992; Bold et al., 1994; Thomas et al., 1992).

GRP-receptor antagonists may be used to slow the growth of cancers other than SCLC. RC-3095 slowed the growth of HT 29 human colon cancers, PC-82 human prostate cancers, NKN45 human gastric cancers (Radulovic et al., 1991), MXT breast cancers in mice (Pinski et al., 1994) and human pancreatic cancers. These data indicate that the growth of many cancers may be slowed by GRP-receptor antagonists.

The process of carcinogenesis may also be altered by GRP receptors. Previously, it was found that BB promotes gastric carcinogenesis induced by Nmethyl-N'-nitro-nitrosoguanidine in Wistar rats (Tatsuta et al., 1989). GRP may function as a promoter of carcinogenesis due to its ability to activate PKC.

III. VIP/PACAP

A. VIP Structure

The VIP family of peptides is more hydrophobic and larger than GRP. VIP, a 28 amino acid peptide initially isolated from porcine stomach (Said and Mutt, 1970), is basic with an essential histidine at the N-terminal (Table 6). The C-terminal is amidated, and VIP has several aromatic amino acids, such as phenylalanine and leucine. A major difference between VIP-related peptides and the GRP-related

Table 6. Structure of VIP-Like Peptides

VIP	His-Ser-Asp-Ala-Val-Phe-Thr-Asp-Asn-Tyr-Thr-Arg-Leu-Arg-Lys-Gln-Met-Ala-Val-Lys-Lys-Tyr-Leu-Asn-Ser-Ile-Leu-Asn-NH_2
PACAP-27	His-Ser-Asp-Gly-Ile-Phe-Thr-Asp-Ser-Tyr-Ser-Arg-Tyr-Arg-Lys-Gln-Met-Ala-Val-Lys-Lys-Tyr-Leu-Ala-Ala-Val-Leu-NH_2
Helodermin	His-Ser-Asp-Ala-Ile-Phe-Thr-Gln-Gln-Tyr-Ser-Lys-Leu-Leu-Ala-Lys-Leu-Ala-Leu-Gln-Lys-Tyr-Leu-Ala-Ser-Ile-Leu-Gly-Ser-Arg-Thr-Ser-Pro-Pro-Pro-NH_2
PHI	His-Ala-Asp-Gly-Val-Phe-Thr-Ser-Asp-Phe-Ser-Arg-Leu-Leu-Gly-Gln-Leu-Ser-Ala-Lys-Lys-Tyr-Leu-Glu-Ser-Leu-Ile-NH_2
Secretin	His-Ser-Asp-Gly-Thr-Phe-Thr-Ser-Glu-Leu-Ser-Arg-Leu-Arg-Asp-Ser-Ala-Arg-Leu-Gln-Arg-Leu-Leu-Gln-Gly-Leu-Val-NH_2
Glucagon	His-Ser-Glu-Gly-Thr-Phe-Thr-Ser-Asp-Tyr-Ser-Lys-Tyr-Met-Asp-Asn-Arg-Arg-Ala-Lys-Asp-Phe-Val-Gln-Trp-Leu-Met-Asn-Thr
VIPhyb	Lys-Pro-Arg-Arg-Pro-Tyr-Thr-Asp-Asn-Tyr-Thr-Arg-Leu-Arg-Lys-Gln-Met-Ala-Val-Lys-Lys-Tyr-Leu-Asn-Ser-Ile-Leu-Asn-NH_2

Note: Sequential homologies relative to VIP are underlined.

peptides, however, is that members of the VIP family are N-terminally active peptides, whereas the GRP-related peptides are COOH-terminally active peptides. As a result the C-terminal of VIP can be extended with retention of biological activity because amidation is not essential (Fahrenkrug, 1992), whereas there is no evidence that the glycine-extended forms of GRP have biologic activity. PACAP is structurally similar to VIP with 19 of 27 amino acid homologies. Nonetheless, distinct genes for VIP and PACAP exist, and there are different receptor subtypes. VIP also has sequential homology to peptide histidine isoleucine (PHI) and a peptide isolated from the Gila monster, helodermin (Hd), having 15 and 13 amino acid homologies, respectively. PHI and Hd bind with moderate affinity to VIP receptors. Secretin and glucagon have sequential homology to VIP with 9 and 7 amino acids, respectively, being identical, but secretin interacts with low affinity with VIP receptors whereas glucagon does not inhibit ^{125}I-VIP binding. Circular dichroism studies provide evidence for a helical structure with amino acid residues 12–20 primarily involved. The Chou–Fosman parameters provide evidence for β-turns, predicting a possible reverse turn involving the NH_2 terminus of VIP, although the probability for a second turn is less strong than for secretin. It was predicted that VIP has a first β-turn in the 1–4 region of the amino terminus and, in addition, a second β-turn in the 6–10 region (Jensen, 1994; Bodansky et al., 1979; Christophe et al., 1988; Coy et al., 1985). It is, therefore, not surprising that VIP fragments which lack the NH_2 terminus have no biological activity. However, the COOH fragments can bind to the receptor, although with low affinity, and can function as VIP receptor antagonists. At present, no potent antagonists for the VIP receptor exist. Three classes of receptor antagonists have been described (Jensen, 1994) including COOH-terminal analogues of VIP, such as VIP(10–28) or VIP(6–28) (Fishbein et al., 1994); D-amino acid substituted analogues of VIP or growth hormone-releasing factor (GRF), such as [4Cl-DPhe^6,Leu^{17}]VIP or [Ac-Tyr^1,DPhe^2]GRF (Waelbroeck et al., 1985; Pandol et al., 1986); and chimeric analogues made up of a VIP fragment and fragment of another peptide, such as neurotensin(6–11)-VIP(7–28) (VIPhyb) (Gozes et al., 1991a; Hill et al., 1991) or VIP(6–28)-PACAP(28–37). Recently, we found that the VIPhyb (neurotensin-VIP chimeric peptide) (Table 6) is a higher affinity antagonist for VIP receptors (Gozes et al., 1991b) than VIP$^{10-28}$.

B. VIP Biosynthesis and Processing

The VIP-PHI gene is expressed widely in the peripheral and central nervous system and in the gastrointestinal tract. The VIP gene contains 8837 base pairs and 6 introns (Gozes et al., 1984). By Northern analysis, VIP mRNA (2 kb) is present in many lung cancer cell lines, such as NCI-H727. VIP mRNA is increased threefold after treatment of NCI-H727 cells with 1-μM 12-o-tetradecanoyl-phorbol-13 acetate (PMA). Previously PMA was found to cause VIP secretion from neuroblastoma cells and a PMA-responsive element was found 4 kb upstream from the VIP gene (Waschek et al,, 1988). Also, a cAMP-responsive element was found 70 base pairs upstream from the transcription initiation site of the VIP gene. These data indicate that the VIP gene is regulated as a result of PKC and PKA activation.

VIP, similar to GRP, is synthesized as a high molecular weight precursor protein (Bloom et al., 1983; Itoh et al., 1983). PreproVIP is a 170 amino acid protein and the N-terminal signal sequence is cleaved by a signal protease. The 149 amino acid proVIP contains VIP (preproVIP125–152) and peptide histidine methionine (PHM, preproVIP81–107), the mammalian equivalent to PHI. ProVIP is metabolized by trypsin-like enzymes to VIPGKR, PHMGKR, and inactive proVIP fragments. VIPGKR and PHMGKR are metabolized by carboxypeptidase B-like enzymes to VIPG and PHMG. VIPG and PHMG are metabolized by PAM enzymes to VIP and PHM. Thus, processing of the VIP precursor protein yields two active peptides, VIP and PHM. Because amidation of VIP is not essential for biological activity (Fahrenkrug, 1992) proVIP may have appreciable biological activity. ProVIP is more abundant than VIP in NCI-H727 extracts, suggesting that VIP may not be efficiently processed in lung cancer cells.

PACAP is synthesized from a different precursor protein which contains 176 amino acid residues (Arimura, 1992). PreproPACAP is metabolized to proPACAP by a signal protease. ProPACAP (152 amino acids) is metabolized by trypsin-like enzymes to PACAP-38GRR. The 41 amino acid PACAP-38GRR is processed by carboxypeptidase B-like enzymes to PACAP-38G, and PACAP-38G is metabolized by PAM enzymes to PACAP-38 which is biologically active. PACAP-38 may be metabolized further to PACAP which contains 27 amino acids and is biologically active. PACAP-38 immunoreactivity is present in many lung cancer cell lines, and both PACAP and PACAP-38 bind with high affinity to lung cancer cells.

C. VIP Receptor Ligands

Numerous VIP and PACAP receptor binding studies have been conducted. PACAP inhibits ^{125}I-VIP and ^{125}I-PACAP binding with high affinity (Table 7). In contrast, VIP inhibits ^{125}I-VIP binding with high affinity but ^{125}I-PACAP binding with low affinity (Moody et al., 1993b). Hd and PHI inhibits both ^{125}I-VIP and ^{125}I-PACAP binding with moderate affinity. Secretin inhibits ^{125}I-VIP and ^{125}I-PACAP binding with low affinity whereas glucagon does not compete (Shaffer et al., 1987; Lee et al., 1990b). VIPhyb inhibits ^{125}I-VIP binding with moderate affinity and PACAP(6-38) inhibits ^{125}I-PACAP binding with high affinity (Moody et al., 1994a, 1993c).

^{125}I-VIP binds with high affinity (Kd = 1 and 50 nM) to 2 classes of sites (Bmax = 100 and 2000 fmol/mg protein) using membranes derived from SCLC cell line NCI-N592 (Shaffer et al., 1987). ^{125}I-VIP was cross-linked to an 80-kDa protein using NCI-N592 membranes and 1-mM disuccinimidylsuberate (DSS). Also, ^{125}I-VIP binding was inhibited by 10-uM GMP(NH)PP. The GTP analogue may interact with a stimulatory guanine nucleotide binding subunit (Gs) resulting in increased dissociation of bound VIP.

Table 7. VIP Receptor Binding

Peptide	^{125}*I-VIP Binding*	^{125}*I-PACAP Binding*
PACAP	20	10
VIP	10	1000
Hd	100	150
PHI	500	700
VIPhyb	700	n.d.
PACAP(6–38)	n.d.	30
Secretin	2000	3000
Glucagon	>10,000	>10,000

Notes: The mean IC_{50}, nM to inhibit ^{125}I-VIP and ^{125}I-PACAP binding to NCI-H838 cells is indicated.
n.d.: not determined.

D. VIP Receptors

VIP and the structurally related mammalian peptides PACAP, GRF, PHI, secretin, and glucagon-like peptide I (GLP-I) can interact with six classes of receptors: the VIP receptor, PACAP receptor, secretin receptor, GLP-I receptor, GRF receptor, and the VIP_2 receptor. VIP interacts with high affinity with two classes of these receptors: the VIP receptor and the VIP_2 receptor. Confusion exists in various classifications because PACAP, which shares a 68% homology with VIP, interacts also with high affinity with the two receptors that have a high affinity for VIP and, in addition, with the PACAP receptor. The VIP receptor (also called the PACAP type II receptor) has a high affinity for VIP, PACAP-27, and PACAP-38, appropriately, a threefold lower affinity for the Gila monster venom peptides, helodermin, helospectin I and II and rGRF, a 50-fold lower affinity for PHI or PHM, a 500-fold lower affinity for secretin and >1000-fold lower affinity for the GLP-I receptor ligand, GLP-I(7-36)NH_2 (Jensen, 1994; Arimura, 1992; Ishihara et al., 1992; Usdin et al., 1994; Sreedharan et al., 1993; Couvineau et al., 1994). This receptor has also been called the PACAP-type II receptor (Arimura et al., 1992; Tatsuno et al., 1990; Shivers et al., 1991) and the VIP_1 receptor. The VIP receptor has recently been cloned from rat (Ishihara et al., 1992; Usdin et al., 1994) and human (Sreedharan et al., 1993; Couvineau et al., 1994) tissues (Figure 10). The rat and human VIP receptors are 459 and 457 amino acid peptides, respectively, and are members of

Figure 10. Amino acid structures of the VIP receptor and various structurally related receptors. Amino acid identities in all receptors are indicated by the boxed areas. Putative transmembrane regions are indicated by lines above the sequences. Filled circles are at 10 amino acid spacing.

```
rVIP1     MRPPSPPHVRWLCVLAGALACALRPAGSQAASPQHECEYLQLIEIQRQQCLEE-------
rVIP2     -MRASVVLTCYCWLLVRV-------------SSIHPECRFHLEIQEEETKCAELL------
rPACAP    --MARVLQLSLTALLLPV------------AIAMMSDCIF----KKEQAMCLERIQRANDL
rSecretin --MLSTMRPRLSLLLLRLLL--LTKAAHTVGVPPRLCDVRRVLLEERAHCLQQLSKEKKC
rGRF      --MDSLLWATWVLCLLNLWG-------VALGHLHLECDFITQLRDDELACLQA------A
rGLPI     -MAVTPSLLRLALLLLGAVG---RAGPRPQGATVSLSETVQKWREYRHQCQRFLT----E

rVIP1     AQLENETTGCSKMWDNLTCWPTTPRGQAVVLDCPLIFQLFAPIHG---------------
rVIP2     SSQMENHRACSGVWDNITCWRPADIGETVTVPCPKVFSNFYSRP----------------
rPACAP    MGLNESSPGCPGMWDNITCWKPAQVGEMVLVSCPEVFRIFNPDQVWMTETIGDSGFADSN
rSecretin ALGPETASGCEGLWDNMSCWPSSAPARTVEVQCPKFLLMLSKNK----------------
rGRF      EGTNNSSMGCPGTWDGLLCWPPTGSGQWVSLPCPEFFSHFGSDP----------------
rGLPI     APLLATGLFCNRTFDDYACWPDGPPGSFVNVSCPWYLPWASSVLQ---------------

rVIP1     -------YNISRSCTEEG-W----SQLEPGPYHIACGLNDRASSLDEQQQTKFYNTVKTG
rVIP2     -------GNISKNCTSDG-W----SETFP-DFIDACGYNDPE----DESKITFYILVKAI
rPACAP    SLEITDMGVVGRNCTEDG-W----SEPFP-HYFDACGFDDYEPE-SGDQDY-YYLSVKAL
rSecretin -------GSLFRNCTQDG-W----SETFPRPDL-ACGNVINNS--FNERRHAYLLKLKVM
rGRF      -------GAVKRDCTITG-W----SDPFP-PYPVACPVPLELL----TEEKSYFSTVKII
rGLPI     -------GHVYRFCTAEGIWLHKDNSSLPWRDLSECEESKQGE---RNSPEEQLLSLYII

          TM I                               TM II
rVIP1     YTIGYSLSLASLLVAMAILSLFRKLHCTRNYIHMHLFMSFILRATAVFIKDMAL---FNS
rVIP2     YTLGYSVSLMSLTTGSIIICLFRKLHCTRNYIHLNLFLSFMLRAISVLVKDSVL---YSS
rPACAP    YTVGYSTSLATLTTAMVILCRFRKLHCTRNFIHMNFLVSFMLRAISVFIKDWIL---YAK
rSecretin YTVGYSSSLAMLLVALSILCSFRRLHCTRNYIHMHLFVSFILRALSNFIKDAVL---FSS
rGRF      YTTCHSISIVALCVAIAILVALRRLHCPRNYIHTQLFATFILKASAVFLKDAAV---FQG
rGLPI     YTVGYALSFSALVIASAILVSFRHLHCTRNYIHLNLFASFILRALSVKIKDAALKWMYST

                                TM III
rVIP1     GEIDH--CSE-AS----VGCKAAVVFFQYCVMANFFWLLVEGLYLYTLLAVSFFSERKYF
rVIP2     SGTLR--CHD-QPG-SWVGCKLSLVFFQYCIMANFYWLLVEGLYLHTLL-VAILPPSRCF
rPACAP    QDSSH--CFV-ST----VECKAVMVFFHYCVVSNYFWLFIEGLYLFTLLVETFFPERRYF
rSecretin DDVTY--CDA-HK----VGCKLVMIFFQYCIMANYAWLLVEGLYLHTLLAISFFSERKYL
rGRF      DSTDH--CSM-ST----ILCKVSVAVSHFATMTNFSWLLAEAVYLSCLLASTSPRSKPAF
rGLPI     AAQQHQWDGL-LSYQDSLGCRLVFLLMQYCVAANYYWLLVEGVYLYTLLAFSVFSEQRIF

          TM IV                                       TM V
rVIP1     WGYILIGWGVPSVFITIWTVVRIYFEDFGCWDTIINSSLWWIIKAPILLSILVNFVLFIC
rVIP2     LAYLLIGWGIPSVCIGAWIATRLSLEDTCCWDTNDHSIPWWVIRMPILISIVVNFALFIS
rPACAP    YWYTIIGWGTPTVCVTVWAVLRLYFDDAGCWDMNDSTALWWVIKGPVVGSIMVNFVLFIG
rSecretin QAFVLLGWGSPAIFVALWAITRHFLENTGCWDINANASVWWVIRGFVILSIIINFIFFIN
rGRF      WWLVLAGWGLPVLCTGTWVGCKLAFEDTACWDLDDSSPYWWIIKGPIVLSVGVNFGLFLN
rGLPI     KLYLSIGWGVPLLFVIPWGIVKYLYEDEGCWTRNSNMNYWLIIRLPILFAIGVNFLVFIR

rVIP1     IIRILVQKLRPPDIGKNDSSPYS-------------------------------------
rVIP2     IVRILLQKLTSPDVGGNDQSQYK-------------------------------------
rPACAP    IIIILVQKLQSPDMGGNESSIYFSCVQKCYCKPQRAQQHSCKMSELSTITL---------
rSecretin ILRILMRKLRTQETRGSETNHYK-------------------------------------
rGRF      IICILLRKLGPAQGGLHTRAQYNYLLPWSCPLPQVPRERTDLGPSSHEITVQESGTRNCQ
rGLPI     VICIVIAKLKANLMCKTDIKC---------------------------------------

                   TM VI                                 TM VII
rVIP1     ---RLAKSTLLLIPLFGIHYVMFAFFPDNFKAQ----VKMVFELVVGSFQGFVVAILYCF
rVIP2     ---RLAKSTLLLIPFLGVHYMVFAAFPIGISST----YQILFELCVGSFQGLVVAVLYCF
rPACAP    ---RLARSTLLLIPLFGIHYTVFÅFSPENVSKR----ERLVFELGLGSFQGFVVAVLYCF
rSecretin ---RLAKSTLLLIPLFGIHYIVFAFSPEDAME-----VQLFFELALGSFQGLVVAVLYCF
rGRF      LPWRLSKSTLLLIPLFGIHYIIFNFLPDSAGLG----IRLPLELGLGSFQGFVVAVLYCF
rGLPI     ---RLAKSTLTLIPLLGTHEVIFAFVMDEHARGTLRFVKLFTELSFTSFQGFMVAVLYCF

rVIP1     LNGEVQAELRRKWRRWHL------QGVLGSSSKSQHPWGGSNGATCSTQVSMLTRVSP--
rVIP2     LNSEVQCELKRRWRGLCL-------TQPGWRDYRLHSWSMSRNGSESALQ--IHRGSR--
rPACAP    LNGEVQAEIKRKWRSWKV------NRYFTMDFKHRHPSLASSGVNGGTQLSILSKSSS--
rSecretin LNGEVQLEVQKKWRQWHL------QEFPLRPVAFNNSFSNATNGPTHS-----TKAST--
rGRF      LNQEVRTEISRKW---------------YGHDPELLPARRTCTE-----------WTTP--
rGLPI     VNNEVQMEFRKSWERWRL------ERLNIQRDSSMKPLKCPTSSVSSG-----ATVGS--

rVIP1     --SARRSSSFQAEVSLV
rVIP2     --TQSFLQSETSVI
rPACAP    --QLRMSSLPADNLAT
rSecretin --EQSRSIPRASII
rGRF      --PRSRVKVLTSEC
rGLPI     --SVYAATCQNSCS
```

the G-protein-coupled seven transmembrane superfamily. The human VIP receptor amino acid sequence shares an 84% identity with the rat VIP receptor, 44% with the rat secretin receptor and 35% with rat GLP-1 receptor (Sreedharan et al., 1993) (Figure 10). An earlier study reported isolating a cDNA clone (G PRN1) from a Nalm 6 leukemic pre-β lymphoblastic cell line and from HT-29 colonic carcinoma cells that was thought to be the human VIP receptor (Sreedharan et al., 1991). It was a 362 amino acid peptide with seven transmembrane-spanning segments by a hydropathy plot. This cDNA is the human equivalent of a previously described cDNA dog clone called RDC1, which encodes a protein containing seven transmembrane domains. The subsequent cloning of the equivalent rat protein, which has a 92% homology to the dog RDC 1 cDNA or human G PRN1, when expressed, demonstrated no VIP binding. A similar result was shown with the canine homologue of the proposed human VIP receptor, which also demonstrated no VIP binding (Cook et al., 1992). These results, as well as the failure of others (Nagata et al., 1992) to show VIP binding when this gene was expressed, have demonstrated that this protein is an orphan receptor.

Recently a second receptor in rat, mouse, and human tissues called the VIP_2 receptor has been cloned which has high affinity for VIP (Usdin et al., 1994; Lutz et al., 1993; Svoboda et al., 1994; Inagaki et al., 1994) (Figure 10). The rat and human VIP_2 receptor have 459 and 457 amino acids, respectively. The human VIP_2 receptor has an 87% identity in amino acid sequence with the rat VIP_2 receptor, a 49% amino acid identity with the human VIP_1 receptor, and 52% with the human PACAP receptor. The expressed human VIP_2 receptor had the highest affinity for PACAP-38 (IC_{50} = 0.8 nM) and equally high affinities for helodermin, PACAP-27, and VIP (2–3 nM), whereas secretin at 1-μM concentration had little effect (Svoboda et al., 1994). Similarly the rat VIP_2 receptor stimulated cAMP and the relative potencies were PACAP-38 = VIP ≈ helodermin > PACAP-27 > PHI >> GRF (Lutz et al., 1993), whereas, in a second study, they were VIP >> PHI > PACAP-27, PACAP-38 > GRF >> secretin (Usdin et al., 1994). In the latter study, the rat VIP_1 and VIP_2 receptors were both expressed in LVIP cells and compared. They differed principally in the EC_{50} of secretin which was 100 times more potent at the VIP, than the VIP_2 receptor, whereas the rank order of potencies for VIP, PHI, PACAP-38, and PACAP-27 for both receptors was similar. In general, significantly greater concentrations of VIP, PHI, and PACAP-38 were required to activate the VIP_2 than the VIP_1 receptor (Usdin et al. 1994).

The mRNA of the two VIP receptors was recently shown to differ markedly using *in situ* hybridization histochemistry (Usdin et al., 1994). VIP_1 mRNA is found predominantly in the lung, small intestine, thymus, and, in the brain, in the cerebral cortex and hippocampus. In contrast, the VIP_2 receptor mRNA was found in the stomach and the testis. In the CNS, VIP_2 receptor mRNA was exclusively in areas associated with neuroendocrine function including several hypothalamic nuclei. In the periphery, it was also found in the pituitary gland and the pancreatic islets (Usdin et al., 1994).

Both the VIP_1 and VIP_2 receptors and the closely related PACAP, GLP-1, GRF, and secretin receptors (Figure 10) contain a signal sequence. The rat VIP_1 and VIP_2 receptors have a signal peptide of 30 and 22 amino acids, respectively (Lutz et al., 1993; Ishihara et al., 1992). Therefore, the native rat VIP_1 and VIP_2 receptors have 429 and 437 amino acids, respectively, with molecular weights of 48,946 and 49,519, respectively.

The PACAP receptor (also called PACAP type I receptor) (Pisegna and Wank, 1993; Spengler et al., 1993; Journot et al., 1994; Hashimoto et al., 1993; Morrow et al., 1993; Svoboda et al., 1993; Ogi et al., 1993; Hosoya et al., 1993; Miyamoto et al., 1994), the GLP1 receptor (Thorens et al., 1993; Thorens 1992), the secretin receptor (Ishihara et al., 1991), and the GRF receptor (Gaylinn et al., 1993; Mayo 1992), which are structurally related to the VIP_1 and VIP_2 receptors, have been cloned (Figure 10) and are structurally related to the PTH and calcitonin receptors. The PACAP receptor which interacts with VIP with low affinity has been reported to have five different subtypes which differ due to alternative splicing in the third intracellular loop (Spengler et al., 1993; Journot et al., 1994). Expression of the different cDNAs revealed altered patterns of adenylyl cyclase and phospholipase C stimulation, suggesting a novel mechanism for finely adjusting intracellular signalling.

E. VIP Second Messengers and Gene Expression

In all tissues examined, activation of the VIP receptor leads to activation of adenylate cyclase. However, in some tissues, it also alters intracellular calcium (Oettling et al., 1990; Wahl et al., 1993; Fatatis et al., 1994). The VIP_2 receptor is also coupled to adenylate cyclase (Inagaki et al., 1994; Svoboda et al., 1994; Lutz et al., 1993). VIP increased the intracellular cAMP 10-fold in a dose-dependent manner using NCI-H838 or N417 cells. The half maximal effective dose for VIP (ED_{50}) was 1 nM. The increase in cAMP caused by VIP was reversed by 10-uM VIPhyb. VIPhyb had no effect on basal cAMP but antagonized the increase in cAMP caused by VIP (Moody et al., 1993c). VIPhyb had an IC_{50} value of 500 nM using NCI-H838 or NCI-N417 cells, but it is reported to be much more potent in the CNS with an IC_{50} value of 1 nM. VIPhyb appears to be two orders of magnitude more potent on rat CNS VIP receptors than human lung cancer VIP receptors.

PACAP can activate adenylate cyclase by interacting with the VIP_1 or VIP_2 receptors or through the PACAP receptor (Figure 11). PACAP or PACAP-38 elevate the cAMP levels 15-fold using NCI-H838 or N417 cells, and the ED_{50} is 1 nM. The increase in cAMP caused by PACAP is reversed by 1-uM PACAP(6–38); PACAP(6–38) had no effect on basal cAMP. PACAP(6–38) has an IC_{50} value of 30 nM and, hence, is an order of magnitude more potent than VIPhyb *in vitro* (Hoshino et al., 1994). As a result of increased cAMP, both VIP and PACAP may activate protein kinase A (PKA) resulting in phosphorylation of protein substrates.

Besides interacting with Gs, PACAP receptors interact with Gp causing PI turnover (Pisegna and Wank, 1993; Spengler et al., 1993; Delporte et al., 1993;

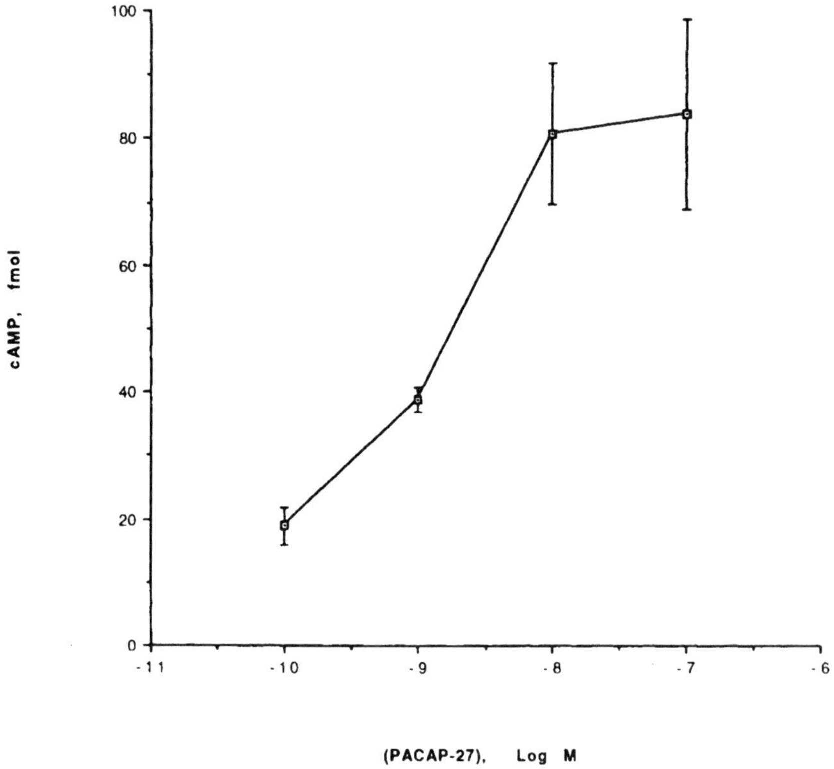

Figure 11. Effect of PACAP on cAMP. PACAP elevates cAMP in a dose-dependent manner using NCI-N417 cells.

Schomerus et al., 1994; Deutsch and Sun, 1992). The IP_3 and DAG released may elevate cytoslic Ca^{2+} and activate PKC, respectively. PACAP elevates cytosolic Ca^{2+} in a dose-dependent manner, and the ED_{50} is 10 nM. Because the ED_{50} for PACAP to cause PI turnover is greater than the ED_{50} to elevate cAMP, PACAP receptors may couple more tightly to Gs than Gp. Because VIP has no effect on cytosolic Ca^{2+} at a 100-nM dose using lung cancer cells, the VIP receptor may not couple to Gp. The increase in cytosolic Ca^{2+} caused by PACAP is reversed by 1-uM PACAP(6–38); PACAP(6–38) has no effect on basal cytosolic Ca^{2+}. As a result of the increase in DAG, PACAP may activate PKC, causing serine and threonine phosphorylation of proteins.

Similar to GRP, PACAP may activate MAP kinase through an indirect mechanism. MAP-kinase phosphorylation results in increased c-fos gene transcription. PACAP causes a sixfold increase in c-fos mRNA after 1 hr, and the ED_{50} is 0.1 nM using NCI-N417 cells. The increase in c-fos mRNA caused by PACAP is reversed by PACAP(6–38) (Moody et al., 1994a). The c-fos mRNA may be translated into c-fos protein which forms a heterodimer with c-jun. The fos/jun heterodimer may activate AP-1 sites resulting in increased transcription of certain growth promoting genes. VIP and PACAP also increase c-myc gene expression in NB-OK-1 cells (Hoshino et al., 1994).

F. VIP Proliferation and Cancer Growth

VIP and PACAP alter the growth of numerous cells. VIP stimulates ^{3}H-thymidine uptake in neuroblastoma cells (Gozes et al., 1994) and lovocolorectal adenocarcinoma cells (Chastre et al., 1994). After 2 weeks, VIP stimulates the clonal growth of NCI-H345 and H1299 cells. The increase in colony size and number caused by 10-nM VIP was reversed by 1-uM VIPhyb (Figure 12).

VIPhybrid decreased NSCLC growth *in vivo* using nude mice. With the nude mice, NSCLC cells (10^7) are injected s.c. in 100 ul of PBS. After a week, a palpable mass forms, and the tumor grows exponentially. Injection of PBS (100 ul, s.c.) has no effect on xenograft growth, whereas, if VIPhyb (10 ug) is daily injected s.c., xenograft growth significantly slows by 75% (Figure 13). The effect was dose-dependent in that 1 ug of VIPhyb also slowed growth by 50% whereas 0.1 ug of VIPhyb had no effect.

Further VIP enhanced survival of spinal cord neurons *in vitro* (Brenneman, and Eiden, 1986). These data suggest that VIP increases proliferation in most assays. In some acute assays for example, ^{3}H-thymidine uptake in NCI-H69 cells, however, VIP decreased ^{3}H-thymidine uptake after 24 hr (Maruno and Said, 1992). This may occur because VIP acutely increases cAMP which may cause a transient differentiation *in vitro* resulting in decreased proliferation. In summary, the effects of VIP on proliferation are complex and depend on the cells and assay utilized.

VIPhyb may slow growth of other tumors as well. Recently, VIPhyb was found to inhibit the growth of MCF-7 xenografts in nude mice. Also, VIP receptors are present in HT29 colon carcinoma cells, and it remains to be determined if VIPhyb slows colon carcinoma growth.

PACAP, similar to VIP, stimulates proliferation of most cells. PACAP increased the labeling index of cerebellar granule neuroblasts *in vitro* (DiCicco–Bloom and Lu, 1994). PACAP increased the colony number and size using NCI-N417 cells. The increase in proliferation caused by 1-nM PACAP was reversed by PACAP(6–38). PACAP(6–38) (0.4 mg/kg daily s.c) slows lung cancer xenograft formation in nude mice.

IV. SUMMARY

GRP and VIP interact with G-protein-coupled receptors which regulate proliferation. In contrast to typical tyrosine kinase growth factor receptor, such as that for EGF, they do not have a catalytic domain in the cytosol tyrosine phosphorylating protein substrates. Rather they cross the plasma membrane seven times and are

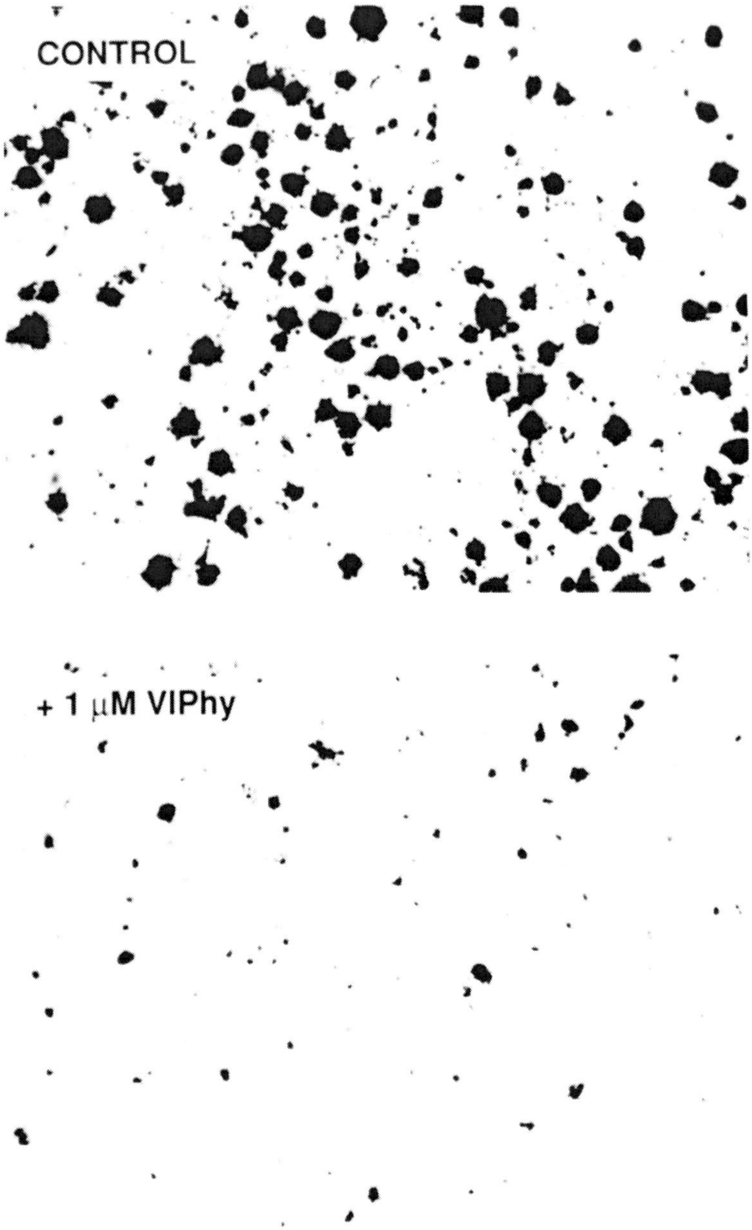

Figure 12. Effect of VIPhyb on NSCLC growth *in vitro*. Numerous NCI-H727 colonies formed and 1-μM VIPhyb decreased the size as well as the number of colonies.

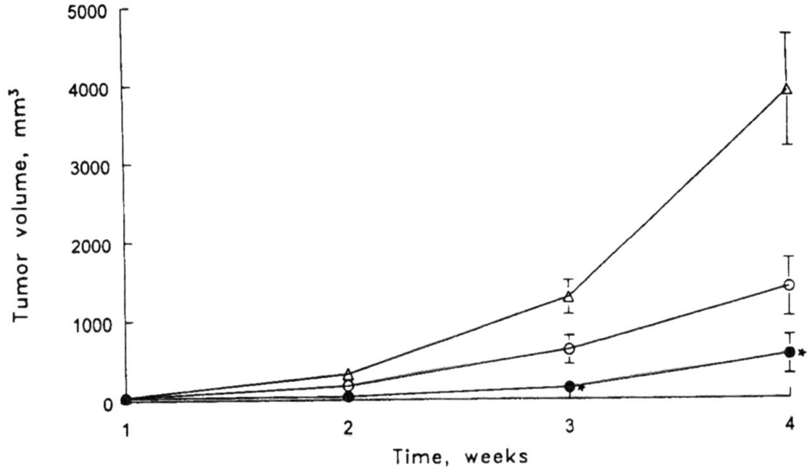

Figure 13. Effect of VIPhyb on NSCLC growth *in vivo*. The effect of 100 μl of PBS (Δ) or PBS containing 0.1-(o) or 10-(•) μg VIPhyb subcutaneously adjacent to the NCI-H157 tumor. The tumor volume (height × width × depth) was determined weekly by calipers and recorded. The mean value ± S.E. of three determinations is indicated; $^*p < 0.05$ using student's t-test.

comprised of approximately 400 amino acid residues. GRP and VIP receptors have an extracellular N-terminal and a short intracellular C-terminal comprised of approximately 60 amino acids. Because the receptors are deeply embedded in the plasma membrane, they interact with G-proteins. Current efforts focus on defining the receptor binding site using receptor chimeras and site-directed mutagenesis techniques.

GRP causes phosphatidylinositol turnover. As a result, the IP_3 released elevates cytosolic Ca^{2+} and the DAG released activates PKC. PKC may phosphorylate various proteins, such as MAP kinase, increasing c-fos gene expression and perhaps growth. GRP receptor activation also increases tyrosine phosphorylation of proteins, such as $p125^{FAK}$ and paxillin, which results in growth. The NMB receptor is structurally similar to the GRP receptor, and its activation increases phospholipase C activity. However, it is currently unknown if it also causes phosphorylation which could contribute to its growth effects.

High-affinity peptide receptor antagonists have been synthesized for the GRP receptor. These inhibit the growth of various cancers especially SCLC which produces and secretes high amounts of GRP. Approximately 40,000 patients die annually from SCLC in the United States. Currently, SCLC is treated with chemo- and radiation therapy but the mean survival time of patients is approximately 1 year.

In the near future, GRP antagonists will be tested in phase I clinical trials of SCLC. An area of concern, however, is the proteolytic degradation of the peptide antagonists *in vivo*. High-affinity NMB receptor antagonists have yet to be developed.

VIP interacts with high affinity with two receptors, the VIP_1, and VIP_2 receptors, both of which activate adenylate cyclase. PACAP also interacts with these receptors with high affinity and with an additional distinct receptor, the PACAP receptor, which has low affinity for VIP. As a result, VIP activates protein kinase A which may phosphorylate various protein substrates, such as MAP kinase. VIP stimulates c-fos gene expression and growth especially of tumors, such as NSCLC, which have VIP mRNA. The PACAP receptor is structurally similar to the VIP receptor and PACAP is an unusual peptide in that it simultaneously elevates cAMP and causes PI turnover.

Peptide antagonists have been synthesized for the VIP and PACAP receptor, however, they are of low affinity, generally. *In vivo* VIPhyb antagonizes the VIP receptor with moderate affinity and inhibits NSCLC growth. NSCLC is currently treated by surgical resection and chemotherapy but unfortunately approximately 120,000 patients in the U.S. die annually and the median survival time is less than 5 years. Current efforts are focused on developing higher affinity VIP and PACAP receptor antagonists. The development of peptide mimetics as GRP and VIP receptor antagonists may greatly improve their ability to inhibit tumor growth *in vivo*.

ACKNOWLEDGMENTS

The authors thank the numerous investigators who have participated in these studies over the past decade including Drs. J. Battey, F. Cuttitta, M. Draoui, S. Mahmoud, J. Mulshine, J. Staley, and F. Zia.

REFERENCES

Alexander, R. W., Upp, J. R., Jr., Poston, G. J., Gupta, V., Townsend, C. M., Jr., & Thompson, J. C. (1988). Effects of bombesin on growth of human small cell lung carcinoma *in vivo*. Cancer Res. 48, 1439–1441.

Anastasi, A., Erspamer, V., & Bucci, M. (1973). Isolation and amino acid sequences of alytesin and bombesin: Two analogous active tetradecapeptides from the skin of European discoglossid frogs. Arch. Biochem. Biophys. 148, 443–446.

Arimura, A. (1992). Pituitary adenylate cyclase activating polypeptide (PACAP): Discovery and current status of research. Reg. Peptides 37, 287–303.

Avis, I. L., Kovacs, T. O. G., Kasprzyk, P. G., Treston, A. M., Bartholomew, R., Walsh, J. H., Cuttitta, F., & Mulshine, J. L. (1991). Preclinical evaluation of an antiautocrine growth factor monoclonal antibody for treatment of patients with small cell lung cancer. JNCI 83, 1470–1476.

Avis, E., Jeff, M., Boyle, T., & Mulshine, J. (1994). The effect of arachidonic acid metabolites and lipoxygenase pathway inhibitors on the *in vitro* growth of human lung cancer cell lines. Cancer Res. Proc. 35, 318.

Battey, J. F., Way, J., Corjay, M. H., Shapira, H., Kusano, K., Harkins, R., Wu, J. M., Slattery, T., Mann, E., & Feldman, R. (1991). Molecular cloning of the bombesin/GRP receptor from Swiss 3T3 cells. Proc. Natl. Acad Sci. 88, 395–399.

Benya, R. V., Wada, E., Battey, J. F., Fathi, Z., Wang, L. H., Mantey, S. A., Coy, D. H., & Jensen, R. T. (1992). Neuromedin B receptors retain functional expression when transfected into BALB 3T3 fibroblasts: Analysis of binding, kinetics, stoichiometry, modulation by guanine nucleotide-binding proteins, and signal transduction and comparison with natively expressed receptors. Mol. Pharmacol. 42(6), 1058–1068.

Benya, R. V., Fathi, Z., Battey, J. F., & Jensen, R. T. (1993). Serines and threonines in the gastrin-releasing peptide receptor carboxyl terminus mediate internalization. J. Biol. Chem. 268(27), 20285–20290.

Benya, R. V., Akeson, M., Mrozinski, J., Jensen, R. T., & Battey, J. F. (1994a). Internalization of the gastrin-releasing peptide receptor is mediated by both phospholipase C-dependent and -independent processes. Mol. Pharmacol. 46(3), 495–501.

Benya, R. V., Fathi, Z., Kusui, T., Pradhan, T., Battey, J. F., & Jensen, R. T. (1994b). Gastrin-releasing peptide receptor-induced internalizaton, downregulation, desensitization, and growth: Possible role for cyclic AMP. Mol. Pharmacol. 46, 235–245.

Benya, R. V., Kusui, T., Shikado, F., Battey, J. F., & Jensen, R. T. (1994c). Desensitization of neuromedin B receptors (NMB-R) on native and NMB-R-transfected cells involves down-regulation and internalization. J. Biol. Chem. 269(16), 11721–11728.

Benya, R. V., Kusji, T., Pradhan, T., Battey, J. F., & Jensen, R. T. (1995a). Expression and characterization of cloned human bombesin receptors. Mol. Pharmacol. 47, 10–20.

Benya, R. V., Kusui, T., Battey, J. F., & Jensen, R. T. (1995b). Chronic desensitization and downregulation of the gastrin-releasing peptide receptor is mediated by a protein kinase C-dependent mechanism. J. Biol. Chem. 270, 3346–3352.

Bishop, J. F., Moody, T. W., & O'Donohue, T. L. (1986). Peptide neurotransmitters of primary sensory neuron: Similar actions of tachykinins and bombesin-like peptides. Peptides 7, 835–842.

Bloom, S. R., Christofides, N. D., Delamarter, J., Buell, G., Kawashima, E., & Polak, J. M. (1983). Diarrhea in vipoma patients associated with cosecretion of a second active peptide (peptide histidine isoleucine) explained by a single coding gene. Lancet 2, 1163–1165.

Bodanszky, M. (1979). Synthetic analogues in the study of hormone receptor interactions. In: Hormone Receptors in Digestion and Nutrition, 15–24 (Rosselin, G., Fromageot, P., & Bonfils S. eds.) Elsevier/North-Holland Biomedical Press, Amsterdam.

Bold, R. J., Lowry, P. S., Ishizuka, J., Battey, J. F., Townsend, C. M., Jr., & Thompson, J. C. (1994). Bombesin stimulates the *in vitro* growth of a human gastric cancer cell line. J. Cell Physiol. 161, 519–525.

Brenneman, D. E. & Eiden, L.E. (1986). Vasoactive intestinal peptide and electrical activity influence neuronal survival. Proc. Natl. Acad. Sci. USA 83, 1159–1162.

Brown, K. D., Laurie, M. S., Littlewood, C. J., Blakeley, D. M., & Corps, A. N. (1988). Characterization of the high-affinity receptors on Swiss 3T3 cells which mediate the binding, internalization and degradation of the mitogenic peptide bombesin. Biochem. J. 252, 227–235.

Bunnett, N. W., Kobayashi, R., Orloff, M. S., Reeve, J. R., Turner, A. J., & Walsh, J. H. (1985). Catabolism of gastrin releasing peptide and substance P by gastric membrane-bound peptidases. Peptides 6, 277–283.

Cai, R. Z., Radulovic, S., Pinski, J., Nagy, A., Redding, T. W., Olsen, D. B., & Schally, A. V. (1992). Pseudononapeptide bombesin antagonists containing C-terminal Trp or Tpi. Peptides 13, 267–271.

Camble, R., Cotton, R., Dutta, A. S., Garner, A., Hayward, C. F., Moore, V. E., & Scholes, P. B. (1989). N-Isobutyryl-His-Trp-Ala-Val-D-Ala-His-Leu-NHMe (ICI 216140), a potent *in vivo* antagonist analogue of bombesin/gastrin releasing peptide (BN/GRP) derived from the C-terminal sequence lacking the final methionine residue. Life Sci. 45, 1521–1527.

Cardona, C., Rabbitts, P. H., Spindel, E. R., Ghatei, M. A., Bleehen, N. M., Bloom, S. R., & Reeve, J. G. (1991). Production of neuromedin B and neuromedin B gene expression in human lung tumor cell lines. Cancer Res. 5205–5211.

Carney, D. N., Cuttitta, F., Moody, T. W., & Minna, J. D. (1987). Selective stimulation of small cell lung cancer clonal growth by bombesin and gastrin-releasing peptide. Cancer Res. 47, 821–825.

Chastre, E., Di Gioia, Y., Djelloul, S., Empereur, S., Louvet, C., Coudray, A. M., Forgue–Lafitte M. E., Bara, J., & Gespach, C. (1994). Pharmacological characterization of VIP/PACAP type II receptors in normal and immortalized rat fetal intestinal epithelial cells. World Scientific, Singapore 568–576.

Christophe, J., Svoboda, M., Waelbroeck, M., Winand, J., & Robberecht, P. (1988). Vasoactive intestinal peptide receptors in pancreas and liver. Structure-function relationship. Ann. NY Acad. Sci. 527, 238–256.

Cirillo, D. M., Gaudino, G., Naldini, L., & Comoglio, P.M. (1986). Receptor for bombesin with associated tyrosine kinase activity. Mol. Cellular Biol. 6, 4641–4649.

Cook, J. S., Wolsing, D. H., Lameh, J., Olson, C. A., Correa, P. E., Sadee, W., Blumenthal, E. M., & Rosenbaum, J. S. (1992). Characterization of the RDC1 gene which encodes the homolog of a proposed human VIP receptor. Expression does not correlate with an increase in VIP binding sites. FEBS Lett. 300, 149–152.

Corjay, M. H., Dobrzanski, D. J., Way, J. M., Viallet, J., Shapira, H., Worland, P., Sausville, E. A., & Battey, J. F. (1991). Two distinct bombesin receptor subtypes are expressed and functional in human lung carcinoma cells. J. Biol. Chem. 266, 18771–18779.

Couvineau, A., Rouyer–Fessard, C., Darmoul, D., Maoret, J. J., Carrero, I., Ogier–Denis, E., & Laburthe, M. (1994). Human intestinal VIP receptor: Cloning and functional expression of two CDNA encoding proteins with different N-terminal domains. Biochem. Biophys. Res. Commun. 200, 769–776.

Coy, D. H., Murphy, W. A., Snejras–Diaz, J., Coy, E. J., & Lance, V. Z. (1985). Structure-activity studies on the N-terminal region of growth hormone releasing factor. J. Med. Chem. 28, 181–185.

Coy, D. H., Heinz–Erian, P., Jiang, N. Y., Sasak, Y., Taylor, J., Moreau, J. P., Wolfrey, W. T., Gardner, J. D., & Jensen, R. T. (1988). Probing peptide backbone function in bombesin: A reduced peptide bond analogue with potent and specific receptor antagonist activity. J. Biol. Chem. 263, 5056–5060.

Cuttitta, F., Carney, D. N., Mulshine, J., Moody, T. W., Fedorko, J., Fischler, A., & Minna, J. D. (1985). Bombesin-like peptides can function as autocrine growth factors in human small cell lung cancer. Nature 316, 823–825.

Davis, T. P., Crowell, S., Taylor, J., Clark, D. L., Coy, D., Staley, J., & Moody, T. W. (1992). Metabolic stability and tumor inhibition of bombesin/GRP receptor antagonists. Peptides 13, 401–407.

Delporte, C., Van Praet, A., Herchuelz, A., Winand, J., & Christophe, J. (1993). Contrasting effects of PACAP and carbachol on $[Ca^{2+}]$ i and inositol phosphates in human neuroblastoma NB-OK-1 cells. Peptides 14, 1111–1118.

Deutsch, P. J. & Sun, Y. (1992). The 38-amino acid form of pituitary adenylate cyclase-activating polypeptide stimulates dual signaling cascades in PC12 cells promotes neurite outgrowth. J. Biol. Chem. 267, 5108–5113.

DiCicco–Bloom, E. & Lu, N. (1994). VIP/PACAP regulate peripheral and central neurogenesis and survival. World Scientific, Singapore 593–594.

Dieterich, K. D., Grigoriadis, D. E., & De Souza, E. B. (1994). Corticotropin-Releasing factor receptors in human small cell lung carcinoma cells: Radioligand binding, second messenger, and northern blot analysis data. Endocrinology 135, 1551–1558.

Draoui, M., Moody, T., Fathi, Z., & Battey, J. (1993). Phorbol esters alter GRP gene expression in SCLC cells. Cell Growth Differentiation 4, 723–729.

Draoui, M., Chung, P., Park, M., Birrer, M., Jakowlew, S., & Moody, T. W. (1995). Bombesin stimulates c-fos and c-jun mRNAs in small cell lung cancer cells. Peptides, in press.

Eden, P. A. & Taylor, J. E. (1993). Somatostatin receptor subtype gene expression in human and rodent tumors. Life Sci. 53, 85–90.

Endo, T., Fukue, H., Kanaya, M., Mizunuma, M., Fujii, M., Yamamoto, H., Tanaka, S., & Hashimoto, M. (1991). Bombesin and bradykinin increase inositol phosphates and cytosolic free Ca^{2+}, and stimulate DNA synthesis in human endometrial stromal cells. J. Endocrinol. 131(2), 313–318.

Erisman, M. R., Linnoila, O., Hernandez, R., DiAugustine, R., & Lazarus, L. (1982). Human lung small cell carcinoma contains bombesin. Proc. Natl. Acad. Sci. USA 79, 2379–2384.

Fahrenkrug, J. (1992). Glycine-extended processing intermediate of proVIP: A new bioactive form of VIP in the rat. Biomed. Res. 13 Sup. 2, 19–23.

Falconieri, E. G., Severini, C., Erspamer, V., Melchiorri, P., Delle Fave, G., & Nakajima, T. (1988). Parallel bioassay of 27 bombesin-like peptides on 9 smooth muscle preparations. Structure-activity relationships and bombesin receptor subtypes. Regul. Pept. 21, 1–11.

Farre, A., Ishizuka, J., Gomez, G., Parekh, D., Koo, J. Y., Townsend, C. M., Jr., & Thompson, J. C. (1993). Bombesin stimulates growth of colon cancer in mice and decreases their survival. Surg. Oncol. 2, 169–173.

Fatatis, A., Holtzclaw, L. A., Avidor, R., Brenneman, D. E., & Russell, J.T. (1994). Vasoactive intestinal peptide increases intracellular calcium in astroglia: Synergism with α-adrenergic receptors. Proc. Natl. Acad. Sci. USA 91, 2036–2040.

Fathi, Z., Benya, R. V., Shapira, H., Jensen, R. T., & Battey, J. F. (1993a). The fifth transmembrane segment of the neuromedin B receptor is critical for high affinity neuromedin B binding. J. Biol. Chem. 268(20), 14622–14626.

Fathi, Z., Corjay, M. H., Shapira, H., Wada, E., Benya, R., Jensen, R., Viallet, J., Sausville, E. A., & Battey, J. F. (1993b). BRS-3: A novel bombesin receptor subtype selectively expressed in testis and lung carcinoma cells. J. Biol. Chem. 268, 5979–5984.

Feldman, R. I., Wu, J. M., Jenson, J. C., & Mann, E. (1990). Purification and characterization of the bombesin/gastrin-releasing peptide receptor from Swiss 3T3 cells. J. Biol. Chem. 265, 17364–17372.

Fishbein, J. A., Coy, D. H., Hocart, S. J., Jiang, N. Y., Mrozinski, J. E., Jr., Mantey, S. A., & Jensen, R. T. (1994). A chimeric VIP-PACAP analogue but not VIP pseudopeptides function as VIP receptor antagonists. Peptides 15, 95–100.

Frankel, A. & Viallet, J. (1993). Protein kinase C is not involved in the homologous desensitization of bombesin responses in small cell lung cancer cells. Exp. Cell Res. 209, 398–401.

Gaylinn, B. D., Harrison, J. K., Zysk, J. R., Lyons, C. E., Lynch, K. R., & Thorner, M. O. (1993). Molecular cloning and expression of a human anterior receptor for growth hormone-releasing hormone. Mol. Endocrinol. 7, 77–84.

Giaccone, G., Battey, J., Gazdar, A. F., Oie, H., Draoui, M., & Moody, T. W. (1991). Neuromedin B is present in lung cancer cell lines. Cancer Res. 52, 2732S–2736S.

Gorbulev, V., Akhundova, A., Buchner, H., & Fahrenholz, F. (1992). Molecular cloning of a new bombesin receptor subtype expressed in uterus during pregnancy. Eur. J. Biochem. 208(2), 405–410.

Gozes, I., Bodner, M., Shwartz, H., Shani, Y., & Fridkin, M. (1984). Studies toward the biosynthesis of vasoactive intestinal peptide (VIP). Peptides 5, 161–166.

Gozes, Y., Brenneman, D. E., Fridkin, M., Asofaky, R., & Gozes, I. (1991a). A VIP antagonist distinguishes VIP receptors on spinal cord cells and lymphocytes. Brain Res. 540, 319–321.

Gozes, I., McCune, S. K., Jacobson, L., Warren, D., Moody, T. W., Fridkin, M., & Brenneman, D. E. (1991b). An antagonist to vasoactive intestinal peptide affects cellular functions in the central nervous system. J. Pharm. Exp. Ther. 257, 959–966.

Gozes, I., Fridkin, M., Westphal, H., Glowa, J., Hill, J. M., Bardea, A., Reshef, A., Zhukovsky, S., Waner, T., Niska, A., Rubinrout, S., Lilling, G., Davidson, A., Glazer–Steiner, R., Moody, T. W., Rostene, W., & Brenneman, D. E. (1994). Neuronal VIP: From gene to sexual behavior, memory and clinical application. World Scientific, Singapore 314–324.

Hashimoto, H., Ishihara, T., Shigemoto, R., Mori, K., & Nagata, S. (1993). Molecular cloning and tissue distribution of a receptor for pituitary adenylate cyclase-activating polypeptide. Neuron 11, 333–342.

Heimbrook, D. C., Boyer, M. E., Garsky, V. M., Balishin, N. L., Kiefer, D. L., Oliff, A., & Riemen, N. W. (1988). Minimal ligand analysis of gastrin-releasing peptide. J. Biol. Chem. 263, 7016–7019.

Heimbrook, D., Saari, W. S., Balishin, N. L., Friedman, A., Moore, K. S., Riemen, M. W., Kiefer, D. M., Rotberg, N. S., Wallen, J. W., & Oliff, A. (1989). Carboxyl-terminal modification of a gastrin releasing peptide derivative generates potent antagonists. J. Biol. Chem. 2640, 11258–11262

Heinz–Erian, P., Coy, D. H., Tamura, M., Jones, S. W., Gardner, J. D., & Jensen, R. T. (1987). (D-Phe^{12}) bombesin analogues: A new class of bombesin receptor antagonists. Am. J. Physiol. 254, G883–G890.

Hill, J. M., Gozes, I., Hill, J. L., Fridkin, M., & Brenneman, D. E. (1991). Vasoactive intestinal peptide antagonist retards the development of neonatal behaviors in the rat. Peptides 12, 187–192.

Hoshino, M., Li, M., Zheng, L. Q., Suzuki, M., Iguchi, K., Yanaihara, C., Mochizuki, T., & Yanihara, N. (1994). Effects of PACAP38, PACAP27 and VIP on human neuroblastoma NB-OK-1 cells. World Scientific, Singapore 577–586.

Hosoya, M., Onda, H., Ogi, K., Masuda, Y., Miyamo, Y., Ohtaki, T., Okazaki, H., Arimura, A., & Fujino, M. (1993). Molecular cloning and functional expression of rat cDNAs encoding the receptor for pituitary adenylate cyclase activating polypeptide (PACAP). Biochem. Biophys. Res. Commun. 194, 133–143.

Huang, N. N., Wang, D. J., & Heppel, L. A. (1994). Role of adenosine 3′, 5′-monophosphate-dependent protein kinase and cAMP levels in ATP-dependent mitogenesis in Swiss 3T3 cells. J. Biol. Chem. 269(1), 548–555.

Inagaki, N., Yoshida, H., Mizuta, M., Mizuno, N., Fujii, Y., Gonoi, T., Miyazaki, J., & Seino, S. (1994). Cloning and functional characterization of a third pituitary adenylate cyclase-activating polypeptide receptor subtype expressed in insulin-secreting cells. Proc. Natl. Acad. Sci. USA 91, 2679–2683.

Ishihara, T., Nakamura, S., Kaziro, Y., Takahashi, T., Takahashi, K., & Nagata, S. (1991). Molecular cloning and expression of a cDNA encoding the secretin receptor. EMBO J. 10, 1635–1641.

Ishihara, T., Shigemoto, R., Mori, K., Takahashi, K., & Nagata, S. (1992). Functional expression and tissue distribution of a novel receptor for vasoactive intestinal polypeptide. Neuron 8, 811–819.

Itoh, N., Obata, K., Yanaihara, N., & Okamoto, H. (1983). Human propro-vasoactive intestinal polypeptide (VIP) mRNA contains the coding sequence for a novel PHI-27-like peptide. Nature 304, 547–549.

Jensen, R. T., Moody, T. W., Pert, C. B., Rivier, J. E., & Gardner, J. D. (1978). Interaction of bombesin and litorin with specific membrane receptors on pancreatic acinar cells. Proc. Natl. Acad. Sci. USA 75, 6139–6143.

Jensen, R. T., Jones, S. W., Folkers, K., & Gardner, J. D. (1984). A synthetic peptide that is a bombesin receptor antagonist. Nature 309, 61–63.

Jensen, R. T., Heinz–Erian, P., Mantey, S., Jones, S. W., & Gardner, J. D. (1988). Characterization of ability of various substance P antagonists to inhibit action of bombesin. Am. J. Physiol. 254, G883–G890.

Jensen, R. T., Mrozinski, J. E., Jr., & Coy, D. H. (1993). Bombesin receptor antagonists: Different classes and cellular basis of action. Cancer Res. 129, 87–113.

Jensen, R. T. (1994). Receptors on pancreatic acinar cells. In: Physiology of the Gastrointestinal Tract, 3rd ed. (Johnson, L. R., Christensen, J., Jacobsen, E. D., Alpers D. H., & Walsh, J. H., eds.) Chapter 38, Vol. 2, pp. 1377–1446, Raven Press Publishing Co., New York.

Journot, L., Spengler, D., Pantaloni, C., Dumuis, A., Sebben, M., & Bockaert, J. (1994). The PACAP receptor: Generation by alternative splicing of functional diversity among G protein-coupled receptors in nerve cells. Cell Biol. 5, 263–272.

Kee, K. A., Finan, T. M., Korman, L. Y., & Moody, T. W. (1988). Somatostatin inhibits the secretion of bombesin-like peptides from small cell lung cancer cells. Peptides 9, 257–261

Kelley, M. J., Avis, I., Linnoila, R. I., Boland, C., Richardson, G. E., Phares, J., Laskin, W. B., Cuttitta, F., Mulshine, J. L., & Johnson, B. E. (1993). Proc. Am. Soc. Clin. Oncol. 12, 339.

Knuhtsen, S., Holst, J. J., Knigge, U., Olesen, M., & Nielsen, O. V. (1984). Radioimmunoassay, pharmacokinetics, and neuronal release of gastrin-releasing peptide in anesthetized pigs. Gastroenterology 87, 372–378.

Korman, L. Y., Carney, D. N., Citron, M. L., & Moody, T. W. (1986). Secretin/VIP stimulated secretion of bombesin-like peptides from human small cell lung cancer. Cancer Res. 46, 1214–1218.

Krane, I. M., Naylor, S. L., Helin–Davis, D., Chin, W. W., & Spindel, E. R. (1988). Molecular cloning of cDNAs encoding human bombesin-like peptide neuromedin B. J. Biol. Chem. 263, 13317–13323.

Kris, R., Hazan, R., Villines, J., Moody, T., & Schlessinger, J. (1987). Identification of the bombesin receptor on murine and human cells by cross-linking experiments. J. Biol. Chem. 262, 11215–11220.

Kull, F. C., Leban, J. J., Landavazo, A., Stewart, K. D., Stockstill, B., & McDermed, J. D. (1992). Conveyance of partial agonism antagonism to bombesin gastrin-releasing peptide analogues on Swiss 3T3 Cells by a carboxyl-terminal leucine insertion. J. Biol. Chem. 267, 21132–21138.

Kusui, T., Benya, R. V., Battey, J. F., & Jensen, R. T. (1994). Glycosylation of bombesin receptors: Characterization, effect on binding, and G-protein coupling. Biochemistry 33(44), 12968–12980.

Ladenheim, E. E., Jensen, R. T., Mantey, S. A., McHugh, P. R., & Moran, T. H. (1990). Receptor heterogeneity for bombesin-like peptides in the rat central nervous system. Brain Res. 537(1–2), 233–240.

Ladenheim, E. E., Jensen, R. T., Mantey, S. A., & Moran, T. H. (1992). Distinct distributions of two bombesin receptor subtypes in the rat central nervous system. Brain Res. 593(2), 168–178.

Ladenheim, E. E., Jensen, R. T., Mantey, S. A., Taylor, J. E., Coy, D. H., & Moran, T. H. (1993). Bombesin receptor antagonists differentiate receptor subtypes in rat brain. Eur. J. Pharmacol. 235(1), 121–125.

Leban, J. J., Kull, F. C., Landavazo, A., Stockstill, B., & McDermed, J. D. (1993). Development of potent gastrin releasing peptide antagonists having a D-Pro-(psi)(CH_2NH)-Phe-NH_2 C-terminus. Proc. Natl. Acad. Sci. USA 90, 1922–1926.

Lee, M., Jensen, R. T., Coy, D. H., & Moody, T. W. (1990a). Autoradiographic localization of neuromedin B binding sites in rat brain. Mol. Cell. Neurosci. 1, 161–167.

Lee, M., Jensen, R. T., Huang, S. C., Bepler, G., Korman, L., & Moody, T. W. (1990b). Vasoactive intestinal polypeptide binds with high affinity to nonsmall cell lung cancer cells and elevates cyclic AMP levels. Peptides 11, 1205–1209.

Lutz, E. M., Sheward, W. J., West, K. M., Morrow, J. A., & Harmar, A. J. (1993). The VIP_2 receptor: Molecular characterization of a cDNA encoding a novel receptor for vasoactive intestinal peptide. FEBS Lett. 334, 3–8.

Mahmet, H., Millar, J. B. A., Lehmann, W., Higgins, T., & Rozengurt, E. (1990). Bombesin stimulation of c-fos expression and mitogenesis in Swiss 3T3 cells: The role of prostaglandin E_2-mediated cyclic AMP accumulation. Expt. Cell Res. 190, 265–270.

Mahmoud, S. M., Palaszynski, E., Fiskum, G., Coy, D. N., & Moody, T. W. (1989). Small cell lung cancer bombesin receptors are antagonized by reduced peptide bond analogues. Life Sci. 44, 367–373.

Mahmoud, S., Staley, J., Taylor, J., Bogden, A., Moreau, J. P., Coy, D., Avis, I., Cuttitta, F., Mulshine, J., & Moody, T. W. (1991). ($Psi^{13,14}$)Bombesin analogues inhibit the growth of small cell lung cancer *in vitro* and *in vivo*. Cancer Res. 51, 1798–1802.

Mantey, S., Frucht, H., Coy, D. H., & Jensen, R. T. (1993). Characterization of bombesin receptors using a novel, potent, radiolabeled antagonist that distinguishes bombesin receptor subtypes. Mol. Pharmacol. 43, 762–774.

Markowska, A., Nussdorfer, G. G., & Malendowicz, L. K. (1993). Effects of bombesin and neuromedin-B on the proliferative activity of the rat adrenal cortex. Histol. Histopathol. 8, 359–362.

Mayo, K. E. (1992). Molecular cloning and expression of a pituitary-specific receptor for growth hormone-releasing hormone. Mol. Endocrinol. 6, 1734–1744.

Maruno, K., Yamaguchi, K., Abe, K., Suzuki, M., Saijo, N., Mishima, Y., Yanaihara, N., & Shimosato, Y. (1989). Immunoreactive gastrin releasing peptide as a specific tumor marker in patients with small cell lung carcinoma. Cancer Res. 49, 629–632.

Maruno, K. & Said, S. I. (1992). Vasoactive intestinal peptide inhibits the proliferation of human small cell lung carcinoma. Biomed. Res. 13, 373–375.

McDonald, T. J., Jornvall, J., Nilsson, G., Vagne, M., Ghatei, M., Bloom, S. R., & Mutt, V. (1979). Characterization of a gastrin releasing peptide from porcine nonantral gastric tissue. Biochem. Biophys. Res. Commun. 90, 227–233.

Mendoza, S. A., Schneider, J. A., Lopez–Rivas, A., Sinnet–Smith, J. W., & Rozengurt, E. (1986). Early events elicited by bombesin and structurally related peptides in quiescent Swiss 3T3 cells. II. Changes in Na^+ and Ca^{2+} fluxes, Na^+, K^+ pump activity and intracellular pH. J. Cell Biol. 102, 2223–2233.

Mervic, M., Moody, T. W., & Komoriya, A. (1991). A structure-function study of bombesin's C-terminal domain. Peptides 12, 1315–1319.

Millar, J. B. & Rozengurt, E. (1988). Bombesin enhancement of cAMP accumulation in Swiss 3T3 cells: Evidence of a dual mechanism of action. J. Cell Physiol. 137, 214–222.

Millar, J. B. & Rozengurt, E. (1990a). Chronic desensitization to bombesin by progressive downregulation of bombesin receptors in Swiss 3T3 cells. Distinction from acute desensitization. J. Biol. Chem. 265, 12052–12058.

Millar, J. B. A. & Rozengurt, E. (1990b). Arachidonic acid release by bombesin. J. Biol. Chem. 265, 19973–19979.

Minamino, N., Kangawa, K., & Matsuo, H. (1983). Neuromedin B: A novel bombesin-like peptide identified in porcine spinal cord. Biochem. Biophys. Res. Commun. 114, 541–548.

Miyamoto, Y., Habata, I., Ohtaki, T., Masuda, Y., Ogi, K., Onda, H., & Fujino, M. (1994). Cloning and expression of a complementary DNA encoding the bovine receptor for pituitary adenylate cyclase-activating polypeptide (PACAP). Biochim. Biophys. Acta 1218(3), 297–307.

Miyata, A., Arimura, A., Dahl, R. R., Minamino, N., Uehara, A., Jiang, L., Culler, M. D., & Coy, D. H. (1989). Isolation of a novel 38 residue hypothalamic polypeptide which stimulates adenylate cyclase in pituitary cells. Biochem. Biophys. Res. Commun. 164, 567–574.

Moody, T. W., Pert, C. B., Rivier, J., & Brown, M. R. (1978). Bombesin: Specific binding to rat brain membranes. Proc. Natl. Acad. Sci. USA 75, 5372–5376.

Moody, T. W., Pert, C. B., Gazdar, A. F., Carney, D. N., & Minna, J. D. (1981). High levels of intracellular bombesin characterize human small cell lung carcinoma. Science 214, 1246–1248.

Moody, T. W., Carney, D. N., Cuttitta, F., Quattrocchi, K., & Minna, J. D. (1985). High-affinity receptors for bombesin/GRP-like peptides on human small cell lung cancer. Life Sci. 37 105–113.

Moody, T. W., Murphy, A., Mahmoud, S., & Fiskum, G. (1987). Bombesin-like peptides elevate cytosolic calcium in small cell lung cancer cells. Biochem. Biophys. Res. Commun. 147, 189–195.

Moody, T. W., Mahmoud, S., Staley, J. S., Naldini, L., Cirillo, D., South, V., Felder, S., and Kris, R. (1989). Human glioblastoma cell lines have neuropeptide receptors for bombesin/GRP. J. Molecular Neurosci. 1, 235–242.

Moody, T. W., Staley, J., Zia, F., Coy, D. H., & Jensen, R. T. (1992). Neuromedin B receptors are present on small cell lung cancer cells. J. Pharmacol. Exp. Ther. 263, 311–317.

Moody, T. W., Zia, F., Goldstein, A., Naylor, P., Sarin, E., Brenneman, D., Koros, A. M. C., Reubi, J. C., Fridkin, M., & Gozes, I. (1993a). VIP analogues inhibit small cell lung cancer growth. Biomed. Res. 13, 131–136.

Moody, T. W., Zia, F., & Makheja, A. (1993b). PACAP elevates cytosolic calcium in small cell lung cancer cell lines. Peptides 14, 241–246.

Moody, T. W., Zia, F., Brenneman, D., Fridkin, M., Davidson, A., & Gozes, I. (1993c). A VIP antagonist inhibits the growth of nonsmall cell lung cancer. Proc. Natl. Acad. Sci. USA 90, 4345–4349.

Moody, T. W., Zia, F., Bitar K., & Coy, D. H. (1994a). PACAP (6–38) is a PACAP type I receptor antagonist on small lung cancer cells. World Scientific, Singapore 537–544.

Moody, T. W., Zia, F., Venugopal, R., Korman, L. Y., Goldstein, A. L., & Fagarasan, M. (1994b). Corticotropin-Releasing factor stimulates cyclic AMP, arachidonic acid release, and growth of lung cancer cells. Peptides 15, 281–285.

Moody, T. W., Venugopal, R., Zia, R., Patierno, S., Leban, J. J., & McDermed, J. (1995). BW2258U89: A GRP receptor antagonist which inhibits small cell lung cancer growth. Life Sci. 5, 521–529.

Morrow, J. A., Lutz, E. M., West, K., Fink, G., & Harmar, A. J. (1993). Molecular cloning and expression of a cDNA encoding a receptor for pituitary adenylate cyclase activating polypeptide. (PACAP). FEBS Lett. 329 (1–2), 99–105.

Nagata, S., Ishihara, T., Robberecht, P., Libert, F., Parmentier, M., Christophe, J., & Vassart, G. (1992). RDC1 may not be VIP receptor [letter]. Trends Pharmacol. Sci. 13(3), 102–103.

Oettling, G., Bruchelt, G., Lohmann, F., Teufel, M., Niethammer, D., Treuner, J., & Drews, U. (1990). Vasoactive intestinal polypeptide (VIP) induces calcium mobilization in the human neuroblastoma cell line SK-N-SH. Cancer Lett. 50(3), 203–207.

Offermanns, S., Heiler, E., Spither, K., & Schultz, G. (1994). Gq and G11 are concurrently activated by bombesin and in Swiss 3T3 cells. FEBS Lett. 349(2), 201–204.

Ogi, K., Miyamoto, Y., Masuda, Y., Habata, Y., Hosoya, M., Ohtaki, T., Masuo, Y., Onda, H., & Fujino, M. (1993). Molecular cloning and functional expression of a cDNA encoding a human pituitary adenylate cyclase activating polypeptide receptor. Biochem. Biophys. Res. Commun. 196(3), 1511–1521.

Orbuch, M., Taylor, J. E., Coy, D. H., Mrozinski, J. E., Jr., Mantey, S. A., Battey, J. F., Moreau, J. P., & Jensen, R. T. (1993). Discovery of a novel class of neuromedin B receptor antagonists, substituted, somatostatin analogues. Mol. Pharmacol. 44, 841–850.

Palumbo, A., Rossino, P., & Comoglio, P. (1986). Bombesin stimulates c-fos gene expression in Swiss 3T3 cells. Expt. Cell Res. 167, 276–280.

Pandol, S. J., Dharmsathaphorn, K., Schoeffield, M. S., Vale, W., & Rivier, J. (1986). Vasoactive intestinal peptide receptor antagonist [4Cl-D-Phe^{6}, Leu^{17}] VIP. Am. J. Physiol. 250 (4 Pt 1), G553–557.

Pang, L., Decker, S.J., & Saltier, A.R. (1993). Bombesin and epidermal growth factor stimulate the mitogen-activated protein kinase through different pathways in Swiss 3T3 cells. J. Biochem. 289, 283–287.

Pert, C. B. & Schumather, U. K. (1982). Plasma bombesin concentrations in patients with extensive small cell carcinoma of the lung. Lancet 1, 509.

Pinski, J., Halmos, G., Yano, T., Szepeshazi, K., Qin, Y., Erte, J., & Schally, A. V. (1994a). Inhibition of growth of MKN human gastric adenocarcinoma xenografts in nude mice by treatment with gastrin-releasing peptide antagonist (RC-3095) and somatostatin analogue RC-160. Int. J. Cancer 57, 54–580.

Pinski, J., Schally, A. V., Halmos, G., Szepeshazi, K., & Groot, K. (1994b). Somatostatin analogues and bombesin/gastrin-releasing peptide antagonist RC-3095 inhibit the growth of human glioblastomas *in vitro* and *in vivo*. Cancer Res. 54, 5895–5901.

Pisegna, J. R. & Wank, S. A. (1993). Molecular cloning and functional expression of the pituitary adenylate cyclase-activating polypeptide type I receptor. Proc. Natl. Acad. Sci. USA 90, 6345–6349.

Profrock, A., Zimmermann, P., & Schulz, I. (1992). Bombesin receptors interact with Gi and p21ras proteins in plasma membranes from rat pancreatic acinar cells. Am. J. Physiol. 263, G240–G247.

Qin, Y., Ertl, T., Cai, R. Z., Halmos, G., & Schally, A. V. (1994a). Inhibitory effect of bombesin receptor antagonist RC-3095 on the growth of human pancreatic cancer cells *in vivo* and *in vitro*. Cancer Res. 54(4), 1035–1041.

Qin, Y., Halmos, G., Cai, R. Z., Szoke, B., Ertl, T., & Schally, A. V. (1994b). Bombesin antagonists inhibit *in vitro* and *in vivo* growth of human gastric cancer and binding of bombesin to its receptors. J. Cancer Res. Clin. Oncol. 120, 519–528.

Radulovic, S., Miller, G., & Schally, A. V. (1991). Inhibition of growth of HT-29 human colon cancer xenografts in nude mice by treatment with bombesin/gastrin-releasing peptide antagonist (RC-3095). Cancer Res. 51, 6006–6009.

Radulovic, S., Schally, A. V., Reile, H., Halmos, G., Szepeshazi, K., Groot, K., Milovanovic, S., Miller, G., & Yano, T. (1994). Inhibitory effects of antagonists of bombesin/gastrin-releasing peptide (GRP) and somatostatin analog (RC-160) on growth of HT-29 human colon cancers in nude mice. Acta Oncol. 33, 693–701.

Rankin, S., Morii, N., Narumiya, S., & Rozengurt, E. (1994). *Botulinum* C3 exoenzyme blocks the tyrosine phosphorylation or p125FAK-and paxillin induced by bombesin and endothelin. FEBS Lett. 354, 315–319.

Rivier, J. & Brown, M. (1978). Bombesin, bombesin analogues and related peptides: Effects on thermoregulation. Biochem. 17, 1766–17710.

Rozengurt, E. (1991). A role for arachidonic acid and its metabolites in the regulation of p21 ras activity. Cancer Cells 3, 397–398.

Rozengurt, E. & Sinett–Smith, J. (1983). Bombesin stimulation of DNA synthesis and cell division in cultures of Swiss 3T3 cells. Proc. Natl. Acad. Sci. 80, 2936–2940.

Said, S. I. & Mutt, V. (1970). Polypeptide with broad biological activity: Isolation from the small intestine. Science 69, 1217–1218.

Sausville, E., Lebacq–Verheyden, A.M., Spindel, E.R., Cuttitta, F., Gazdar, A. F., & Battey, J. (1986). Expression of the gastrin-releasing peptide gene in human small cell lung cancer. J. Biol. Chem. 261, 2451–2459.

Sausville, E. A., Moyer, J. D., Heikkila, R., Neckers, L. M., & Trepel, J. B. (1988). A correlation of bombesin-responsiveness with myc-family gene expression in small cell lung carcinoma cell lines. Ann. NY Acad. Sci. 547, 310–321.

Schomerus, E., Poch, A., Bunting, R., Mason, W. T., & McArdle, C. A. (1994). Effects of pituitary adenylate cyclase-activating polypeptide in the pituitary: Activation of two signal transduction pathways in the gonadotrope-derived α T3-1 cell line. Endocrinology 134, 315–323.

Seckl, M. & Rozengurt, E. (1993). Tyrphostin inhibits bombesin stimulation of tyrosine phosphorylation, c-fos expression, and DNA synthesis in Swiss 3T3 cells. J. Biol. Chem. 268, 9548–9554.

Sethi, T., Langdon, S., Smyth, J., & Rozengurt, E. (1992). Growth of small cell lung cancer cells: Stimulation by multiple neuropeptides and inhibition by broad spectrum antagonists *in vitro* and *in vivo*. Cancer Res. 52, 2737s–2742s.

Seva, C., Dickinson, C. J., & Yamada, T. (1994). Growth-promoting effects of glycine-extended progastrin. Science 265, 410–412.

Shaffer, M. M., Carney, D. N., Korman, L. Y., Lebovic, G. S., & Moody, T. W. (1987). High affinity binding of VIP to human lung cancer cell lines. Peptides 8, 1101–1106.

Shapira, H., Way, J., Lipinsky, D., Oron, Y., & Battey, J. F. (1994). Neuromedin B receptor, expressed in *Xenopus laevis* oocytes, selectively couples to G alpha q and not G alpha 11. FEBS Lett. 348, 89–92.

Shivers, B. D., Gorcs, T. J., Gottschall, P. E., & Arimura, A. (1991). Two high affinity binding sites for pituitary adenylate cyclase-activating polypeptide have different tissue distributions. Endocrinology 128, 3055–3065.

Singh, P., Guo, Y. S., Kull, F. C., Jr., & Leban, J. J. (1992). A novel bombesin receptor antagonist (BW-10) potently inhibits bombesin-evoked release of gastrointestinal hormones from rats and dogs *in vitro* and *in vivo*. Regul. Peptides 40, 75–86.

Sinnett–Smith, J., Zachary, I., Valverde, A. M., & Rozengurt, E. (1993). Bombesin stimulation of p125 focal adhesion kinase tyrosine phosphorylation. J. Biol. Chem. 268, 14261–14268.

Slice, L. W., Wong, H. C., Sternini, C., Grady, E. F., Bunnett, N. W., & Walsh, J. H. (1994). The conserved NPXnY motif present in the gastrin-releasing peptide receptor is not a general sequestration sequence. J. Biol. Chem. 269, 21755–21761.

Sorenson, G. D., Pettengill, O. S., Cate, C. C., Ghatei, M. A., Molyneaux, K. E., Gosselin, E. J., & Bloom, S. R. (1982). Bombesin and calcitonin secretion by pulmonary carcinoma is modulated by cholinergic receptors. Life Sci. 33, 1939–1946.

Spengler, D., Waeber, C., Pantaloni, C., Holsboer, F., Bockaert, J., Seeburg, P. H., & Journot, L. (1993). Differential signal transduction by five splice variants of the PACAP receptor. Nature 365, 170–175.

Spindel, E. R., Chin, W. W., Price, J., Besser, L. H., & Habener, J. F. (1984). Cloning and characterization of cDNAs encoding human gastrin-releasing peptide. Proc. Natl. Acad. Sci. 81, 5699–5703.

Spindel, E. R., Zilberberg, M. D., & Chin, W. W. (1987). mRNAs encoding human gastrin-releasing peptide: An alternative RNA splicing occurs in neural and endocrine tissues. Mol. Endocrinol. 1, 224–232.

Spindel, E. R., Giladi, E., Brehm, T. P., Goodman, R. H., & Segerson, T. P. (1990). Cloning and functional characterization of a cDNA encoding the murine fibroblast bombesin/GRP receptor. Mol. Endocrinol. 4, 1956–1963.

Spindel, E. R., Giladi, E., Segerson, T. P., & Nagalla, S. (1993). Bombesin-like peptides: Of ligands and receptors. Recent Prog. Horm. Res. 48, 365–391.

Sreedharan, S. P., Patel, D. R., Huang, J. X., & Goetzl, E. J. (1993). Cloning and functional expression of a human neuroendocrine vasoactive intestinal peptide receptor. Biochem. Biophys. Res. Commun. 193(2), 546–553.

Sreedharan, S. P., Robichon, A., Peterson, K.E., & Goetzl, E. J. Cloning and expression of the human vasoactive intestinal peptide receptor [published erratum] Proc. Natl. Acad. Sci. USA 88, 4986–4990.

Svoboda, M., Tastenoy, M., Van Rampelbergh, J., Goossens, J. F., De Neef, P., Waelbroeck, M., & Robberecht, P. (1994). Molecular cloning and functional characterization of a human VIP receptor from SUP-T1 lymphoblasts. Biochem. Biophys. Res. Commun. 205, 1617–1624.

Svoboda, M., Tastenoy, M., Ciccarelli, E., Stievenart, M., & Christophe, J. (1993). Cloning of a splice variant of the pituitary adenylate cyclase-activating polypeptide (PACAP) type I receptor. Biochem. Biophys. Res. Commun. 195, 881–888.

Swope, S. L. & Schonbrunn, A. (1990). Desensitization of islet cells to bombesin involves both receptor downmodulation and inhibition of receptor function. Mol. Pharmacol. 37, 758–766.

Tatsuno, I., Gottschall, P. E., Koves, K., & Arimura, A. (1990). Demonstration of specific binding sites for pituitary adenylate cyclase-activating polypeptide (PACAP) in rat astrocytes. Biochem. Biophys. Res. Commun. 168, 1027–1033.

Tatsuta, M., Iishi, M., Baba, A., Nakaizumc, M., Ichi, I., & Taniguchi, H. (1989). Promotion by bombesin of gastric carcinogenesis induced by N-methyl-N'-nitro-N-nitrosoguanidine in Wistar rats. Cancer Res. 49, 5254–5257.

Taylor, J. E., Moreau, J. P., Baptiste, L., & Moody, T. W. (1991). Octapeptide analogues of somatostatin inhibit the clonal growth and vasoactive intestinal peptide stimulated cyclic AMP formation in human small cell lung cancer cells. Peptides 12, 839–844.

Thomas, F., Arvelo, F., Antoine, E., Jacrot, M., & Poupon, M.F. (1992). Antitumoral activity of bombesin analogues on small cell lung cancer xenografts: Relationship with bombesin receptor expression. Cancer Res. 52, 4872–4877.

Thorens, B., Porret, A., Buhler, L., Deng, S. P., Morel, P., & Widmann, C. (1993). Cloning and functional expression of the human islet GLP-1 receptor. Demonstration that exendin-4 is an agonist and exendin-(9-39) an antagonist of the receptor. Diabetes 42, 1678–1682.

Thorens, B. (1992). Expression cloning of the pancreatic β-cell receptor for the gluco-incretin hormone glucagon-like peptide 1. Proc. Natl. Acad. Sci. USA 89, 8641–8645.

Trepel, J. B., Moyer, J. D., Cuttitta, F., Frucht, H., Coy, D. H., Natale, R. B., Mulshine, J. L., Jensen, R. T., & Sausville, E. A. (1988). A novel bombesin receptor antagonist inhibits autocrine signals in a small cell lung carcinoma cell line. Biochem. Biophys. Res. Commun. 156, 1383–1389.

Usdin, T. B., Borner, T. I., & Mezey, E. (1994). Two receptors for vasoactive intestinal polypeptide with similar specificity and complementary distributions. Endocrinology 135, 2662–2680.

Van Lint, J., Agostinis, P., Merlevede, W., & Vandenheede, J. R. (1993). Early responses in mitogenic signaling, bombesin induced protein phosphorylations in Swiss 3T3 cells. Adv. Enzyme Regul. 33, 143–155.

Von Schrenck, T., Heinz–Erian, P., Moran, T., Mantey, S. A., Gardner, J. D., & Jensen, R. T. (1989). Characterization of a neuromedin B-preferring receptor in esophagus muscle: Evidence for subtypes of bombesin receptors. Am. J. Physiol. 256, G747–G758.

Von Schrenck, T., Wang, L. H., Coy, D. H., Villanueva, M. L., Mantey, S., & Jensen, R. T. (1990). Potent bombesin receptor antagonists distinguish receptor subtypes. Am. J. Physiol. 256, G747–G758.

Wada, E., Way, J., Shapira, H., Kusano, K., Lebacq–Verheyden, A.M., Coy, D., Jensen, R., & Battey, J. (1991). cDNA Cloning, characterization and brain region-specific expression of a neuromedin-B preferring bombesin receptor. Neuron 6, 421–430.

Waelbroeck, M., Robberecht, P., Coy, D. H., Camus, J. C., De Neef, P., & Christophe, J. (1985). Interaction of growth hormone-releasing factor (GRF) and 14 GRF analogs with vasoactive intestinal peptide (VIP) receptors of rat pancreas. Discovery of (N-Ac-Tyr^1, D-Phe^2)-GRF (1-29)-NH_2 as a VIP antagonist. Endocrinology 116, 2643–2649.

Wahl, M. A., Straub, S. G., & Ammon, H. P. (1993). Vasoactive intestinal polypeptide-augmented insulin release: Actions on ionic fluxes and electrical activity of mouse islets. Diabetologia 36, 920–925.

Walsh, J. H., Bouzyk, M., & Rozengurt, E. (1993). Homologous desensitization of bombesin-induced increases in intracellular Ca^{2+} in quiescent Swiss 3T3 cells involves a protein kinase C independent mechanism. Cell Physiol. 156, 333–340.

Wang, L. H., Coy, D. H., Taylor, J. E., Jiang, N. Y., Kim, S. H., Moreau, J. P., Huang, S. C., Mantey, S. A., Frucht, H., Gardner, J. D., & Jensen, R. T. (1990a). Des-met alkylamide bombesin analogues: A new class of bombesin receptor antagonists with potent antisecretory activity in pancreatic acini and antimitotic activity in Swiss 3T3 cells. Biochem. 29, 616–622.

Wang, L. H., Coy, D. H., Taylor, J. E., Jiang, N. Y., Moreau, J. P., Huang, S. C., Frucht, H., Haffar, B. M., & Jensen, R. T. (1990b). Des-met carboxyl terminally modified analogues of bombesin function as potent bombesin receptor antagonists, partial agonists or agonists. J. Biol. Chem. 265, 15695–15703.

Wang, L. H., Battey, J. F., Wada, E., Lin, J. T., Mantey, S., Coy, D. H., & Jensen, R. T. (1992). Activation of neuromedin B-preferring bombesin receptors on rat gliablastoma C-6 cells increases cellular Ca^{2+} and phosphoinositides. Biochem. J. 286, 641–648.

Wang, L. H., Mantey, S. A., Lin, J. T., Frucht, H., & Jensen, R. T. (1993). Ligand binding, internalization, degradation and regulation by guanine nucleotides of bombesin receptor subtypes: A comparative study. Biochem. Biophys. Acta 1175, 232–242.

Waschek, J. A., Hsu, C.–M., & Eiden, L. E. (1988). Lineage-specific regulation of the vasoactive intestinal peptide gene in neuroblastoma cells is conferred by 5.2 kilobases of 5′ -flanking sequence. Proc. Natl. Acad. Sci. USA 85, 9547–9551.

Westendorf, J. & Schonbrun, A. (1983). Characterization of bombesin receptors in a rat pituitary cell line. J. Biol. Chem. 258, 7527–7535.

Willey, J. C., Lechner, J. F., & Harris, C. C. (1984). Bombesin and the C-terminal tetradecapeptide of gastrin-releasing peptide are growth factors for normal human bronchial epithelial cells. Exp. Cell Res. 153, 245–248.

Wood, S., Wood, J., Ghatei, M., Lee, U., O'Shaghnessy, D., & Bloom, S. R. (1981). Bombesin, somatostatin and neurotensin-like immunoreactivity in bronchial carcinoma. J. Clin. Endocrinol. 53, 1310–1312.

Wood, S. M., Wood, J. R., Ghatei, M. A., Sorenson, G. D., & Bloom, S. R. (1982). Is bombesin a tumor marker for small cell carcinoma? Lancet 1, 690.

Yano, T., Pinski, J., Groot, K., & Schally, A. V., (1992). Stimulation by bombesin and inhibition by bombesin/gastrin-releasing peptide antagonist RC-3095 of growth of human breast cancer cell lines. Cancer Res. 52, 4545–4547.

Yano, T., Pinski, J., Szepeshaz, K., Halmos, G., Radulovic, S., Groot, K., & Schally, A. V. (1994). Inhibitory effect of bombesin/gastrin-releasing peptide antagonist RC-3095 and luteinizing hormone-releasing hormone antagonist SB-75 on the growth of MCF-7 MIII human breast xenografts in athymic nude mice. Cancer 73, 1229–1238.

Zachary, I. & Rozengurt, E. (1985). High-affinity receptors for peptides of the bombesin family in Swiss 3T3 cells. Proc. Natl. Acad. Sci. USA 82, 7616–7620.

Zachary, I., Sinnett–Smith, J. W., & Rozengurt, E. (1986). Early events elicited by bombesin and structurally related peptides in quiescent swiss 3T3 cells. Activation of protein kinase C and inhibition of epidermal growth factor binding. J. Cell Biol. 102, 2211–2222.

Zachary, I. & Rozengurt, E. (1987a). Identification of a receptor for peptides of the bombesin family in Swiss 3T3 cells by affinity cross-linking. J. Biol. Chem. 262, 3947–3950.

Zachary, I. & Rozengurt, E. (1987b). Internalization and degradation of peptides of the bombesin family in Swiss 3T3 cells occurs without ligand-induced receptor downregulation. EMBO J. 6, 2233–2239.

Zachary, I., Millar, J., Nonberg, E., Higgins, T., & Rozengurt, E. (1987). Inhibition of bombesin-induced mitogenesis by pertussis toxin: Dissociation from phospholipase C pathway. Biochem. Biophys. Res. Commun. 146, 456–463.

Zhu, W. Y., Goke, B., & Williams, J. A. (1991). Binding, internalization, and processing of bombesin by rat pancreatic acini. Am. J. Physiol. 261, G57–64.

ROLE OF SPHINGOLIPID METABOLITES IN THE ACTIONS OF GROWTH FACTORS AND CYTOKINES

Sarah Spiegel and Sheldon Milstien

	Abstract	538
I.	Introduction	539
II.	Sphingolipid Metabolism	540
III.	Ceramide, a Signal for Arresting Cell Growth	541
IV.	The Role of Ceramide in TNFα and Interleukin-1 Signal Transduction	543
V.	Downstream Signaling of Ceramide	544
VI.	Sphingosine and Sphingosine-1-phosphate, Positive Regulators of Cell Growth	545
VII.	Downstream Signaling of Sphingosine and Sphingosine-1-phosphate	547
	A. Protein Kinase C and Other Kinases	547
	B. Phosphatidic Acid	547
	C. Calcium and Inositol Trisphosphate	550
	D. cAMP	551
VIII.	Growth Factors Modulate Sphingosine and Sphingosine 1-phosphate Levels	551

Growth Factors and Cytokines in Health and Disease
Volume 1B, pages 537–563.

ISBN: 0-7623-0091-4

IX. Transcription Factors: Sphingosine and Sphingosine-1-phosphate Stimulate DNA Binding Activity of AP-1 553
X. Sphingosinephosphorylcholine Is a Remarkably Potent Mitogen 554
A. Sphingosinephosphorylcholine Signaling Pathways 555
B. Does Endogenous Sphingosylphosphorylcholine Regulate Mitogenesis? . 555
XI. Concluding Remarks 556
Abbreviations 556
Acknowledgments 557
References 557

ABSTRACT

There is convincing evidence that breakdown products of membrane glycerophospholipids are involved in transmitting information from external stimuli to internal cellular machinery. Less is known of the second messenger roles of the metabolites derived from another major class of membrane lipids—sphingolipids. Analogous to a glycerophospholipid cycle providing intermediates with second-messenger functions, it has been suggested that a sphingolipid metabolic cycle also generates its own intermediates which have important biological functions. This review summarizes current knowledge regarding the role of sphingolipid metabolites in the actions of growth factors and cytokines, and, especially, the second-messenger roles of ceramide and sphingosine and its metabolite, sphingosine-1-phosphate, in regulating cellular processes and, also, discusses possible mechanisms underlying their action. Recent evidence has suggested that branching pathways of sphingolipid metabolism may mediate either mitogenic or apoptotic effects depending on the cell type and the nature of the stimulus. Ceramide has been reported to induce apoptosis and cell arrest in several mammalian cell lines whereas sphingosine and sphingosine-1-phosphate generally provide positive mitogenic signals.

Several cytokines, including TNFα, vitamin D_3, γ-interferon, and IL-1 stimulate the activity of sphingomyelinase which catalyzes sphingomyelin degradation resulting in the formation of ceramide. Ceramide, in turn, has been shown to play previously unrecognized roles in both cellular differentiation and apoptosis. The cellular targets mediating the action of ceramide are not yet known, but the most likely candidates appear to be a ceramide-stimulated protein kinase and a ceramide-stimulated protein phosphatase.

A tantalizing link between sphingolipids and cellular proliferation has emerged from the discovery that sphingosine and sphingosine-1-phosphate stimulate growth of quiescent Swiss 3T3 fibroblasts via a protein kinase C-independent pathway. We proposed that sphingosine and SPP have many appropriate properties that make them suitable candidates to function as intracellular second messengers: they elicit diverse cellular responses; their turnover is extremely rapid; their levels in cells are very low and increase rapidly and transiently in response to growth factors; they can release calcium from internal sources in a previously unknown, inositol trisphosphate-independent manner; finally, they could also act in a positive feedback loop to amplify the cascade of events following receptor stimulation via their effect on phosphatidic acid levels, which might link growth signaling to cellular Ras activity. Sphingosine-

and sphingosine-1-phosphate-induced activation of the transcription factor, activator protein-1 (AP-1), one of the transcription factors activated by the Ras-MAPK pathway, is the first link between the effects of sphingolipid metabolites on cellular proliferation and gene expression. AP-1 may be a convergence point in the nucleus where signals carried by sphingolipid-derived messengers are integrated with the well-known intracellular second messengers derived from glycerophospholipids, for example, diacylglycerol.

I. INTRODUCTION

Lipid second messengers derived from glycerophospholipids, such as diacylglycerol, inositol 1,4,5-trisphosphate ($InsP_3$), phosphatidic acid, and arachidonic acid, participate in signaling pathways mediated by many growth factors and cytokines (Berridge, 1993; Exton, 1990; Irvine, 1982; Nishizuka, 1992). Recently, the importance of another class of lipid second messengers derived from membrane sphingolipids has begun to be appreciated. Sphingolipids are all composed of a long-chain sphingoid base backbone of which sphingosine is the most prominent, an amide-linked fatty acid of varying chain length, and various polar head groups (hydroxyl for ceramide, phosphorylcholine for sphingomyelin, or carbohydrate residues for glycosphingolipids) (Figure 1). Sphingolipids and, in particular, glycosphingolipids (for example, cerebrosides, gangliosides, and sulfatides), have much greater structural diversity than glycerophospholipids. It has long been known that these membrane lipids are important for cell growth and differentiation, oncogenesis, and cell–cell contact (reviewed in (Hakomori, 1990; Olivera and Spiegel, 1992). Furthermore, sphingolipids have been shown to be cell-surface receptors for some viruses and bacterial toxins. Recently, it has been suggested that they can modulate the functions of certain growth factors (Hakomori, 1990). However, it is now clear that not only do the more complex species have biological activities, but their metabolic products, such as ceramide (Hannun, 1994; Heller and Kronke, 1994; Kolesnick and Golde, 1994), sphingosine (Hannun and Bell, 1989; Merrill, 1991; Spiegel et al., 1993; Spiegel et al., 1994; Zhang et al., 1990a; Zhang et al., 1990b), sphingosine-1-phosphate (SPP) (Mattie et al., 1994; Olivera and Spiegel, 1993; Spiegel et al., 1994; Zhang et al., 1991), lysosphingomyelin or sphingosylphosphorylcholine (SPC) (Desai et al., 1993; Desai and Spiegel, 1991), and di- and trimethylsphingosine (Hakomori and Igarashi, 1993; Igarashi et al., 1989; Igarashi et al., 1990), also have important activities. Analogous to the glycerophospholipid cycle providing intermediates with second-messenger functions, the sphingolipid metabolic cycle also generates its own intermediates with important biological functions. This review summarizes current knowledge regarding the role of sphingolipid metabolites in the actions of growth factors and cytokines, and, especially, the second-messenger roles of ceramide, and sphingosine and its metabolites, SPP and SPC, in regulating cellular processes.

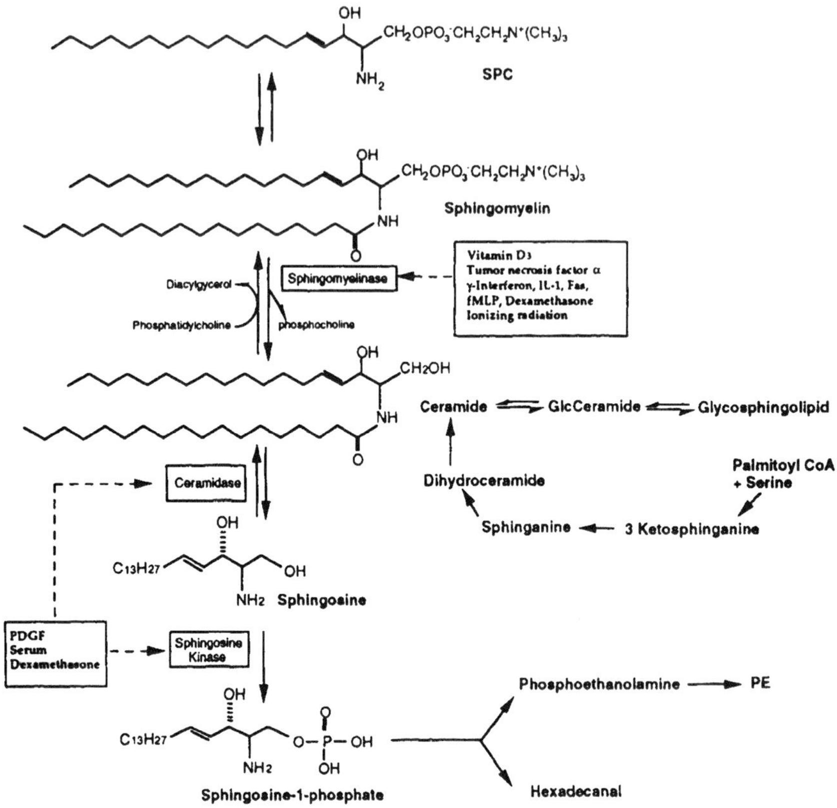

Figure 1. Sphingolipid metabolic pathways leading to the formation of sphingosine, SPP, and ceramide, putative intracellular second messengers. The influences of growth factors and cytokines are indicated. For more detailed information, see text and Table 1. (Modified with permission from Spiegel et al., 1994).

II. SPHINGOLIPID METABOLISM

Characterization of the molecular defects in human sphingolipid storage diseases has led to the elucidation of the metabolic pathways for sphingolipid synthesis and degradation. The de novo sphingolipid biosynthetic pathway begins with the condensation of serine with palmitoyl-CoA to form 3-ketodihydrosphingosine (3-ketosphinganine), a reaction catalyzed by serine palmitoyl transferase. 3-Keto-dihydrosphingosine is then reduced by an NADPH-dependent reductase to dihydrosphingosine (sphinganine) followed by acylation with fatty acyl-CoA to dihydroceramide (see Figure 1). In an FAD-dependent reaction, dihydroceramide is converted directly to ceramide, the precursor of all sphingolipids (Merrill and Wang, 1992; Rother et al., 1992). Addition of various polar head groups to ceramide yields the complex sphingolipids and glycosphingolipids. The cellular localization and topology of the enzymes that are involved in sphingolipid biosynthesis have been extensively investigated by Pagano and colleagues using fluorescent analogues and subcellular fractionation (Rosenwald and Pagano, 1993). However, only serine palmitoyltransferase, the first enzyme in this pathway, has been cloned and sequenced from yeast (Pinto et al., 1992; Zhao et al., 1994).

Metabolism of sphingolipids requires sequential removal of the saccharide groups by exoglyosidases or endoglycosidases and the hydrolysis of phosphorylcholine from sphingomyelin, catalyzed by at least three distinct sphingomyelinases, to yield ceramide. The best characterized sphingomyelinase is an acidic lysosomal phospholipase which is deficient in Nieman–Pick disease (Spence, 1993). Cleavage of the amide-linked fatty acid from ceramide to form sphingosine is catalyzed by ceramidase. Three distinct ceramidases have been identified which differ in their pH optima (reviewed in Merrill and Jones, 1990). Sphingosine can either be converted back to ceramide by reacylation or metabolized further via a pathway involving sphingosine kinase-catalyzed phosphorylation to SPP (Stoffel et al., 1973; Stoffel et al., 1968). SPP is cleaved by a specific pyridoxal phosphate-dependent lyase to ethanolamine phosphate (required for the synthesis of phosphatidyl ethanolamine) and *trans*-2-hexadecanal (Stoffel and Assmann, 1970; Van Veldhoven and Mannaerts, 1991). This lyase has recently been shown to be located in the endoplasmic reticulum (Van Veldhoven and Mannaerts, 1991). The phosphate group on SPP can also be removed by a specific phosphatase (Van Veldhoven and Mannaerts, 1994). Hakomori and colleagues have suggested that sphingosine may also be metabolized by methylation to form di- and trimethyl sphingosine (Hakomori and Igarashi, 1993; Igarashi et al., 1989; Igarashi et al., 1990). The metabolism of sphingolipids resembles other metabolic cycles which produce intracellular messengers, and the intermediates of sphingolipid turnover may function in previously unappreciated modes to regulate cellular processes.

III. CERAMIDE, A SIGNAL FOR ARRESTING CELL GROWTH

Much attention has been focused recently on the sphingomyelin metabolite, ceramide, which, it has been proposed, is pivotal in a variety of differentiation processes (Hannun, 1994; Heller and Kronke, 1994; Kolesnick and Golde, 1994). Increased turnover of sphingomyelin has been demonstrated in human polymorphonuclear leukocytes (Nelson et al., 1982) and in rat epididymal fat cell ghosts in response to corticosteroids (Murray et al., 1979). Differentiation of 3T3-L1 preadipocyte cells in response to dexamethasone is accompanied by activation of membrane-bound neutral sphingomyelinase (Ramachandran et al., 1990), indicating that sphingomyelin turnover might be involved in the action of corticosteroid hormones (Nelson et al., 1982; Ramachandran et al., 1990). Furthermore, other studies convincingly show an important role for sphingomyelin hydrolysis in monocytic differentiation of human promyelocytic leukemia HL60 cells induced

Table 1. Increases of Cellular Levels of Sphingolipids Metabolites Modulated by Growth Factors and Cytokines

Stimuli	*Sphingolipid*	*Cell Type*
1α,25-dihydroxyvitamin D_3	ceramide	HL-60
TNFα	ceramide	HL-60, U937, Jurkat
TNFα	ceramide/sphingosine	HL-60
γ-interferon	ceramide	HL-60
IL-1 β	ceramide	EL4 thymoma
dexamethasone	ceramide/sphingosine	T3-L1 preadipocyte fibroblasts
complement	ceramide	JY25 B
anti CD95 (Fas/Apo-1 antigen)	ceramide	U937, Jurkat, HUT78
DAG	ceramide	GH_3 pituitary
HIV	ceramide	HIV-infected CEM
fMLP	ceramide	human neutrophils
anti IgG/dexamethasone/ irradiation	ceramide	WEHI 231 murine B lymphoma
ionizing radiation	ceramide	bovine aortic endothelial
serum/plasma	sphingosine	human neutrophils
serum/PDGF	sphingosine/SPP	Swiss 3T3 fibroblasts
PDGF	sphingosine	vascular smooth muscle
phorbol dibutyrate	sphingosine	Epstein–Barr transformed human B-lymphocytes
dexamethasone	sphingosine	Epstein–Barr transformed human B-lymphocytes
partial hepatectomy	sphingosine	rat liver nuclei
fumonisin B_1	sphingosine sphinganine	many cell types, cattle, horses

by 1α-25-dihydroxyvitamin D_3 (Okazaki et al., 1989; Okazaki et al., 1990), tumor necrosis factor α (TNFα), or interferon γ (Kim et al., 1991) (Table 1). Other differentiation agents, including retinoic acid or phorbol esters, which induce granulocytic or macrophage-like differentiation, respectively, fail to induce sphingomyelin hydrolysis (Kim et al., 1991). Addition of short-chain, cell-permeable ceramide analogues mimics the effects of vitamin D_3 and TNFα on differentiation of HL-60 cells (Hannun, 1994). Using N-acyl derivatives of phenylaminoalcohols, Hannun and co-workers showed that neither the primary nor secondary hydroxyl group of ceramide was necessary but that the amide-linked fatty acid was essential (reviewed in Hannun, 1994). Thus, it has been suggested that sphingomyelin hydrolysis and its immediate breakdown product, ceramide, may play previously unrecognized roles in cellular differentiation (Hannun, 1994; Heller and Kronke, 1994; Kolesnick and Golde, 1994). Other studies raise the possibility that ceramide may play a role in cellular differentiation and may also be involved in cell growth regulation (Hauser et al., 1994; Olivera and Spiegel, 1992). In confluent and quiescent fibroblasts, ceramide enhances thymidine uptake and cellular proliferation (Olivera and Spiegel, 1992), whereas it inhibits growth of normal, logarithimically growing fibroblasts (Hannun, 1994; Venable et al., 1994). Thus, ceramide has a bimodal growth effect similar to that of TNFα. In addition to cytotoxic/cytostatic action on many cell types, TNFα can also stimulate the growth of other cells (Vilcek and Lee, 1991). Furthermore, TNFα inhibits the growth of rapidly dividing 3T3 fibroblasts, although TNFα stimulates the growth of the same cells when they are confluent and quiescent (Palombella and Vilcek, 1989).

Initial studies with ceramide in different cell types were hampered by its cytotoxicity. The finding that short-chain ceramides mimic the cytotoxicity of TNFα together with the structural specificity of the ceramide effect raised the possibility that ceramide mediates the effects of TNFα on activation of programmed cell death or apoptosis. In lymphoid and myeloid cells, cell-permeable ceramide analogues induced intranucleosomal DNA fragmentation (Obeid et al., 1993), induced loss of integrity of bulk DNA, and elicited expression of apoptotic morphology (Jarvis et al., 1994). In addition, ceramide levels are significantly increased in HIV-infected T-lymphocytes known to undergo apoptosis, suggesting that ceramide may mediate the cytopathic effects of TNFα in AIDS (Van Veldhoven et al., 1992).

IV. THE ROLE OF CERAMIDE IN TNFα AND INTERLEUKIN-1 SIGNAL TRANSDUCTION

Both TNFα and interleukin-1 (IL-1) are cytokines with pleiotrophic biological activities, exerting broad-range and common biological actions. The redundancy of IL-1 and TNFα activities may be due to activation of similar or the same signaling pathways. Both TNFα and IL-1 increase the level of diacylglycerol, leading to activation of protein kinase C, and, also, stimulate sphingomyelin hydrolysis (Heller and Kronke, 1994).

TNFα binds to two different molecular weight cell-surface receptors; one is the p55 TNFR and the second is the p75 TNFR. Most of the effects of TNFα are mediated through the p55 TNFR. Transfection of p55 TNFR into pre-B cells which lack TNFα receptors, sensitizes the cells to TNFα which then stimulates sphingomyelin hydrolysis (Weigmann et al., 1992). The mechanisms coupling TNFR to activation of sphingomyelinase are poorly understood but several provocative explanations for this coupling have been proposed. Hannun and co-workers suggested that a cytosolic neutral sphingomyelinase is activated in response to TNFα by arachidonic acid generated via TNFα -mediated activation of phospholipase A_2 in an intervening step (Jayadev et al., 1994). In this case, the sphingomyelinase must be acting on sphingomyelin located in the inner leaflet of the bilayer. The studies of Kolesnick and his co-workers indicated tight coupling of TNFR to sphingomyelinase. They found that, in HL-60 cells, both TNF-α and IL-1 stimulated sphingomyelin hydrolysis to ceramide within seconds and, furthermore, that the effects could also be demonstrated in cell-free microsomal fractions (Dressler et al., 1992). The ceramide produced in this manner can then in turn activate a membrane-bound 97 kD kinase called the ceramide-activated protein kinase, a member of the family of proline-directed serine/threonine protein kinases (Kolesnick and Golde, 1994). However, in contrast to other members of this family, ceramide-activated protein kinase recognizes the motif Leu-Thr-Pro. Kronke and his colleagues have suggested that, in addition to neutral plasma, membrane-associated sphingomyelinase, TNFα also activates an acidic sphingomyelinase (Heller and Kronke, 1994). They demonstrated that diacylglycerol generated by phosphatidylcholine-specific phospholipase C leads to the activation of acidic sphingomyelinase, which presumably resides in an acidic intracellular compartment (lysosomes or endosomes). The significance of p55 TNFR internalization for this activation is not clear (Heller and Kronke, 1994).

V. DOWNSTREAM SIGNALING OF CERAMIDE

Several potential targets for ceramide action have been proposed. Ceramide, similar to TNFα and IL-1, can activate a unique protein kinase that might be involved in signal transduction during cellular differentiation of HL60 and A431 cells (Goldkorn et al., 1991; Mathias et al., 1991). It has been suggested that activation of ceramide-activated protein kinase will in turn lead to activation of the mitogen-activated protein kinase (MAPK) cascade (Raines et al., 1993). MAPK activation may connect the ceramide signaling pathway to other intracellular signaling molecules, such as cytoplasmic phospholipase A_2, cyclin B, S6 kinase, and the transcription factors AP-1, NF-IL-6, Elk-1, and ATF-2. On the other hand, others have identified a phosphatase as the molecular target for the action of ceramide, defining a different intracellular signaling pathway (Dobrowsky and Hannun, 1992). Ceramide-activated protein phosphatase (CAPP) belongs to the heterotrimeric subfamily of protein phosphatase 2A which is inhibited by okadaic acid (Hannun, 1994). In agreement, okadaic acid inhibits ceramide-induced apoptosis and c-Myc downregulation which seems to be important in the differentiation of HL-60 cells. CAPP is likely to be an important enzyme because it is conserved in evolution (reviewed in Hannun, 1994).

Ceramide has been shown to activate the nuclear factor NF-κB in permeabilized Jurkat T-cells (Schutze et al., 1992). Kronke and co-workers demonstrated that both exogenous sphingomyelinase and ceramide analogues induced degradation of IκB, the physiological inhibitor of NF-κB (Schutze et al., 1992). They proposed that ceramide generated by TNFα-responsive phosphatidylcholine specific-phospholipase C stimulates an acidic sphingomyelinase resulting in the generation of ceramide which triggers induction of nuclear NF-κb activity (Heller and Kronke, 1994; Schutze et al., 1992). In contrast, TNFα in intact HL-60 cells, similar to ceramide, stimulates increased nuclear NF-κB whereas diacylglycerol has no effect (Yang et al., 1993). To add more confusion, addition of ceramide by itself is not sufficient to activate nuclear NF-κb in Jurkat T-cells (Dbaibo et al., 1993), raising the possibility that ceramide may be necessary but not sufficient to activate this transcription factor. Furthermore, although both the ceramide and diacylglycerol pathways may contribute to TNFα activation of NF-κb, impediments of these pathways do not block the activation of NF-κB by TNFα (Johns et al., 1994). This adds to the inherent difficulties of dissecting the interaction and involvement between these two classes of lipid second messengers, especially, because many agonists that stimulate sphingomyelinase also stimulate phospholipase C activity, leading to increases in ceramide and diacylglycerol, respectively.

VI. SPHINGOSINE AND SPHINGOSINE-1-PHOSPHATE, POSITIVE REGULATORS OF CELL GROWTH

Although sphingosine was originally proposed as a negative growth regulator functioning as an endogenous inhibitor of protein kinase C (Hannun and Bell, 1987; Hannun and Bell, 1989; Hannun et al., 1986), we and others have discovered a more versatile role for sphingosine independent of any effects on protein kinase C (Merrill, 1991; Spiegel et al., 1993; Zhang et al., 1990a). We found that low concentrations of sphingosine stimulated the proliferation of quiescent Swiss 3T3 fibroblasts and also potentiated the mitogenic responses to other growth factors, including insulin and EGF, via a protein kinase C-independent pathway (Zhang et al., 1990a). In contrast to the lack of stereospecificity in the inhibition of protein kinase C, the mitogenic effect of sphingosine was stereospecific, because only the D-*erthyro*-isomer of sphingosine, the natural isomer, stimulated DNA synthesis. Structurally related analogues of sphingosine or L-*threo*-stereoisomers were completely inactive (Olivera et al., 1994). The mitogenic effects of sphingosine have subsequently been shown to be widespread. Sphingosine stimulated proliferation of rodent fibroblasts (Gomez et al., 1994; Schroeder et al., 1994) and reinitiated DNA synthesis in HL-60 cells after phorbol-ester-induced, cell-growth arrest (Merrill et al., 1986). In contrast, sphingosine had no effect on the proliferation of A431 human epidermoid carcinoma cells (Faucher et al., 1988) and WI38 human fetal lung fibroblasts (Davis et al., 1988), despite sphingosine-induced stimulation of EGF receptor tyrosine kinase activity.

Fumonisin B, a fungal toxin with a structural resemblance to long-chain sphingoid bases, causes an increase in levels of sphinganine and sphingosine in Swiss 3T3 fibroblasts by inhibiting sphinganine-N-acyltransferase and, also, stimulates DNA synthesis (Schroeder et al., 1994). These effects on sphingolipid metabolism and cell growth may be responsible for the cancer-inducing properties of fumonisins (Schroeder et al., 1994).

The inhibition of cell growth by exogenous sphingosine that was originally observed (Stoffel and Bister, 1973) could be partly due to cytotoxicity as a consequence of its detergent properties (Merrill and Stevens, 1989), although it has since been proposed that inhibition of protein kinase C activity may mediate this effect (Merrill and Stevens, 1989; Stevens et al., 1990). It is not unprecedented for the same effector to stimulate and inhibit the growth of cells. Transforming growth factor β is a bifunctional regulator of cell growth, stimulating growth of non-neoplastic fibroblasts while inhibiting growth of many tumor cells (Sporn and Roberts, 1988). Similarly, TNFα can also stimulate the growth of some cells although it has well-known cytotoxic/cytostatic actions on many cell types (Vilcek and Lee, 1991). TNFα and the B subunit of cholera toxin stimulate proliferation of quiescent 3T3 fibroblasts, and they inhibit proliferation of the same fibroblasts when they are rapidly growing (Buckley et al., 1990; Palombella and Vilcek, 1989; Spiegel and Fishman, 1987). Similarly, sphingosine also has bimodal effects on cell growth depending on cell type and the nature of the stimuli (Chao et al., 1992; Spiegel et al., 1993; Stevens et al., 1990; Zhang, et al., 1990a).

Recently, we have shown that a metabolite of sphingosine, SPP, which is rapidly produced in response to mitogenic concentrations of sphingosine, may mediate the mitogenic activity of sphingosine and also may be a new intracellular second messenger (Spiegel et al., 1993; Zhang et al., 1991). SPP increased DNA synthesis and cell division in quiescent cultures of Swiss 3T3 fibroblasts similarly to sphingosine, albeit effectively at lower concentrations (Zhang et al., 1991). Both sphingosine and SPP acted synergistically with a wide variety of growth factors. However, there were no additive or synergistic effects in response to maximally effective doses of both sphingosine and SPP, suggesting either that they act through common pathways or that SPP mediates the growth promoting activity of sphingosine.

In addition to its ability to simulate cell growth, it was recently found that at very low concentrations (10 nM), SPP effectively and specifically inhibited chemotactic motility and invasiveness of several tumor cell lines, including mouse and human melanoma and human osteosarcoma cells and was more potent than sphingosine (Sadahira et al., 1992). SPP also inhibited extracellular matrix protein-induced haptotactic motility but not adhesion of B16 mouse melanoma cells, and other related compounds were much less effective (Sadahira et al., 1994). SPP significantly reduced invasion of MDA-MB-231 cells in Matrigel-lined Boyden chamber assays (Spiegel et al., 1994). The MDA-MB-231 cell line lacks estrogen receptors, is unresponsive to antiestrogens and provides a model for the more aggressive, hormone-independent class of breast cancers (Lippman and Dickson, 1989). In contrast to other effects of SPP, sphingosine was relatively inactive in chemoinvasion and motility assays (Sadahira et al., 1992; Spiegel et al., 1994). These results suggest that SPP may inhibit the cellular enzymatic activity responsible for degrading cell matrices, a process required for invasion. The potent inhibitory effect of SPP on invasion, particularly in certain aggressive tumor cell lines, suggests that it could be a useful agent for suppressing metastasis, the major cause of mortality in human cancers.

VII. DOWNSTREAM SIGNALING OF SPHINGOSINE AND SPHINGOSINE-1-PHOSPHATE

In the following sections, we discuss the signaling pathways which are activated by the sphingolipid metabolites, sphingosine and SPP, and possible mechanisms by which they may exert their diverse effects. These sphingolipid metabolites have been implicated in regulating calcium, glycerophospholipid signaling, cAMP, and protein kinases. Little emphasis will be placed here on protein kinase C because it has been the subject of several excellent reviews (Hannun and Bell, 1987; Hannun and Bell, 1989; Merrill, 1991).

A. Protein Kinase C and Other Kinases

The discovery that sphingosine is an inhibitor of protein kinase C attracted much attention (Hannun and Bell, 1987; Hannun and Bell, 1989; Merrill, 1991). There are now numerous examples of sphingosine-regulated pathways in cells in which the effects of sphingosine are mediated by protein kinase C inhibition (reviewed in Hannun and Bell, 1987; Hannun and Bell, 1989; Merrill, 1991). However, other roles for sphingosine have since been discovered that are independent of its effect on protein kinase C (reviewed in Merrill, 1991; Spiegel, 1993b; Spiegel et al., 1993). The mitogenic effect of sphingosine on Swiss 3T3 fibroblasts was independent of protein kinase C (Zhang et al., 1990a), and SPP, in contrast to sphingosine, has no effect on protein kinase C activity *in vitro* (Van Veldhoven et al., 1989).

Sphingosine also inhibits the activities of several other kinases, including insulin receptor tyrosine kinase (Arnold and Newton, 1991) and calmodulin-dependent kinases (Jefferson and Schulman, 1988). Sphingosine also induced dephosphorylation of the retinoblastoma gene product, a nuclear phosphoprotein that functions as a tumor suppresser, which was correlated with its ability to inhibit growth and lock lymphoblastic leukemic cells in the Go/G1 phase of the cell cycle (Chao et al., 1992). On the other hand, sphingosine stimulated the activities of several other protein kinases, such as casein kinase II (McDonald et al., 1991) and the tyrosine kinase of the EGF receptor (Davis et al., 1988), and also induced *in vitro* phosphorylation of cytosolic proteins in Jurkat T-cells (Pushkareva et al., 1992). However, it is not clear whether the ability of sphingosine to stimulate these protein kinases is related to its mitogenic effect.

B. Phosphatidic Acid

The mitogenic effect of sphingosine in Swiss 3T3 fibroblasts was accompanied by an increase in the levels of phosphatidic acid (Zhang et al., 1990b), a phospholipid implicated in regulating cell growth (Moolenaar et al., 1986). Sphingosine stimulates phospholipase D activity leading to an increase in phosphatidic acid in many types of cells, including 3T3 fibroblasts (Kiss and Anderson, 1990; Lavie and Liscovitch, 1990; Zhang et al., 1990b). It is interesting to note that, in a wide variety of cell types with different pathways for formation of phosphatidic acid, sphingosine and SPP, uniformly, have a stimulatory effect on phosphatidic acid levels (Spiegel, 1993b). Recently, it has been suggested that an increase in the concentration of phosphatidic acid is essential for the mitogenic signal-transduction cascades triggered by several growth factors, including platelet-derived growth factor (PDGF) (Ben-Av and Liscovitch, 1989; Exton, 1990). Furthermore, phosphatidic acid may be involved in regulating the biological action of cellular Ras (Tsai et al., 1989; Tsai et al., 1991; Tsai et al., 1990). Ras is one of several key oncoproteins that plays a central role as a molecular switch in the early steps of signal transduction associated with cell growth, differentiation, and neoplasia (Barbacid, 1987). This small GTP-binding protein is active only when in its GTP-complexed form. A cytoplasmic GTPase activating protein (GAP) stimulates the conversion to the inactive form. Activation of Ras is an initial and key step of the signaling cascade initiated by binding of growth factors to receptors and subsequent tyrosine phosphorylation, leading to the recruitment of several proteins containing src homology 2 domains (SH2 proteins), including phosphatidyinositol-3 kinase (PI-3 kinase), GRB2/Sos, Shc, Nck, SH-PTP-2, and GAP. GRB-2 links the receptor to the guanine exchange factor Sos, which appears to activate Ras by stimulating binding of GTP to Ras. Ras then activates other proteins, including PI-3 kinase and Raf, which, in turn, activate the rest of the MAP kinase cascade, culminating in the activation of $p90^{rsk}$ and other important transcriptional factors, including c-Myc, NF-IL-6, Elk-1, ATF-2 and AP-1 (Pelech and Sanghera, 1992) (Figure 2). In *in vitro* studies, phosphatidic acid has been shown to stabilize Ras-GTP by inhibiting Ras-GAP (GTPase activating protein) and activating a cytosolic protein which inhibits Ras-GTPase activity (Tsai et al., 1989; Tsai et al., 1991; Tsai et al., 1990). Thus, it has been suggested that phosphatidic acid might, thereby, activate Ras *in vivo*. In view of the potential roles of phosphatidic acid in cellular proliferation, signal transduction, and activation of the Ras cascade, a connection between long-chain sphingoid bases and phosphatidic acid has important implications for the biological actions of sphingolipid metabolites. Therefore, it is not suprising that many studies have been directed toward examining sphingosine-mediated regulation of phosphatidic acid metabolism (reviewed in Spiegel, 1993a; Spiegel, 1993b; Spiegel and Milstien, 1995).

In contrast to the stimulatory effects of long-chain sphingoid bases on phosphatidic acid levels, ceramide decreases phosphatidic acid levels by inhibiting phospholipase D activation (Gomez et al., 1994). Short chain, cell-permeable ceramide analogs not only inhibit the stimulation of DNA synthesis in Rat-1 fibroblasts but also inhibit phospholipase D activity induced by many growth promoting agents, including phorbol ester, thrombin, serum, phosphatidate and lysophosphatidate. In agreement with our results, sphingosine had opposite effects to ceramides on DNA synthesis and phospholipase D activation in these cells (Gomez et al., 1994). Thus, it has been proposed that ceramides may produce part of their growth-arresting effects (discussed above) by inhibiting the phospholipase D signaling pathway (Gomez et al., 1994). Further support for this proposal emerged from the observa-

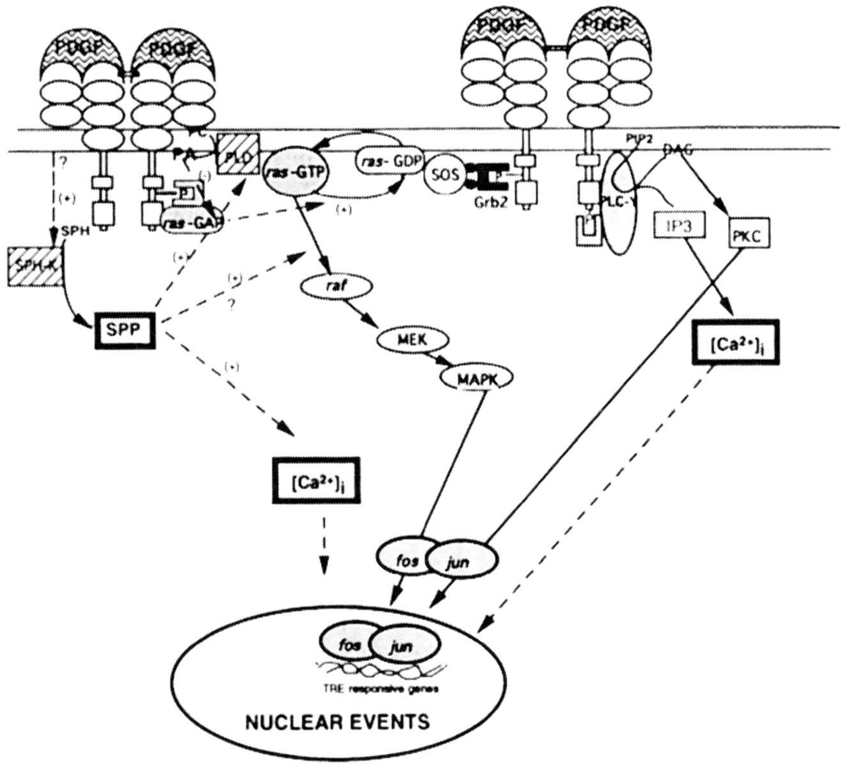

Figure 2. Platelet derived growth factor-mediated signaling pathways and possible sites of action of sphingolipid metabolites. PDGF receptors are linked to and activate other effector systems, including phosphatidylinositol 3-hydroxy kinase (PI3K), phospholipase D (PLD), phospholipase C-γ (PLCγ1) GTPase activating protein (GAP), and the adaptor Grb2-SOS complex. This leads to activation of the signaling cascade involving Ras and Raf-1, which phosphorylates and activates MEK, that, then, activates mitogen-activated protein kinases (MAPKs) and culminates in the phosphorylation of transcription factors (such as Jun/Fos) and other cellular factors that bring about the cellular responses. For more detailed information, see text. Recently, it has been shown that PDGF stimulates sphingosine kinase (SPH-K), leading to an increase in SPP. SPP has complex intracellular actions which may contribute to its mitogenic effect, including mobilization of intracellular calcium, activation of phospholipase D, and activation of Raf/MEK/MAPK. Dashed arrows indicate influences on pathways. Stimulatory and inhibitory effects are indicated by (+) or (–), respectively.

tion that one of the defects in mitogenic signaling pathways in cellular senescence is caused by an elevation of endogenous ceramide levels which, in turn, inhibits proliferation by inhibiting phospholipase D activation (Venable et al., 1994).

Previously, it has been suggested that the glycerophospholipid cycle could influence the sphingolipid cycle (Koval and Pagano, 1991). As discussed here, the increase of phosphatidic acid levels induced by sphingosine and SPP, and the decrease by ceramide, indicates that the sphingolipid cycle could regulate the glycerophospholipid cycle. Cross talk between these pathways adds another level of complexity to signaling pathways leading to cellular proliferation and additional sites of regulation.

C. Calcium and Inositol Trisphosphate

Sphingosine induces a profound release of calcium from inositol trisphosphate-sensitive and -insensitive intracellular pools in permeabilized smooth muscle cells (Ghosh et al., 1990) and in viable 3T3 fibroblasts (Zhang et al., 1991). It is important to note that calcium release is the only early event in sphingosine-induced mitogenesis that has been demonstrated to be stereospecific. Only the natural D-(+)-*erythro* sphingosine stereoisomers (*cis* and *trans*) stimulated DNA synthesis and mobilized calcium from internal sources, whereas all of the stereoisomers induced similar increases in phosphatidic acid (Olivera et al., 1994). Calcium release was time- and temperature-dependent, implicating an intervening enzymatic step, consistent with the conversion of sphingosine to SPP (Ghosh et al., 1990). Similar to sphingosine, SPP can also release calcium from internal sources in viable Swiss 3T3 fibroblasts (Mattie et al., 1994; Zhang et al., 1991). Recently, SPP has been shown to mobilize calcium from internal sources in human foreskin fibroblasts (Chao et al., 1994) and in REF52 fibroblasts and astrocytoma cells (Randriamampita and Tsien, 1993). However, in rat pancreatic acinar cells, although sphingosine released calcium from internal sources, SPP was inactive (Yule et al., 1993).

Sphingosine also enhanced phosphatidylinositol turnover by stimulating phospholipase C activity, and the activation of this process was modulated by a G-protein that is a substrate for pertussis toxin (Chao et al., 1994; Ritchie et al., 1992; Zhang et al., 1990b). However, SPP stimulated a more rapid release of intracellular calcium than sphingosine in human skin fibroblasts, without any effects on phosphoinositol turnover (Chao et al., 1994). In agreement, mobilization of calcium by SPP proceeded by a mechanism which was independent of phosphoinositides hydrolysis and $InsP_3$ formation (Ghosh et al., 1990; Mattie et al., 1994). Although SPP increased $InsP_3$ levels in Swiss 3T3 fibroblasts, complete inhibition of $InsP_3$ formation by brief treatment with the phorbol ester TPA did not inhibit SPP-mediated calcium responses, indicating that formation of $InsP_3$ is not required for release of calcium by SPP (Mattie et al., 1994). Furthermore, in permeabilized cells (Mattie et al., 1994), and in endoplasmic reticulum from smooth muscle cells (Ghosh et al., 1994), heparin, an $InsP_3$ antagonist, inhibited calcium release induced by exogenous $InsP_3$, but did not affect calcium release induced by SPP.

The calcium response to SPP has many of the hallmarks of a receptor-mediated event. The rapidity, reversibility, and specificity of the effect of SPP all closely resemble the calcium-releasing effect of $InsP_3$ which is mediated by direct calcium-channel activation. SPP may interact directly with calcium channels within the membranes of intracellular stores, such as the $InsP_3$ receptor or the ryanodine receptor. Although these calcium channels share sequential homology and tertiary structural features, they can be distinguished physiologically by their responses to ligands and regulators (Taylor and Marshall, 1992). It seems reasonable to assume that SPP may interact directly with calcium channels within the membranes of some intracellular stores which are under the control of the ryanodine receptor or $InsP_3$ receptors. Nevertheless, it is possible that SPP may affect another, yet unknown, calcium channel. Sphingosine may have differential effects on different types of calcium channels. In contrast to its effects on permeabilized smooth muscle cells and on Swiss 3T3 fibroblasts (Ghosh et al., 1990; Mattie et al., 1994; Zhang et al., 1991), sphingosine inhibited the release of calcium from skeletal muscle sarcoplasmic reticulum and directly inhibited sarcoplasmic reticulum ryanodine receptors (Sabbadini et al., 1992). However, ryanodine channels have not been detected in Swiss 3T3 fibroblasts (Mattie et al., 1994).

D. cAMP

Both sphingosine (Zhang et al., 1990b) and SPP (Goodemote et al., 1995) rapidly and drastically decreased cellular cAMP levels in Swiss 3T3 fibroblasts. Adrenaline-, propranolol,- and forskolin-stimulated increases in cAMP levels were also lowered in S49 lymphoma cells by sphingosine treatment (Johnson and Clark, 1990). Pertussis toxin (PT) prevented the inhibitory effect of sphingosine and SPP on cAMP accumulation, suggesting that a PT-sensitive G_i protein may be involved. PT also inhibited DNA synthesis induced by sphingosine and SPP (Goodemote et al., 1995). However, the inhibitory effects on cAMP accumulation may not be associated with G_i regulation of mitogenesis. In Swiss 3T3 cells, cAMP unequivocally acts as a positive effector of proliferation (Rozengurt, 1986); thus, it is unlikely that sphingosine- or SPP-induced decreases in cAMP play a significant role in inducing proliferation. However, in several other cell types including Rat-1 and human foreskin fibroblasts, cAMP is most likely a negative effector of mitogenesis, and, therefore, it seems reasonable to assume that reduction of cAMP may have growth-promoting effects in these cells (Gomez et al., 1994). The relevance of changes in cAMP levels to mitogenic effects of sphingosine or SPP in these cell merits further study.

VIII. GROWTH FACTORS MODULATE SPHINGOSINE AND SPHINGOSINE 1-PHOSPHATE LEVELS

If sphingosine and SPP function as intracellular second messengers, like other second messengers, their levels should be relatively low and regulated by external stimuli. Levels of free sphingosine have been found in the range of 10–100 pmol/10^6 cells in a wide variety of cells (Merrill, 1989; Merrill et al., 1988), and can be regulated by several agonists (Merrill, 1991; Olivera and Spiegel, 1993; Ramachandran et al., 1990; Wilson et al., 1988). Dexamethasone induced a modest increase in free sphingosine in 3T3-L1 fibroblasts (Ramachandran et al., 1990). Furthermore, increases in sphingosine levels were observed in nuclei of regenerating rat liver cells during RNA and DNA synthesis, suggesting a connection between sphingosine levels in the nucleus and nucleic acid synthesis (Alesenko et al., 1992). In addition, TPA decreased the level of free sphingosine in human neutrophils, whereas serum, plasma, or serum lipoproteins induced increases (Wilson et al., 1988; Table 1).

Recently, we found that both PDGF and serum induced rapid and transient increases in sphingosine and SPP levels in Swiss 3T3 fibroblasts (Olivera and Spiegel, 1993). Basal levels of SPP in quiescent Swiss 3T3 fibroblasts were 16 ± 2 pmol/10^6 cells and increased by 2.1 and threefold in response to PDGF and serum, respectively. PDGF also increased sphingosine levels by 1.5-fold in vascular smooth muscle cells with a concomitant decrease in ceramide levels (Jacobs and Kester, 1993). These responses may be specific for certain growth promoting agents, because EGF did not induce significant changes (Olivera and Spiegel, 1993).

In Swiss 3T3 fibroblasts, serum and PDGF also induced transient increases in sphingosine kinase, the enzyme that catalyzes the phosphorylation of the primary hydroxyl group of sphingosine to form SPP (Olivera and Spiegel, 1993). Other mitogens, such as the B subunit of cholera toxin and TPA, also increased the activity of this kinase, whereas bombesin, bradykinin, insulin, and EGF had no effect. These results further demonstrate that effects on sphingoid base levels are restricted to certain growth factors. Recently, it was found that brief treatment of Balb/c 3T3 fibroblasts with TPA also markedly stimulated sphingosine kinase activity (Mazurek et al., 1994).

The fact that SPP is a potent mitogenic agent when added exogenously to cells, combined with the observation that PDGF and serum can activate sphingosine kinase and consequently regulate endogenous levels of SPP, suggests that SPP has an important role in the signal-transduction pathways activated by these growth factors. To further substantiate a potential role for SPP in cellular proliferation induced by PDGF, we utilized a known competitive inhibitor of sphingosine kinase, DL-*threo*-dihydro-sphingosine. This compound inhibited sphingosine kinase activity from 3T3 fibroblasts *in vitro* and also, inhibited the production of SPP, induced by a mitogenic concentration of sphingosine by 63 ± 2%, and the initiation of DNA synthesis in response to sphingosine by more than 55%. Most importantly, this sphingosine kinase inhibitor completely eliminated the production of SPP elicited by PDGF and, also, reduced DNA synthesis induced by PDGF and serum by 40 and 44%, respectively, consistent with the multiplicity signal transduction systems known to be stimulated by these growth factors. In contrast, this inhibitor did not abrogate cellular proliferation induced by EGF, which seems to act independently of SPP formation. In agreement, L-cycloserine, an inhibitor of sphingolipid synthesis, inhibited PDGF-stimulated DNA synthesis in vascular smooth muscle cells (Jacobs and Kester, 1993).

In sum, these results led us to suggest that sphingosine and SPP are second messengers in signal-transduction pathways leading to cellular proliferation. Sphingosine and SPP have many appropriate properties that make them suitable candidates to function as intracellular second messengers: they elicit diverse cellular responses; their turnover is extremely rapid; their levels in cells are very low and increase rapidly and transiently in response to growth factors; they can release calcium from internal sources in a previously unknown, $InsP_3$-independent manner; finally, they could also act in a positive feedback loop amplifying the cascade of events following (or subsequent to) receptor stimulation via their effect on phosphatidic acid levels, which might link growth signaling to cellular Ras activity.

IX. TRANSCRIPTION FACTORS: SPHINGOSINE AND SPHINGOSINE-1-PHOSPHATE STIMULATE DNA BINDING ACTIVITY OF AP-1

Activation of Ras is a key step in the signaling cascade to the nucleus initiated by binding of a growth factor to its receptor and subsequent tyrosine phosphorylation. Activation of Ras results in activation of the mitogen-activated protein kinase (MAP kinases) also known as extracellular-signal-regulated kinases. A 42-kDa MAP kinase (p42mapk/ERK2) and a 44 kDa MAP kinase (p44mapk/ERK2) are the two best characterized MAP kinases. Both enzymes are activated by dual specificity MAP kinase kinases (MKKs or MEKs) which are, in turn, regulated by serine phosphorylation by several kinases, including MEK kinase and Raf-1 (Figure 2). Recent studies have demonstrated that SPP, similar to PDGF, rapidly and transiently activates Raf, MEK, and MAP kinases (Wu et al., 1995). *In vitro* experiments have not revealed any direct effects of SPP on Raf, MEK, or MAP kinase, implying action(s) upstream of Raf. The ability of SPP to stimulate phospholipase D, leading to an increase in cellular phosphatidic acid, could be important because phosphatidic acid can inhibit Ras-GTPase activating protein and might thereby activate Ras.

Many studies have focused on the transcription factor AP-1 which appears to function as a major convergence point, coupling growth factor receptors to long term cellular phenotypic responses by regulating expression of specific target genes. AP-1, which is also one of the transcription factors activated by the Ras-MAPK pathway, consists of homo- and/or heterodimers of the fos and jun gene products and controls genes required for cell growth by binding to AP-1 consensus sequences, named TREs (TPA-responsive elements). Inhibition of either Fos synthesis with a vector expressing antisense c-fos RNA (Holt et al., 1986), or Fos function by microinjection of Fos-specific antibodies (Riabowol et al., 1988), inhibited cell-cycle progression, confirming the function of AP-1 protein in controlling cell proliferation. Furthermore, microinjection of oligonucleotides containing consensus AP-1-binding sites blocked DNA synthesis in response to growth factors (Riabowol et al., 1992).

As established for growth factors, sphingosine and SPP, also, markedly simulated DNA binding activity of AP-1 (Su et al., 1994). The induced AP-1 complex contained both c-Fos and c-Jun proteins. This is an important finding because heterodimer formation of any Jun protein with c-Fos results in the generation of potent transactivators, and part of this increased activity is due to the increased stability of these heterodimers (Karin and Smeal, 1992). Several experiments suggested that the stimulation of DNA synthesis in Swiss 3T3 fibroblasts by SPP was mediated, at least in part, by AP-1 (Su et al., 1994). First, the dose response for activation of AP-1 by SPP correlated closely with the induction of DNA synthesis. Second, although both SPP and sphingosine induced DNA-binding activity of AP-1, DL-*threo*-dihydrosphingosine, which inhibits sphingosine-induced DNA synthesis, only inhibited sphingosine-stimulated AP-1 DNA-binding activity. This result lends further support to our proposal that SPP mediates the mitogenic effect of sphingosine. Finally, SPP induced AP-1 activity within 30 min after stimulation, and the activity remained elevated for at least 21 h. This finding is consistent with the notion that AP-1 may be required for the transcriptional activation of a cascade of genes necessary for the entry of the cells into the S-phase by 18 h after stimulation. The activation of AP-1 by sphingosine and SPP is the first evidence linking the effects of sphingolipid metabolites on cellular proliferation to gene expression (Su et al., 1994). AP-1 may be a convergence point where signals carried by sphingolipid-derived messengers are integrated with the well-known intracellular second messages derived from glycerophospholipids, for example, diacylglycerol.

X. SPHINGOSINEPHOSPHORYLCHOLINE IS A REMARKABLY POTENT MITOGEN

To identify the structural components of SPP required for inducing proliferation, many analogs of SPP were tested for mitogenic activity. Of these, only lysosphingomyelin or sphingosinephosphorylcholine (SPC), had greater mitogenicity than sphingosine or SPP (Desai et al., 1993; Desai and Spiegel, 1991). SPC-induced proliferation was equal to that of serum and was much greater than all other fibroblast growth factors, including platelet activating factor, which is structurally similar to SPC. Unlike most peptide mitogens which require long exposure to induce proliferation in fibroblasts, treatment with SPC for only 10 min was sufficient to induce significant DNA synthesis. SPC also stimulated proliferation of many diverse cell types, including preadipocytes, HeLa carcinoma cells, C6 rat glioma cells, human astrocytes, and appears to be a new type of potent, wide-spectrum, growth-promoting agent (Desai and Spiegel, 1991).

A. Sphingosinephosphorylcholine Signaling Pathways

SPC induced rapid and transient increases in cytosolic free calcium in Swiss 3T3 fibroblasts (Desai et al., 1993). The rapid rise in $[Ca^{2+}]_i$ was independent of calcium in the external medium, indicating that the response was mainly due to mobilization of calcium from internal stores. Thus, SPC is also a calcium-mobilizing agonist in viable 3T3 fibroblasts, acting apparently similarly to sphingosine and SPP (Zhang et al., 1991). In agreement, SPC and sphingosine also mediated Ca^{2+} oscillations in pancreatic acinar cells. This response was not, however, mimicked by the addition of SPP, ceramide, or sphingomyelin (Yule et al., 1993). Recently, a novel intracellular SPC-gated calcium channel with unique pharmacological and electrophysiological properties which differed from other voltage-gated and ligand-gated calcium channels (Kindman et al., 1994), was detected in rat basophilic leukemia cells.

Previously, we found that sphingosine and SPP stimulated formation of inositol phosphates (Mattie et al., 1994; Zhang et al., 1990b). Thus, despite shared structural features, sphingosine and SPP do not share a common mechanism of mitogenic stimulation with SPC (Desai et al., 1993). Although SPC, SPP, and sphingosine all mobilize calcium in fibroblasts, SPC did not mimic the other changes in intracellular signaling induced by sphingosine or SPP, which included activation of phospholipase D leading to phosphatidic acid accumulation, stimulation of inositol phosphate formation, and inhibition of cAMP production. Moreover, in contrast to sphingosine and SPP, SPC induced an increase in the release of arachidonic acid and may act via both protein kinase C-dependent and independent signaling pathways whereas the mitogenic activities of sphingosine and SPP in Swiss 3T3 cells were clearly independent of protein kinase C (Zhang et al., 1990a; Zhang et al., 1990b). It is intriguing to note that protein kinase Cα, the predominant protein kinase C isoform in Swiss 3T3 cells, can be activated by free arachidonic acid in the presence of calcium (Nishizuka, 1992). The increase in $[Ca^{2+}]_i$ induced by SPC, together with the increase in arachidonic acid levels, might activate protein kinase Cα and could account for the partial protein kinase C dependency observed in these studies (Desai et al., 1993). Recently, in contrast to the effect of sphingosine, SPC was found to have dual effects on the phosphorylation of cytosolic proteins in Jurkat T-cells (Pushkareva and Hannun, 1994). Thus, it seems likely that SPC acts as an extremely potent mitogen due to activation of signaling pathway(s) distinct from those utilized by sphingosine or SPP.

B. Does Endogenous Sphingosylphosphorylcholine Regulate Mitogenesis?

Although SPC has not yet been detected in normal cells, it has been found in the spleens of patients with Niemann–Pick disease (Strasberg and Callahan, 1988), a form of sphingolipidosis resulting from a deficiency of sphingomyelinase activity

(Strasberg and Callahan, 1988). The potential role of SPC in the pathophysiology of this disease was suggested by its deleterious effects on mitochondrial function and calcium uptake (Strasberg and Callahan, 1988). Earlier, SPC had been proposed as a potential precursor of sphingomyelin (Brady et al., 1965). It was recently reported that SPC stimulates an increase in sphingomyelin levels in neuroblastoma cells (Sugiyama et al., 1993) and in human leukemia HL-60 cells (Dressler et al., 1991). In both cases, the increases in sphingomyelin were associated with differentiation. Nevertheless, exogenous SPC in 3T3 cells has not been found to be significantly converted to sphingomyelin (Desai et al., 1993). Furthermore, other studies have convincingly demonstrated that most, if not all, sphingomyelin is synthesized in mammalian cells by direct transfer of phosphorylcholine from phosphatidylcholine to ceramide, suggesting that SPC is probably not a precursor of sphingomyelin (Futerman and Pagano, 1991; Pagano, 1988). Although there is no indication that external stimuli mediate the formation of SPC, yet, analogous to receptor-linked phospholipase A_2, which catalyzes liberation of arachidonic acid from phospholipids, a receptor-linked "sphingomyelin deacylase" might function to increase endogenous levels of SPC, which potentially could act as a modulator of cellular proliferation via distinct signaling pathways.

XI. CONCLUDING REMARKS

Several studies have demonstrated that intracellular ceramide and/or sphinogsine and SPP levels are increased by treating various types of cells with growth factors or cytokines (see Table 1). Furthermore, as reviewed here, abundant evidence has been accumulated to suggest strongly that sphingolipid metabolites may constitute a new class of intracellular second messengers mediating some of the effects of growth factors and cytokines. However, many questions remain to be answered. Are sphingosine, SPP, SPC, and ceramide the only active sphingolipids or are there still others to be discovered? Are the effects of these compounds the same or similar in all cells or are they limited to specific cells and tissues? More complete descriptions of the mechanisms of regulation of cell growth and other cellular processes by cytokines and growth factors should enable us to answer some of these questions.

ABBREVIATIONS

The following abbreviations are used:

SPP, sphingosine-1-phosphate
SPC, sphingosylphosphorylcholine
TPA, 12-O-tetradecanoylphorbol 13-acetate
PDGF, platelet-derived growth factor
BSA, bovine serum albumin
FCS, fetal calf serum

FAD, flavine adenine dinucleotide
NADPH, nicotinamide adenine dinucleotide phosphate, reduced

ACKNOWLEDGMENTS

The authors gratefully acknowledge the contributions of collaborators and co-workers, especially, Drs. Ana Olivera, Alvin Berger, and Hong Zhang. This work was supported by research grants 1RO1 GM43880 and CA61774 from the National Institutes of Health.

REFERENCES

Alessenko, A. V., Pantaz, E. A., Pushkareva, M. Y., Rusakov, S. A., & Fillipova, G. N. (1992). Changes in sphingosine content in regenerating rat liver cell nuclei. Biokhimiya 58, 298–305.

Arnold, R. S. & Newton, A. C. (1991). Inhibition of the insulin receptor tyrosine kinase by sphingosine. Biochemistry 30, 7747–7754.

Barbacid, M. (1987). ras genes. Annu. Rev. Biochem. 56, 779–827.

Ben–Av, P. & Liscovitch, M. (1989). Phospholipase D activation by the mitogens platelet-derived growth factor and 12-O-tetradecanoylphorbol 13-acetate in NIH-3T3 cells. FEBS Lett. 259, 64–66.

Berridge, M. J. (1993). Inositol trisphosphates and calcium signaling. Nature 361, 315–325.

Brady, R. O., Bradley, R. M., Young, O. M., & Kaller, H. (1965). An alternative pathway for the enzymatic synthesis of sphingomyelin. J. Biol. Chem. 240, 3693–3694.

Buckley, N. E., Matyas, G. R., & Spiegel, S. (1990). The bimodal growth response of Swiss 3T3 cells to the B subunit of cholera toxin is independent of the density of its receptor, ganglioside GM1. Exp. Cell. Res. 189, 13–21.

Chao, C. P., Laulederkind, S. J., & Ballou, L. R. (1994). Sphingosine-mediated phosphatidylinositol metabolism and calcium mobilization. J. Biol. Chem. 269, 5849–5856.

Chao, R., Khan, W., & Hannun, Y. A. (1992). Retinoblastoma protein dephosphorylation induced by D-erythro-sphingosine. J. Biol. Chem. 267, 23459–23462.

Davis, R. J., Girones, N., & Faucher, M. (1988). Two alternative mechanisms control the interconversion of functional states of the epidermal growth factor receptor. J. Biol. Chem 263, 5373–5379.

Dbaibo, G. S., Obeid, L. M., & Hannun, Y. A. (1993). Tumor necrosis factor α (TNF-α) signal transduction through ceramide. Dissociation of growth inhibitory effects of TNFα from activation of nuclear factor κB. J. Biol. Chem. 268, 17762–17766.

Desai, N. N., Carlson, R. O., Mattie, M. E., Olivera, A., Buckley, N. E., Seki, T., Brooker, G., & Spiegel, S. (1993). Signaling pathways for sphingosylphosphorylcholine-mediated mitogenesis in Swiss 3T3 fibroblasts. J. Cell Biol. 121, 1385–1395.

Desai, N. N. & Spiegel, S. (1991). Sphingosylphosphorylcholine is a remarkably potent mitogen for a variety of cell lines. Biochem. Biophys. Res. Comm. 181, 361–366.

Dobrowsky, R. T. & Hannun, Y. A. (1992). Ceramide stimulates a cytosolic protein phosphatase. J. Biol. Chem. 267, 5048–5051.

Dressler, K. A., Kan, C. C., & Kolesnick, R. N. (1991). Sphingomyelin synthesis is involved in adherence during macrophage differentiation of HL-60 cells. J. Biol. Chem. 266, 11522–11527.

Dressler, K. A., Mathias, S., & Kolesnick, R. N. (1992). Tumor necrosis factor α activates the sphingomyelin signal transduction pathway in a cell-free system. Science 255, 1715–1717.

Egan, S. E., Giddings, B. W., Brooks, M. W., Buday, L., Sizeland, A. M., & Weinberg, R. A. (1993). Association of Sos Ras exchange protein with Grb2 is implicated in tyrosine kinase signal transduction and transformation. Nature 363, 45–51.

Exton, J. H. (1990). Signaling through phosphatidylcholine breakdown. J. Biol. Chem. 265, 1–4.

Faucher, M., Girones, N., Hannun, Y. A., Bell, R. M., & Davis, R. J. (1988). Regulation of the epidermal growth factor receptor phosphorylation state by sphingosine in A431 human epidermoid carcinoma cells. J. Biol. Chem. 263, 5319–5327.

Futerman, A. H. & Pagano, R. E. (1991). Determination of the intracellular sites and topology of glucosylceramide synthesis in rat liver. Biochem. J. 280, 295–302.

Gale, N. W., Kaplan, S., Lowenstein, E. J., Schlessinger, J., & Bar–Sagi, D. (1993). Grb2 mediates the EGF-dependent activation of guanine nucleotide exchange on Ras. Nature 363, 88–92.

Ghosh, T. K., Bian, J., & Gill, D. L. (1990). Intracellular calcium release mediated by sphingosine derivatives generated in cells. Science 248, 1653–1656.

Ghosh, T. K., Bian, J., & Gill, D. L. (1994). Sphingosine 1-phosphate generated in the endoplasmic reticulum membrane activates release of stored calcium. J. Biol. Chem. 269, 22628–22635.

Goldkorn, T., Dressler, K. A., Muindi, J., Radin, N. S., Mendelsohn, J., Menaldino, D., Liotta, D., & Kolesnick, R. N. (1991). Ceramide stimulates epidermal growth factor receptor phosphorylation in A431 human epidermoid carcinoma cells. J. Biol. Chem. 266, 16092–16097.

Gomez, M. A., Martin, A., O'Brien, L., & Brindley, D. N. (1994). Cell-permeable ceramides inhibit the stimulation of DNA synthesis and phospholipase D activity by phosphatidate and lysophosphatidate in rat fibroblasts. J. Biol. Chem. 269, 8937–8943.

Goodemote, K. A., Mattie, M. E., Berger, A., & Spiegel, S. (1995). Involvement of a pertussis toxin sensitive G protein in the mitogenic signaling pathways of sphingosine-1-phosphate. J. Biol. Chem. 270, 10272–10277.

Hakomori, S. (1990). Bifunctional role of glycosphingolipids. Modulators for transmembrane signaling and mediators for cellular interactions. J. Biol. Chem. 265, 18713–18716.

Hakomori, S. & Igarashi, Y. (1993). Gangliosides and glycosphingolipids as modulators of cell growth, adhesion, and transmembrane signaling. Adv. Lipid. Res 25, 147–162.

Hannun, Y. A. (1994). The sphingomylein cycle and the second messenger function of ceramide. J. Biol. Chem. 269, 3125–3128.

Hannun, Y. A. & Bell, R. M. (1987). Lysosphingolipids inhibit protein kinase C: Implications for the sphingolipidoses. Science 235, 670–674.

Hannun, Y. A. & Bell, R. M. (1989). Functions of sphingolipids and sphingolipid breakdown products in cellular regulation. Science 243, 500–507.

Hannun, Y. A., Loomis, C. R., Merrill, A. H., & Bell, R. M. (1986). Sphingosine inhibition of protein kinase C activity and of phorbol dibutyrate binding *in vitro* and in human platelets. J. Biol. Chem. 261, 12604–12609.

Hauser, J. M., Buehrer, B. M., & Bell, R. M. (1994). Role of ceramide in mitogenesis induced by exogenous sphingoid bases. J. Biol. Chem. 269, 6803–6809.

Heller, R. A. & Kronke, M. (1994). Tumor necrosis factor receptor-mediated signaling pathways. J. Cell Biol. 126, 5–9.

Holt, J. T., Gopal, T. V., Moulton, A. D., & Nienhuis, A. W. (1986). Inducible production of c-fos antisense RNA inhibits 3T3 cell proliferation. Proc. Natl. Acad. Sci. USA 83, 4794–4798.

Igarashi, Y., Hakomori, S., Toyokuni, T., Dean, B., Fujita, S., Sugimoto, M., Ogawa, T., el Ghendy, K., & Racker, E. (1989). Effect of chemically well-defined sphingosine and its N-methyl derivatives on protein kinase C and src kinase activities. J. Biol. Chem. 264, 14354–14360.

Igarashi, Y., Kitamura, K., Toyokuni, T., Dean, B., Fenderson, B., Ogawass, T., & Hakomori, S. (1990). A specific enhancing effect of N,N-dimethylsphingosine on epidermal growth factor receptor autophosphorylation. Demonstration of its endogenous occurrence (and the virtual absence of unsubstituted sphingosine) in human epidermoid carcinoma A431 cells. J. Biol. Chem. 265, 5385–5389.

Irvine, R. (1982). How is the level of free arachidonic acid controlled in mammalian cells? Biochem. J. 204, 3–16.

Jacobs, L. S. & Kester, M. (1993). Sphingolipids as mediators of effects of platelet-derived growth factor in vascular smooth muscle cells. Am. J. Physiol. 265, c740–c747.

Jarvis, W. D., Kolesnick, R. N., Fornari, F. A., Traylor, R. S., Gewitz, D. A., & Grant, S. (1994). Induction of apoptotic DNA damage and cell death by activation of the sphingomyelin pathway. Proc. Natl. Acad. Sci. USA 91, 73–77.

Jayadev, S., Linardic, C. M., & Hannun, Y. A. (1994). Identification of arachidonic acid as a mediator of sphingomyelin hydrolysis in response to tumor necrosis factor α. J. Biol. Chem. 269, 5757–5763.

Jefferson, A. B. & Schulman, H. (1988). Sphingosine inhibits calmodulin-dependent enzymes. J. Biol. Chem. 263, 15241–15244.

Johns, L. D., Sarr, T., & Ranges, G. E. (1994). Inhibition of ceramide pathway does not affect ability of TNF-α to activate nuclear factor-κB. J. Immunol. 152, 5877–5882.

Johnson, J. A. & Clark, R. B. (1990). Multiple nonspecific effects of sphingosine on adenylate cyclase and cyclic AMP accumulation in S49 lymphoma cells preclude its use as a specific inhibitor of protein kinase C. J. Biol. Chem 265, 9333–9339.

Karin, M. & Smeal, T. (1992). Control of transcription factors by signal transduction pathways: the beginning of the end. TIBS 17, 418–422.

Kim, M. Y., Linardic, C., Obeid, L., & Hannun, Y. (1991). Identification of sphingomyelin turnover as an effector mechanism for the action of tumor necrosis factor alpha and gamma-interferon. Specific role in cell differentiation. J. Biol. Chem. 266, 484–489.

Kindman, L. A., Kim, S., McDonald, T. V., & Gardner, P. (1994). Characterization of a novel intracellular sphingolipid-gated Ca^{2+}-permeable channel from rat basophilic leukemia cells. J. Biol. Chem. 269, 13088–13091.

Kiss, Z. & Anderson, W. B. (1990). ATP stimulates the hydrolysis of phosphatidylethanolamine in NIH 3T3 cells. Potentiating effects of guanosine triphosphates and sphingosine. J. Biol. Chem. 265, 7188–7194.

Kolesnick, R. & Golde, D. W. (1994). The sphingomyelin pathway in tumor necrosis factor and interleukin-1 signaling. Cell 77, 325–328.

Koval, M. & Pagano, R. E. (1991). Intracellular transport and metabolism of sphingomyelin. Biochim. Biophys. Acta 1082, 113–125.

Lavie, Y. & Liscovitch, M. (1990). Activation of phospholipase D by sphingoid bases in NG108-15 neural-derived cells. Biochem. Biophys. Res. Commun. 167, 607–613.

Lippman, M. E. & Dickson, R. B. (1989). Mechanisms of growth control in normal and malignant breast epithelium. Recent Prog. Horm. Res. 45, 383–439.

Mathias, S., Dressler, K. A., & Kolesnick, R. N. (1991). Characterization of a ceramide-activated protein kinase: Stimulation by tumor necrosis factor α. Proc. Natl. Acad. Sci. USA 88, 10009–10013.

Mattie, M. E., Brooker, G., & Spiegel, S. (1994). Sphingosine-1-phosphate, a putative second messenger, mobilizes calcium from internal stores via an inositol trisphosphate-independent pathway. J. Biol. Chem. 269, 3181–3188.

Mazurek, N., Megidish, T., Hakomori, S., & Igarashi, Y. (1994). Regulatory effect of phorbol esters on sphingosine kinase in BALB/C 3T3 fibroblasts (variant A31): Demonstration of cell type-specific response—a preliminary note. Biochem. Biophys. Res. Commun. 198, 1–9.

McDonald, O. B., Hannun, Y. A., Reynolds, C. H., & Sahyoun, N. (1991). Activation of casein kinase II by sphingosine. J. Biol. Chem. 266, 21773–21776.

Merrill, A. H. (1989). Lipid modulators of cell function. Cancer Res. 49, 3229–3234.

Merrill, A. H. (1991). Cell regulation by sphingosine and more complex sphingolipids. J. Bioenerg. Biomembr. 23, 83–104.

Merrill, A. H. & Jones, D. D. (1990). An update of the enzymology and regulation of sphingomyelin metabolism. Biochim. Biophys. Acta 1044, 1–12.

Merrill, A. H., Sereni, A. M., Stevens, V. L., Hannun, Y. A., Bell, R. M., & Kinkade, J. M. (1986). Inhibition of phorbol ester-dependent differentiation of human promyelocytic leukemic (HL-60) cells by sphinganine and other long-chain bases. J. Biol. Chem. 261, 12610–12615.

Merrill, A. H. & Stevens, V. L. (1989). Modulation of protein kinase C and diverse cell functions by sphingosine—a pharmacologically interesting compound linking sphingolipids and signal transduction. Biochim. Biophys. Acta 1010, 131–139.

Merrill, A. H. & Wang, E. (1992). Enzymes of ceramide biosynthesis. Meth. Enzymol. 209, 427–437.

Merrill, A. H., Wang, E., Mullins, R. E., Jamison, W. C., Nimkar, S., & Liotta, D. C. (1988). Quantitation of free sphingosine in liver by high-performance liquid chromatography. Anal. Biochem. 171, 373–381.

Moolenaar, W. H., Krujer, W., Tilly, B. C., Verlaan, I., Bierman, A. J., & deLaat, S. W. (1986). Growth factor-like action of phospatidic acid. Nature 323, 171–173.

Murray, D. K., Ruhmann–Wennhold, A., & Nelson, D. H. (1979). Dexamethasone effect on the phospholipid content of isolated fat cell ghosts from adrenalectomized rats. Endocrinology 105, 774–777.

Nelson, D. H., Murray, D. K., & Brady, R. O. (1982). Dexamethasone-induced change in the sphingomyelin content of human polymorphonuclear leukocytes *in vitro*. J. Clin. Endocrinol. Metab. 54, 292–295.

Nishizuka, Y. (1992). Intracellular signaling by hydrolysis of phospholipids and activation of protein kinase C. Science 607, 607–614.

Obeid, L. M., Lenardic, C. M., Karolak, L. A., & Hannun, Y. A. (1993). Programmed cell death induced by ceramide. Science 259, 1769–1771.

Okazaki, T., Bell, R. M., & Hannun, Y. A. (1989). Sphingomyelin turnover induced by vitamin D_3 in HL-60 cells. Role in cell differentiation. J. Biol. Chem. 264, 19076–19080.

Okazaki, T., Bielawska, A., Bell, R. M., & Hannun, Y. A. (1990). Role of ceramide as a lipid mediator of 1 alpha,25-dihydroxyvitamin D_3-induced HL-60 cell differentiation. J. Biol. Chem. 265, 15823–15831.

Olivera, A. & Spiegel, S. (1992). Sphingomyelinase and cell-permeable ceramide analogs stimulate cellular proliferation in quiescent Swiss 3T3 fibroblasts. J. Biol. Chem. 267, 26121–26127.

Olivera, A. & Spiegel, S. (1993). Sphingosine-1-phosphate as second messenger in cell proliferation induced by PDGF and FCS mitogens. Nature 365, 557–560.

Olivera, A., Zhang, H., Carlson, R. O., Mattie, M. E., Schmidt, R. R., & Spiegel, S. (1994). Stereospecificity of sphingosine-induced intracellular calcium mobilization and cellular proliferation. J. Biol. Chem. 289, 17924–17930.

Pagano, R. E. (1988). What is the fate of diacylglycerol produced at the Golgi apparatus. TIBS 13, 202–205.

Palombella, V. J. & Vilcek, J. (1989). Mitogenic and cytotoxic actions of tumor necrosis factor in BALB/c 3T3 cells. Role of phospholipase activation. J. Biol. Chem. 264, 18128–18136.

Pelech, S. L. & Sanghera, J. S. (1992). MAP kinases: Charting the regulatory pathways. Science 257, 1355–1356.

Pinto, W. J., Wells, G. W., & Lester, R. L. (1992). Characterization of enzymatic synthesis of sphingolipid long-chain bases in *Saccharomyces cerevisiae*: Mutant strains exhibiting long-chain base auxotrophy are deficient in serine palmitoyl transferase activity. J. Bact. 174, 2575–2581.

Pushkareva, M. Y. & Hannun, Y. A. (1994). Modulation of cytosolic protein phosphorylation by sphingosylphosphorylcholine. Biochim. Biophys. Acta 1221, 54–60.

Pushkareva, M. Y., Khan, W. A., Alessenko, A. V., Sahyoun, N., & Hannun, Y. A. (1992). Sphingosine activation of protein kinases in Jurkat T cells. *In vitro* phosphorylation of endogenous protein substrates and specificity of action. J. Biol. Chem. 267, 15246–15251.

Raines, M. A., Kolesnick, R. N., & Golde, D. W. (1993). Sphingomyelinase and ceramide activate mitogen-activated protein kinase in myeloid HL-60 cells. J. Cell Biol. 268, 14572–14575.

Ramachandran, C. K., Murray, D. K., & Nelson, D. H. (1990). Dexamethasone increases neutral sphingomyelinase activity and sphingosine levels in 3T3-L1 fibroblasts. Biochem. Biophys. Res. Commun. 167, 607–613.

Randriamampita, C. & Tsien, R. Y. (1993). Emptying of intracellular Ca^{2+} stores releases a novel small messenger that stimulates Ca^{2+} influx. Nature 364, 809–814.

Riabowol, K. T., Schiff, J., & Gilman, M. Z. (1992). Transcription factor AP-1 activity is required for initiation of DNA synthesis and is lost during cellular aging. Proc. Natl. Acad. Sci. USA 89, 157–161.

Riabowol, K. T., Vosatka, R. J., Ziff, E. B., Lamb, N. J., & Feramisco, J. R. (1988). Microinjection of *fos*-specific antibodies blocks DNA synthesis in fibroblast cells. Mol. Cell. Biol. 8, 1670–1676.

Ritchie, T., Rosenberg, A., & Noble, E. P. (1992). Regulation of phosphoinositide hydrolysis in cultured astrocytes by sphingosine and psychosine. Biochem. Biophys. Res. Commun. 186, 790–795.

Rosenwald, A. G. & Pagano, R. E. (1993). Inhibition of glycoprotein traffic through the secretory pathway by ceramide. J. Biol. Chem. 268, 4577–4579.

Rother, J., van Echten, G., Schwarzmann, G., & Sandhoff, K. (1992). Biosynthesis of sphingolipids: dihydroceramide and not sphinganine is desaturated by cultured cells. Biochem. Biophys. Res. Commun. 189, 14–20.

Rozakis–Adcock, M., Fernley, R., Wade, J., Pawson, T., & Bowtell, D. (1993). The SH2 and SH3 domains of mammalian Grb2 couple the EGF receptor to the Ras activator mSos1. Nature 363, 83–87.

Rozengurt, E. (1986). Early signals in the mitogenic response. Science 234, 161–166.

Sabbadini, R. A., Betto, R., Teresi, A., Fachechi–Cassano, G., & Salviati, G. (1992). The effects of sphingosine on sarcoplasmic reticulum membrane calcium release. J. Biol. Chem. 267, 15475–15484.

Sadahira, Y., Ruan, F., Hakomori, S., & Igarashi, Y. (1992). Sphingosine 1-phosphate, a specific endogenous signaling molecule controlling cell motility and tumor cell invasiveness. Proc. Natl. Acad. Sci. USA 89, 9686–9690.

Sadahira, Y., Zheng, M., Ruan, F., Hakomori, S., & Igarashi, Y. (1994). Sphingosine-1-phosphate inhibits extracellular matrix protein-induced haptotactic motility but not adhesion of B16 mouse melanoma cells. FEBS Lett. 340, 99–103.

Schroeder, J. J., Crane, H. M., Xia, J., Liotta, D. C., & Merrill, A. H. (1994). Disruption of sphingolipid metabolism and stimulation of DNA synthesis by fumonisin B1. A molecular mechanism for carcinogenesis associated with *Fusarium moniliforme*. J. Biol. Chem. 269, 3475–3481.

Schutze, S., Potthoff, K., Machleidt, T., Berkovic, D., Wiegmann, D., & Kronke, M. (1992). TNF activates NF-κB by phosphatidylcholine-specific phospholipase C-induced "acidic" sphingomyelin breakdown. Cell 71, 765–776.

Spence, M. W. (1993). Sphingomyelinases. Adv. Lipid Res. 26, 3–23.

Spiegel, S. (1993a). Sphingosine and sphingosine-1-phosphate in cellular proliferation: Relationship with protein kinase C and phosphatidic acid. J. Lipid Med. 8, 169–175.

Spiegel, S. (1993b). Sphingosine, a breakdown product of cellular sphingolipids in cellular proliferation and phospholipid metablolism. In: Phospholipids and Signal Transmission (Massarelli, R., Horrocks, L. A., Kanfer, J. N. & Loffelholz, K., eds.), pp. 190–203, Springer-Verlag, Berlin.

Spiegel, S. & Fishman, P. H. (1987). Gangliosides as bimodal regulators of cell growth. Proc. Natl. Acad. Sci. USA 84, 141–145.

Spiegel, S. & Milstien, S. (1995). Sphingoid bases and phospholipase D activation. Chem. Phys. Lipids, in press.

Spiegel, S., Olivera, A., & Carlson, R. O., eds. (1993). The role of sphingosine in cell growth and transmembrane signaling. Advances in Lipid Research: Sphingolipids in Signaling, Part A, (Bell, R. M., Hannun, Y. A. & Merrill, A. H., eds.), pp. 105–129, Academic Press Inc., Orlando, FL.

Spiegel, S., Olivera, A., Zhang, H., Thompson, E., Su, Y., & Berger, A. (1994). Sphingosine-1-phosphate, a novel second messenger involved in cell growth regulation and signal transduction, affects growth and invasiveness of human breast cancer cells. Breast Cancer Res. Treat., in press.

Sporn, M. B. & Roberts, A. B. (1988). Peptide growth factors are multifunctional. Nature 332, 217–221.

Stevens, V. L., Nimkar, S., Jamison, W. C., Liotta, D. C., & Merrill, A. H. (1990). Characteristics of the growth inhibition and cytotoxicity of long-chain (sphingoid) bases for Chinese hamster ovary cells: evidence for an involvement of protein kinase C. Biochim. Biophys. Acta 1051, 37–45.

Stoffel, W. & Assmann, G. (1970). Enzymatic degradation of 4t-sphigenine 1-phosphate (sphingosine-1-phosphate) to 2t-hexadecen-1-al and ethanolamine phosphate. Hoppe–Seyler's Z. Physiol. Chem. 351, 1041–1049.

Stoffel, W. & Bister, K. (1973). Stereospecificities in the metabolic reactions of the four isomeric sphinganines (dihydrosphingosines) in rat liver. Hoppe–Seyler's Z. Physiol. Chem. 354, 169–181.

Stoffel, W., Heimann, G., & Hellenbroich, B. (1973). Sphingosine kinase in blood platelets. Hoppe–Seyler's Z. Physiol. Chem. 354, 562–566.

Stoffel, W., Sticht, G., & LeKim, D. (1968). Synthesis and degradation of sphingosine bases in *Hansenula ciferrii*. Hoppe–Seyler's Z. Physiol. Chem. 349, 1149–1156.

Strasberg, P. M. & Callahan, J. W. (1988). Lysosphingolipids and mitochondrial function. II. Deleterious effects of sphingosylphosphorylcholine. Biochem. Cell Biol. 66, 1322–1332.

Su, Y., Rosenthal, D., Smulson, M., & Spiegel, S. (1994). Sphingosine-1-phosphate, a novel signaling molecule, stimulates DNA binding activity of AP-1 in quiescent 3T3 fibroblasts. J. Biol. Chem. 269, 16512–16517.

Sugiyama, E., Uemura, K., Hara, A., & Taketomi, T. (1993). Metabolism and neurite promoting effect of exogenous sphingosylphosphocholine in cultured murine neuroblastoma cells. J. Biochem. (Tokyo) 113, 467–472.

Taylor, W. & Marshall (1992). Calcium and inositol 1,4,5-trisphosphate receptors: A complex relationship. TIBS 17, 403–407.

Tsai, M., Hall, A., & Stacey, D. W. (1989). Inhibition by phospholipids of the interaction between R-Ras, rho, and their GTPase-activating proteins. Mol. Cell Biol. 9, 5260–5264.

Tsai, M., Roudebush, M., Dobrowolski, Yu, C., Gibbs, J. B., & Stacey, D. W. (1991). Ras GTPase-activating protein physically associates with mitogenically active phospholipids. Mol. Cell. Biol. 11, 2785–2793.

Tsai, M., Yu, C., & Stacey, D. W. (1990). A cytoplasmic protein inhibits the GTPase activity of H-Ras in a phospholipid-dependent manner. Science 250, 982–985.

Van Veldhoven, P. P., Foglesong, R. J., & Bell, R. M. (1989). A facile enzymatic synthesis of sphingosine-1-phosphate and dihydrosphingosine-1-phosphate. J. Lipid Res. 30, 611–616.

Van Veldhoven, P. P. & Mannaerts, G. P. (1991). Subcellular localization and membrane topology of sphingosine-1-phosphate lyase in rat liver. J. Biol. Chem. 266, 12502–12507.

Van Veldhoven, P. P. & Mannaerts, G. P. (1994). Sphingosine 1-phosphate metabolism in cultured skin fibroblasts: evidence for the existence of a sphingosine phosphatase. Biochem. J. 299, 567–601.

Van Veldhoven, P. P., Matthews, T. J., Bolognesi, D. P., & Bell, R. M. (1992). Changes in bioactive lipids, alkylacylglycerol and ceramide, occur in HIV-infected cells. Biochem. Biophys. Res. Commun. 187, 209–216.

Venable, M. E., Blobe, G. C., & Obeid, L. M. (1994). Identification of a defect in the phospholipase D/diacylglycerol pathway in cellular senescence. J. Biol. Chem. 269, 26040–26044.

Vilcek, J. & Lee, T. H. (1991). Tumor necrosis factor. New insights into the molecular mechanisms of its multiple actions. J. Biol. Chem. 266, 7313–7316.

Weigmann, K., Schutze, S., Kampen, E., Himmler, A., Machleidt, T., & Kronke, M. (1992). Human 55-kDa receptor for tumor necrosis factor coupled to signal transduction cascades. J. Biol. Chem. 267, 7313–7316.

Wilson, E., Wang, E., Mullins, R. E., Uhlinger, D. J., Liotta, D. C., Lambeth, J. D., & Merrill, A. H. (1988). Modulation of the free sphingosine levels in human neutrophils by phorbol esters and other factors. J. Biol. Chem. 263, 9304–9309.

Wu, J., Spiegel, S., & Sturgill, T. W. (1995). Sphingosine-1-phosphate rapidly activates the MAP kinase pathway by a G-protein dependent mechanism. J. Biol. Chem. 11484–11488.

Yang, Z., Costanzo, M., Golde, D. W., & Kolesnick, R. N. (1993). Tumor necrosis factor activation of the sphingomyelin pathway signals nuclear factor kB translocation in intact HL-60 cells. J. Biol. Chem. 268, 20520–20523.

Yule, D. I., Wu, D., Essington, T. E., Shayman, J. A., & Williams, J. A. (1993). Sphingosine metabolism induces Ca^{2+} oscillations in rat pancreatic acinar cells. J. Biol. Chem. 268, 12353–12358.

Zhang, H., Buckley, N. E., Gibson, K., & Spiegel, S. (1990a). Sphingosine stimulates cellular proliferation via a protein kinase C-independent pathway. J. Biol. Chem. 265, 76–81.

Zhang, H., Desai, N. N., Murphey, J. M., & Spiegel, S. (1990b). Increases in phosphatidic acid levels accompany sphingosine-stimulated proliferation of quiescent Swiss 3T3 cells. J. Biol. Chem. 265, 21309–21316.

Zhang, H., Desai, N. N., Olivera, A., Seki, T., Brooker, G., & Spiegel, S. (1991). Sphingosine-1-phosphate, a novel lipid, involved in cellular proliferation. J. Cell. Biol. 114, 155–167.

Zhao, C., Beler, & Dunn, T. (1994). Suppressors of the Ca^{2+}-sensitive yeast mutant (*csg2*) identify genes involved in sphingolipid biosynthesis. Cloning and characterization of SCS1, a gene required for serine palmitoyltransferase activity. J. Biol. Chem. 269, 21480–21488.

INDEX

A-chain gene, 126
Achondroplasia, 162, 206
Acid box, 184
Acoustic neurinoma, 240
Acromegaly, 28
Activin, 371-375
Aging, 17
AIDS, 17, 543
Alzheimer's disease, 164-165, 241-243
Amphiregulin, 280
Amyotrophic lateral sclerosis, 164, 241-243
Angiogenesis, 150, 203, 279, 342, 434, 443, 450
Angioplasty, 133
Antinerve growth factor antibody, 235
AP-1, 419-420, 441, 553-554
Apert's syndrome, 206
Arginine-glycine-aspartate (RDG) sequence (Arg-Gly-Asp), 11
ASH (abundant Src homology), 189
Atherosclerosis, 133
AUG codon, 150-152
Autocrine growth factor, 5, 129
Autophosphorylation, 185, 403

B-chain gene, 126
B-sheet conformation, 103
Balb/c 3T3, 451
Basic fibroblast growth factor (bFGF), 440
BB gene *in vivo*, 375
Beckwith-Wiedeman syndrome, 12
Betacellulin, 280
Betaglycan, 399-400, 410
Bidentate site, 476
BMPs *see* "Bone morphogenetic proteins..."
Bombesin (BB)/GRP, 492-513
  GRP biosynthesis and processing, 496-498
  GRP cancer growth, 512-513
  GRP gene expression, 508-510
  GRP proliferation, 510-512
  GRP receptor ligands, 499
  GRP receptors, 499-504
  GRP second messengers, 504-508
  GRP secretion, 498-499
  GRP structure, 492-496
Bond remodeling, 279
Bone formation, 32, 44, 201, 373, 376-378
Bone morphogenetic proteins (BMPs), 358, 376-380
Brachypodism (*bp*) mouse, 378-380
Brain-derived neurotrophic factor (BDNF), 242-244
Breast cancer/tumors, 20, 167, 513, 546

c-*fos*, 63-66, 95-96, 239, 419
c-*jun*, 63, 95-96, 419
c-*jun*-N-terminal kinase, 194
c-MET, 470, 472, 479-481
c-*Myc*, 95-96, 418
c-Ras, 158
c-Ron, 470, 474, 478
c-terminal amino acids, 98, 102-103, 107, 126, 156-156, 168, 186-187, 232, 265, 287, 359-365, 372, 476, 493-495
CAAT/enhancer binding protein (C/EBP), 5
CAG donor sequence, 296
cAMP-dependent protein kinase pathway, 441
Cancer
  breast cancer/tumors, 20, 167, 513, 546
  colon cancer, 291, 421, 513
  gastric cancer cells, 421
  GRP and, 512-513
  liver tumors, 294-295
  lung carcinomas, 292
  mammary tumors, 291-294, 299
  prostate cancer, 205, 233
  skin cancer, 342
  small-cell lung cancer, 509-512
Carcinogenesis, 341-342
Cardiac development, 201
Catabolic states, 15
cC-Sea, 470, 474, 478
CCAAt boxes, 4, 5, 7
CdkIs (cyclin-dependent kinase inhibitors), 415-417
Cell adhesion molecules (CAMs), 164
Cell motility, 128
Cell-cell adhesion, 403
Central nervous system (CNS), 163-164, 269-271
Ceramide
  arresting cell growth signaling, 541-543
  downstream signaling, 544-545
  TNFα and interleukin-1 signal transduction, 543-544
Chemotaxis, 131
Chimeric TGFα proteins, as therapeutic agents, 295-296
Chondrogenesis, 162
Chromaffin-like cell, 234-235
Chronic renal failure, 17, 34, 55
Ciliary neurotrophic factor (CNTF), 48
Circulating IGFs, 3, 13
Cis, 150, 442
Contractile phenotype, 207
Corneal trauma, 88-89
Corticotropin releasing factor (CRF), 498
Craniosynostosis, 162
Crouzon syndrome, 206
Cysteine, 90, 98, 323
Cystic kidney disease, 90
Cytokine/hematopoietien receptor subfamily, 48-52

Decubitus ulcers, 131
Depot storage system, 159
Digital abnormalities, 206
Disulfide bonds, 87, 91, 126, 280, 364
DNase hypersensitivity site, 62
Dwarfism, 12, 17, 28, 34-36, 55-56
  sex-linked dwarfism, 56

E-peptides, 4
ECM deposition, 321, 327
EGF (epidermal growth factor)
  biological activity, 88-89
  gene expression response, 95-96
  gene organization, 89-90
  historical overview, 86-87
  mutant proteins
    acidic residues substitution, 106-107

aromatic residues substitution, 104-105
basic residues, substitution, 107-108
hydrophobic residues substitution, 102-104
kinase stimulation, 97
mitogenicity, 97-98
polar residues mutation, 105-106
receptor affinity, 97-98
receptor binding mutations (cumulative effect), 108-109
single-site mutations on receptor ligand, 98-108
structure of, 96-97
precursor of, 90
primary sequences of family members, 90-91
receptor, 92-94
mechanism of activation, 94-95
solution structure, 91
wound healing and, 88-89
EGF-R, 96, 280, 295-296
Elastase enzymes, 286
Embryogenesis, 443
inhibin and, 373-374
Embryonic development
embryonal stem (ES) cells, 334
embryonic lethality, 331
VEGF/VPF and, 442
Endocrine function, 129
negative feedback loop, 15
Endoglin, 399-400, 403, 422-423
Epidermal growth factor *see* "EGF..."
Epiregulin, 280
Erythroid differentiation factor (EDF), 372
Erythropoiesis, 442
Erythropoietin (EPO), 48-49
Estrogens, 441
Extracellular matrix (ECM), 321, 327
Familial dysautonomia, 240
Fasting, 55
FGF (fibroblast growth factor)
bioactivity regulation, 159-160
dimers, 158
localization, 155
molecular properties of family, 153
physiology and pathology, 160-168
angiogenic factors, 165-167
limb development, 161-162
lung development, 162-163
nervous system development, 163-164
neurodegenerative conditions, 164-165
tumorigenesis, 167-168
secretion, 154
signaling, 157-159
target cells and distribution, 149-150
FGF genes, 148-149
isoforms, 148
molecular regulation of transcription, 150-153
FGF-1, 148-151
FGF-2, 148, 151-152, 162-163, 205
FGF-3, 148, 152-153, 162-163, 205
FGF-5, 148, 153, 162
FGF-7, 153, 162, 187
peptide localization, 149
FGF ligands and biological effects, 183
FGF receptors, 184-185
heparin sulfate proteoglycans, 196-197
high-affinity receptors, 156-157
Fibroblast growth factor *see* "FGF..."
Fibronectin, 131

Fibrosis, 134
Fk506, 412
FkBP12, 412-413
Follicle-stimulating hormone (FSH), 371-374
Freemartin calf, 369-370

Gastric ulcers, 89
Gastrin, 95
Gastrin releasing peptide (GRP), 492
GH (growth hormone)
  binding protein (GHBP) and receptor dimerization, 28, 34, 57
  bioinactive GH, 37
  biological actions, 32-33
  calcium response, 67-68
  deficiency and insufficiency, 35-36
  excess GH conditions, 37
  immune system and, 32
  induction of c-*fos* and other genes, 63
  induction of IGF-1 synthesis and secretion, 62
  IRS-1 and IRS-2 tyrosyl phosphorylation, 67
  JAK2 tyrosine kinase activation, 58-61
  lactogenic activity, 32-33
  MAP kinases activation, 61-62
  nutrition and, 34
  plasma transport and metabolism, 34-35
  posttranslational modifications, 31
  pregnancy and, 33-34
  protein kinase C and G, 66
  secretion, regulation of, 33-34
  signal transduction via GHR, 56
  stats 1, 3, and, 5, 64-65
  structural properties, 31-32
GH genes
  bovine GH genes, 30-31
  gene expression and products, 30-31
  genomic organization, 29-30
GH resistance-Laron syndrome, 15
GH-releasing hormone (GHRH), 28
GHR (growth hormone receptor)
  development regulation of, 53
  differential expression between sexes, 53-54
  dwarfism and, 55-56
  fetal development and, 53
  glycosylation, 57
  growth and, 55-56
  internalization, 57-58
  localization of, 52-53
  regulation by GH, 54
  regulation by insulin and glucocorticoids, 54-55
  signal transduction, 56-69
  signaling pathways, 58-60
  tyrosyl phosphorylation, 58
GHR gene
  alternative splicing of GHR/mRNA, 52
  cloning of, 45-48
  cytokine/hematopoietien receptor subfamily, 48-52
Gigantism, 13, 28
Glia cell line-derived neurotrophic factor (GDNF), 358, 380-382
Gliomas, 204, 234
Glucocorticoids, 54-55, 55
Glumerulonephritis, 134-135
Glycine, 98
Golgi cells, 154
Granulocyte-colony stimulating factor (G-CSF), 48
Granulocyte-macrophage colony stimulating factor (GM-CSF), 48
Growth hormone *see* "GH..."

Growth hormone receptor *see* "GHR..."
GRP *see* "Bombesin (BB)/GRP..."

Heat shock response, 154
Hematopoiesis, 372
Heparan sulfate proteoglyeans (HSPG), 148, 155-156
Heparin, 436, 440
Heparin-binding EGF, 280
Hepatic bGH-binding sites, 54
Hepatocellular necrosis, 374
Hepatocyte growth factor *see* "HGF..."
HER2/neu, 280
HER3, 279
Hereditary hemorrhagic telangiectasia (HHT), 422-423
  mutations in endoglin, 422-423
Heregulin/neu differentiation factor (NDF), 91, 280
Hetero-FGFR dimers, 158
Heterodimers, 125, 181, 280, 360, 364, 473
HGF (hepatocyte growth factor)
  GF activated by extra cellular cleavage, 467-469
  low- and high-affinity receptors, 469-470
    subcellular distribution of p190 *MET*, 470-471
  mitogen, motogen and morphogen, 466-467
  posttranslational processing generates dominant-negative truncated HGF receptor isoforms, 472-473
  receptor kinase activity, positive and negative regulation, 473-474
  structure (heterodimeric molecule), 467
HGF receptor
  signal transduction, 474-478
  tissue-specific regulation, 478-479
HGF receptor gene (MET)
  in vitro cell transformation, 479-481
  in vivo harmful oncogene, 481-482
High-affinity FGF receptors (FGFRs), 156
High-performance liquid chromatography (HPLC), 278
Hind limb ischemia model, 447
Hinge region of EGF, 105
Homo-oligomers, 408, 409
Homodimers, 125, 181
Housekeeping genes, 297
Human gastric ulcers, 164
Human glioblastoma, 132
Huntington's chorea, 164-165
Hydrophobic pocket, 186
Hypoglycemia, 36, 240
Hypopituitarism, 36
Hypoxia, 441, 449
  VEGF/VPF and, 442

Iduronic acid (IdoA), 156
IGFs (insulin-like growth factors)
  binding proteins, 2, 9-11
    cell surface membrane and, 11
    modulation of IGF action, 10-11
    molecular structure, 10
  IGF-I, 2, 28, 44
    bone growth and, 13
    intermediary metabolism and, 13-15
    molecular and cellular aspects, 3-4
    neurological disorders and, 17
    transcription, 4
  IGF-I and intermediary metabolism, 13-15

IGF-I receptor, 2, 7-8
human receptor, 7
organ development and, 9
IGF-II, 2, 17-18
human, 17
molecular and cellular aspects, 4-5
promoters, 5
transcription, 5
IGF-II/M-6-P receptor, 8-9
molecular and cellular aspects, 2-11
neoplasia and, 18
physiological and clinical aspects, 11-18
clinical uses, 15-17
embryonic growth and development, 12
in vitro studies, 11
postnatal growth, 12-13
Immune system, growth hormone (GH) and, 32
Immunoglobulin (Ig) superfamily, 156-157
Immunosuppressor, 329, 412
Immunosympathectomy, 230
Inhibins/activins, 371-375
inhibin A, 372
inhibin B, 372
inhibin BA, 365
inhibin-deficient animals, 374-375
Insulin, 13-15, 54-55
insulin receptor-related-receptor (IRR), 5, 7
insulin-dependent diabetes mellitus (IDDM), 14-15, 34
insulin-resistance disorders, 16
Interleukin-1 (IL-1), 543
Intrauterine growth retardation, 12
Ischemia, 228, 240

Janus kinase/signal transducer, 195

Kaposi's sarcoma, 167
KDR-flk-1, 452
Keratinocytes, 418
Kidney defect, 197
Knockout BDNF, 268, 270
Knockout of receptor, 197, 199, 206, 236, 243

Lactogenic activity, 32-33, 65
Laron-type dwarfism, 55
Larson syndrome, 32-34
Leu-Thr-Pro motif, 544
Leucine-rich motifs (LRM), 261
Leucine-zipper, 204, 419
Leydig cell, 370, 372
Ligand binding, 59
Ligand-induced dimerization, 136, 158
Limb abnormalities, 270
Limb formation and FGFs, 199
Lipid second messenger, 539
Liver, 52, 54, 291
cirrhosis of, 134
liver-enriched activator protein (LAP), 5
regeneration of, 279
tumors of, 294-295
Longitudinal bone growth, 32, 44, *see also* "Bone formation..."
Low-affinity NGF receptor, 258
Lung carcinomas, 292
Lung development, MIS and, 371
Lung and FGFs, 201
Luteinizing hormone (LH), 441-442

Macrophages, 131
Malignancies, 132-133
Mammary gland, 65, 290
ectopic TRFB1 expression in, 333-334
Mammary tumorigenesis, 291-294, 299
MAP, 158, 193-194, 265, 279

MAPK, 190, 544
Marfan syndrome, 325
MCF-7 cells, 421-422
MEK, 193
Mesodermal markers, 198
Met receptor, 468, 470, 474
Metastatic activity, 152
MIS *see* "Müllerian inhibiting substance..."
Mitogen, HGF, 466
Mitogen-activated protein (MAP/MAPK), 61, 544
Mitogenic effect, 128
Monocytes, 452
Morphogen, HGF, 466
Motogen, HGF, 466
MRNA expression, 52-54
Müllerian inhibiting substance (MIS), 365, 369-371
  reproductive development and, 369-371
Multifunctional docking site, 470, 477
Multifunctional super site, 477, 478
Myelofibrosis, 134

N-sulfated glucasamine (GleNSO$_3$), 156
N-terminal, 103, 107, 126, 150, 151, 153, 232, 359-360, 467, 493
Naf (not a functional raf), 198
Neomycin, 136
Neoplasia, 18
Nerve growth factor (NGF), 228
  biosynthesis, 232-233
  clinical perspectives, 240-242
  early experiments, 229-230
  gene/gene expression, 231-232
  mechanism of action, 236-240
  occurrence, 233
  receptors, 258-259
  signaling, 264
  structure, 230-231
  target tissues, 233-236
Neural tissue, inhibin/activin and, 373
Neuroblastoma, 240
Neurofibromatosis, 240
Neurogenesis, 203
Neuromedin B (NMB), 492, 493, 499, 501
Neurotrophins, 258
  NT-3, 244-245
  NT-4/5 and, 6, 245-246
  signaling, 267
NGF *see* "Nerve growth factor..."
Niemann-Pick disease, 555
Nodal, GDNF, and other growth/differentiation factors related to TBFB, 380-382
  glia cell line-derived neurotrophic factor (GDNF), 358
Nodose-petrosal complex, 269
Northern blot, 52, 436, 439, 443, 449
Notch lipoprotein receptor, 282

Ocular disease, 442, 447-448
Oncofetal growth factor, 290
One-way species specificity, 28
Open reading frame (ORF), 153
Organogenesis, 161, 198
Osteoporosis, 17
Ovarian hyperstimulation syndrome (OHSS), 445

P27, 417
P75 receptor
  role *in vivo*, 270-271
  structural features, 264
P190*MET*, 473, 474
Parkinson's disease, 164-165, 241-243
Patch mutation, 197
PDGF (platelet derived growth factor)
  antagonists, 135-137
  atherosclerosis, 133

autocrine loop, 204
cell culture and, 128-129
disease and, 132-135
embryonal development, 130
FGF (fibroblast growth factor), 181-183
FGF receptors
embryonic development, 197-204
genomic organization and structural features, 183-185
human disease, 204-207
fibrosis, 134
glumerulonephritis, 134-135
in vivo function, 130-132
embryogenesis, 130-131
inhibition of platelet aggregation, 131-132
wound healing, 131
isoforms, 181
ligands and biological effects, 181-182
malignancies, 132-133
production, 129
structure of, 125
PDGF genes, 125-126
receptor binding, 127-128
sequence of, 126
three-dimensional structure, 126
Persistent Müllerian duct syndrome (PMDS), 370
Pfeiffer's syndrome, 162, 206
Phosphoinsitide pathways, 191
Phospholipase C (enzyme) (PLC-*Y*), 191
Pituitary dwarfism, 12
Pituitary gland, 27, 37
Placenta, 27, 30, 31, 130
placenta growth factor (PIGF), 126, 436
placental GH (hGH-V), 33-34, 36
Postaglandins (PGs), 452
Pox viruses, 282
Prepro EGF, 90
Prepro GRP, human prepro GRP, 496-497
Pro EGF, 90
Pro-HGF, 468
Pro-region, 365
Prolactin, 95
Prolactin receptor (PRLR), 48
Proline, 98
Promoters, 5
murine species, 5, 12
promoter-specific genomic imprinting, 5
Prostate cancer, 205, 233
Prostate NGF-like material, 233
Protamine, 440
Protein binding site, 62
Protein kinase C (PKC), 66
Pseudoacromegalic state, 34
Psoriasis, 450-451
Pulmonary fibrosis, 134

Rabson-Mendenhall syndrome, 14
Raf, 93, 158, 192, 198, 265, 279
Rapamycin, 412
Ras, 93, 188-190, 191-192, 265, 279, 475, 477, 553
Rb (retinoblastoma susceptibility gene product Rb), 413-414, 419
Receptors, 5-9
activation and signal transduction, 185-190
dimerization, transphosphorylation and signaling complex formation, 185
p21 ras and its downstream targets, 188-190
PH domains, 187
PI-3 kinase, role in signaling, 190-192
PTB domains, 186
Raf downstream of Ras, 192-194

Ras-independent pathways, 195-196
SH2 domains, 185-186
SH3 domains, 186-187
signaling pathways used by PDGF and FGF receptors, 187-188
dimerization, 94
IGF-I receptor, 7-8
IGF-II/M-6-P receptor, 8-9
receptor tyrosine kinases (RTKs), 180, 439
Recombinant human IGF-I (rhIGF-I), 2
Response-to-injury hypothesis, 133
Rheumatoid arthritis (RA), 134, 450-451
Rhombomeres, 164

Salivary gland (mouse), 231
Sarcoma growth factor (SGF), 278
Sarcoma necrosis factor, 466
Sarcoma viruses, 279
Scatter factor, 466, 467
SCLC (small-cell lung cancer), 509-512
Scleroderma (systemic sclerosis), 134
Secreted variant, 159
Sensory defects, 270
Sertoli cells, 369, 372
SH2 (SRC homology 2) domains, 93
She, 475
Short ear (*se*) mouse, 378-380
Signal transduction, 264-267
pathway, 135-136, 266
receptors, 405-406
p75 receptor, 258, 267
TrkA tyrosine kinase receptors, 264-265
TrkB and TrkC tyrosine kinase receptors, 265-267
signaling complexes, 185
signaling molecules, 44-45, 61
STAT (signal transducers and activators of transcription), 96, 109
activation, 65
proteins, 279
stat1, 64
stat3, 65
stat5, 65
transcription factor, 195
Simian sarcoma virus (SSV), 204
Sis oncogene, 132
Skin cancer, 342
Skin cells and FGFs, 200
Small latent complex, 325
Somatic growth, 62
Son of sevenless (SOS) of Ras pathway, 93, 189
Sphingolipid metabolism, 540-541
Sphingosine and sphingosine-1-phosphate
calcium and inositol triphosphate, 550-551
downstream signaling, 547-551
growth factors modulate levels, 551-553
phosphatidic acid, 547-550
positive regulators of cell growth, 545-546
protein kinase C and other kinases, 547
transcription factors stimulate DNA binding activity of AP-1, 553-554
Sphingosinephosphorylcholine
as mitogen, 554
as mitogenesis regulator, 555-556
signaling pathways, 555
Src family, 94, 110
Src homology 2 (SH2) domain, 58, 189, 475, 477, 478
Stats *see* "Signal transduction..."
Streptozotocin-induced diabetes, 55
Stress activated protein kinase, 194

Stress fibers, 128
Stroma formation, 133
Supersite, 477, 478
Suramin, 136, 440
Swiss 3T3 cells, 504-511, 545, 552
Synthetic phenotype, 207
System-specific anabolic effect, 16-17

T-cell receptor, 266
Tamm-Horsfall protein, 90
TATA box, 4, 5, 7, 151, 153, 239
TGFα, 434
  digestive track and, 290
  EGF-R ligands and biological activity, 284
  EGF-R and related proteins, 279-280
  historical background, 278-279
  integument and, 289
  mammalian EGF superfamily, 280-282
  midgestation tissues, 289
  neoplastic cells and tissues, 291-292
  proTGFα
    membrane-anchored growth factors, 287-288
    normal tissues, 288-290
    processing of, 284-287
  structure/function analyses of mature TGFα, 283
  transformation, 292-295
  viral and nonvertebrate EGF family homologues, 282-283
TGFα gene, 296-297
  gene expression, 298-300
  knockout mice, 200-303
  promoter, 297-298
TGFB, 320-321, 434
  benign skin tumors to malignant conversion, 341-342
  biochemistry/molecular biology, 321-326
    activation of latent TGFB, 325-326
    molecular structure, 321-323
    transcriptional/posttranscriptional control of gene expression, 323-325
  bone morphogenetic proteins, 376-380
  in vitro effects, 326-330
    apoptosis induction, 329-330
    cell proliferation control, 326-327
    cellular differentiation, 328
    extracellular matrix modulation, 327
    immune system, 329
  in vivo
    biological activities, 331-343
    ectopic expression in transgenic mice, 331-333
    keratin promoter TGFB1 in transgenic mice, 332-333
    TGFB gene knockout mice, 334-337
    TGFB in multistage epithelia carcinogenesis, 340
    TGFB wound healing and tissue repair, 338-340
  inhibins/activins, 371-375
  isoforms, differential functions, 330
  ligands, 397-398
  Müllerian inhibiting substance (MIS), 369-371
  nodal, GDNF, and other growth/differentiation factors related to TGFB, 380-383
  oncogenesis and, 422
  overview, 358, 396-398
  pro-regions, 364
  sequences and structures, 358-369

signal transduction, 404-420
biological effects mediated by multiple signaling pathways, 404
cell-cycle machinery, 413-419
functional properties of receptors, 406-408
gene regulation, 419-420
identification of receptor interacting proteins, 411-413
mechanism of receptor activation, 406-410
pathways in carcinogenesis, 420-422
signaling receptors, 404-406
TGFB regulates expression of genes with diverse functions, 419-420
as stimulator of malignant progression, 342-343
as tumor suppressing factor, 340-341
TGFB receptors, 398-404
cloning and structural features of TGFB receptors, 399-404
identification of cell-surface TGFB-binding proteins, 398-399
signaling pathways, 396-397
introduction/overview, 396-397
Type I TGFB receptor, 401-403
functional interaction, 408-410
Type II TGFB receptor, 400-401
functional interaction, 408-410
Type III receptor, 399-400
betaglycan presents TGFB to the Type II receptor, 410-411
Type V TGFB receptor, 403-404
TGFB1, 365
TGFB1 null embryo lethality, 336
Thyroid papillary carcinomas, 262
Tissue regeneration, 88-89
Tme mutation, 9
TNFα, 543
Tooth-lid factor, 86
Trans acting elements, 150
Transforming growth factor *see* "TGF..."
Transmembrane (TM) domain V, 504
TRK receptors
role *in vivo*, 268-270
structural features, 259-264
TrkA defective mice, 268-269
TrkA receptors, 258, 262
TrkB defective mice, 269-270
TrkB receptors, 258, 262-263
TrkC defective mice, 270
TrkC receptors, 258, 263-264
Tumor angiogenesis, 448-450
Tumor biology, 167, 204
Tumor necrosis factor (TNF), 264, 434
Tumorigenesis, 164-165, 279, 291, 293
Tumors, 18, 28, 37, *see also* "Cancer..."
benign, 341-342
Tyrosine autophosphorylation, 473-474
Tyrosine kinase, 58, 94

Untranslated regions (UTRs), 325
Urogastrone protein, 87

V-*erB* gene product, 94
VEGF/VPF (vascular endothelia cell growth factor/vascular permeability factor)
binding sites, 437-438
disease and, 447-451
ocular diseases, 447-448
psoriasis and rheumatoid arthritis, 450-451
tumor angiogenesis, 448-450

induction of, 440-442
differentiating agents, 441
hormonal agents, 441-442
hypoxia, 442
isoforms and related growth factors, 434-437
receptors, 437-440
cloning of VEGF/VPF receptors, 438-440
embryonic development, 443-444
female reproductive system, 444-446
modulation of VEGF/VPF binding to its receptors, 440
VEGF/VPF binding sites, 437-438
structural characteristics, 434-437
target cells and tissues other than vascular endothelium response, 451-453
therapeutic angiogenesis, 446-447
vascular endothelia cell growth factor, 126, 435
VIP/PA/CAP, 513-521
VIP biosynthesis and processing, 514-515
VIP proliferation and cancer growth, 521
VIP receptor ligands, 515
VIP receptors, 516-519
VIP second messengers and gene expression, 519-521
VIP structure, 513-514
VIP_2 receptor, 518-519

Wilms' tumor, 234
Wound healing, 17, 88-89, 131, 160, 166, 279, 293
fetal wound healing, 339
FGFs and, 200
WSXWS motif, 49

Yolk sac blood islands, 336-337

CPSIA information can be obtained at www.ICGtesting.com
Printed in the USA
LVOW05*0355020114

367657LV00008B/74/P

9 780762 300914